COMPUTERS
IN HOTELS
Concepts and Applications

PARTHO PRATIM SEAL

Assistant Professor
Welcomgroup Graduate School of Hotel Administration
Manipal University

OXFORD
UNIVERSITY PRESS

OXFORD
UNIVERSITY PRESS

Oxford University Press is a department of the University of Oxford.
It furthers the University's objective of excellence in research, scholarship,
and education by publishing worldwide. Oxford is a registered trade mark of
Oxford University Press in the UK and in certain other countries.

Published in India by
Oxford University Press
22 Workspace, 2nd Floor, 1/22 Asaf Ali Road, New Delhi 110 002

First published in 2013
Sixth impression 2021

ISBN-13: 978-0-19-808400-6
ISBN-10: 0-19-808400-5

Typeset in Garamond
by iPlus Knowledge Solutions Private Limited, Chennai
Printed in India by Repro India Ltd., Surat

For product information and current price, please visit www.india.oup.com

In honour of my father,
whose guidance and support
have given meaning to my existence.

The lessons that I learnt in my life were
mostly by emulating the way he lived.

Preface

Computers play a major role in our lives, so much so that we cannot even think of a life without them. The use of computers in various establishments and organizations including hotels began in the late 20th century. Hotels used to have ledger books, registers, and files, which had to be maintained and updated regularly. The process used to be tedious with much manpower and man-hours being used for its upkeep.

Computerization not only makes work easy for the hotelier but also fast. It reduces a lot of paper work and interaction with other departments. These systems have also re-modelled the guest reservation system, and food and beverage (F&B) management system in restaurants and food outlets. The billing process has also undergone tremendous change, and now a guest can even pay the bill from the room and opt for an express check-out. This has led to better guest facilities and comfort, faster bill settlement, lesser interaction with hotel staff, and improved guest services.

Individuals can specialize in any of the four major departments in the hotel industry, viz. front office, housekeeping, food production, and F&B service. However, their learning would be incomplete without proper knowledge of computers as all these departments are now linked with computers; information pertaining to both guests and non-guests is shared amongst them.

The syllabi of hotel management programmes and courses cover the basics of computers such as hardware, software, operating systems, MS Office, Internet, database, and programming. Therefore, the books in use also cover only these basics. However, a need has been felt for books that explain not just the fundamentals of computers but also their applications in the hotel industry, such as hotel information system (HIS), computerized reservation system (CRS), property management system (PMS), and F&B management.

ABOUT THIS BOOK

Computers in Hotels: Concepts and Applications has been prepared keeping in mind the needs of hotel management students. It provides an elementary view about the origin and history of computers, as well as an overview of the software and hardware requirements, the uses of the Internet, and some basic programming. The applicability of computers to the hospitality industry has been illustrated throughout the book.

The book also explains the various PMSs, and their features and uses in the hospitality sector. The different kinds of hardware, user-friendly systems, and commonly used PMSs in hotels have also been discussed in the book. The book has been prepared bearing in mind the latest information technology used in the industry in the current scenario.

PEDAGOGICAL FEATURES

The book has been written in an engaging manner so as to make it student-friendly. Some of the salient features are as follows:

- Each chapter begins with a brief story that brings out the essence of the topic being covered.
- Learning objectives have been added at the beginning of every chapter.

- The chapters are well-supported by images, screenshots, tables, and diagrams.
- All the chapters end with a summary, which gives a brief overview of the chapter contents.
- Key terms with their definitions, at the end of each chapter, to help readers understand the important words and abbreviations used in the chapter, are provided.
- Concept review questions and multiple choice questions are provided in view of the various examination formats.
- Project work at the end of every chapter gives hands-on practical understanding of the topics discussed.
- Case studies at the end of almost all the chapters help students analyse and discuss real-time problems with peer groups and mentors.

COVERAGE AND STRUCTURE

The book comprising 11 chapters is divided into two parts.

The first part, *Computer Fundamentals,* provides an overview of the concepts and applications of computer, software, MS Office, Internet and Internet security, DBMS, and FoxPro programming.

Chapter 1 deals with the history and classification of computers. Hardware devices, data, and information are also briefly described.

Chapter 2 briefly describes software, types of software, operating systems and their types such as MS DOS, Windows, and Linux, as well as programming languages.

Chapter 3 provides an overview of the various types of MS Office applications such as word packages, spreadsheet, presentation, and database.

Chapter 4 describes the Internet and its applications, types of networking, latest wireless technology, e-commerce, and antivirus software.

Chapter 5 covers data and information and gives a brief idea of database models. It describes the database software, MS Visual FoxPro 6.0, and also discusses programming.

The second part, *Computer Applications in the Hotel Industry,* helps the students understand the usage of computers in hotels.

Chapter 6 provides an overview of the different hotel departments. It also gives an overview of the revenue- and non-revenue-generating departments, and why computers are now necessary in all departments. Three major PMSs—Micros, IDS, and ShawMan—are also discussed.

Chapter 7 describes the ways in which reservation of rooms is carried out with the help of computers. The room management module discusses assignment of rooms to guests, organization of housekeeping activities, and generation of reports.

Chapter 8 covers both the guest accounting system and the hotel accounting system. It also discusses inventory management systems, including purchase and payroll modules.

Chapter 9 discusses the point-of-sale systems used in the F&B system as well as the equipment used in the processes. It also talks about the recipe management, menu management, sales analysis, and automated beverage management systems.

Chapter 10 describes common PMS interfaces such as call accounting system (CAS), electronic locking system (ELS), energy management system (EMS), and auxiliary guest service devices.

Chapter 11 describes management information system (MIS) including its definition, concepts, design, functions, and evaluation.

ACKNOWLEDGEMENTS

I acknowledge the part played by certain individuals and organizations for their help and support in the completion of this book.

My father, the late Mr Sarat Kumar Seal, and my mother, Mrs Anjali Seal, have been a constant source of strength in my endeavours to write this book. Their blessings have been my guiding force.

I thank my wife, Mrs Samita Seal, without whose help, inspiration, and encouragement, the first part of the book would not have been so intricately prepared. I am also grateful to the Chairman and Secretary of DSMS Group of Institutions for allowing me to use their resources.

I am deeply indebted to Mr Sumit Sharma, Assistant Vice-President, Corporate Marketing Division of Oberoi Hotels & Resorts, for allowing me to use their screenshots, and Mr James Nicol, Vice-President of Marketing, Easybar Beverage Dispensing Systems, for providing me with an image of the automated beverage dispensing system. My heartfelt gratitude to Mr Sameer Kriplani, Owner at RCH Supply Co., for permitting me to use the auto check-in images and Mr Steve Driessens of Resort Software for his help on recipe management.

I express my sincere appreciation to Mr Sagun Shawney, Country Manager of Micros-Fidelio, Mr Jimmy P. Shaw, Managing Director of ShawMan Software Pvt. Ltd, Ms Sangita S. Mani, Senior Manager Marketing of IDS–Next, Mr Kamlesh Jain, CEO, Dataman Computer Systems, Mr Prabhas Bhatnagar, Founder, HMS Infotech (Hotelogix), and Mr Sunil Sondhi of Hotel Management Systems for giving me access to various software features, screenshots, reports, and images.

My genuine thanks are also due to Mr Rajive P. Sood, Proprietor, Real Power Software and Mr O.V. Krishna Reddy, CEO, Samudra Technologies, for their assistance.

I am much obliged to the academicians and reviewers of the book for their feedback, which helped in improving the presentation of the contents of the book. I express my gratitude to the editorial team at Oxford University Press India, for their constant feedback and guidance, which led to the timely completion of the project.

I thank my colleagues, friends, and industry professionals who helped me prepare this manuscript. In addition, I owe a lot to the students who inspired me to write this comprehensive title.

Partho Pratim Seal

Brief Contents

Detailed Contents

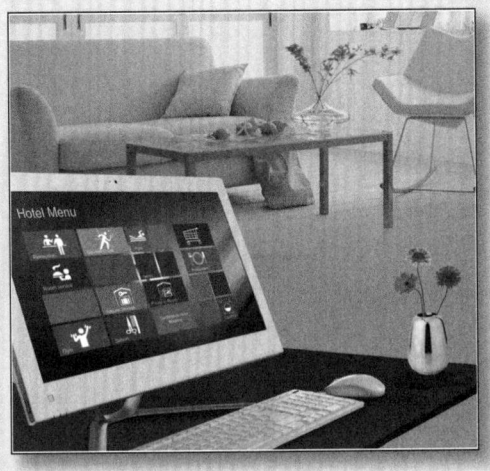

PART I

Computer Fundamentals

- Introduction to Computers

- Software Concepts and Operating Systems

- MS-Office

- Internet

- Introduction to DBMS

CHAPTER **1**

Introduction to Computers

LEARNING OBJECTIVES

After reading this chapter, you will be able to understand the following:

- Concepts of data, information, and data processing
- Components of a computer system
- Data entry, data output, and storage device
- Classification and characteristics of a computer system

The word 'computer' comes from the Latin word *compute*, which means 'to calculate'. People usually think that a computer is just a calculating device that can perform arithmetic operations at a high speed with great accuracy. However, more than 80 per cent of the work done by computers is non-arithmetic.

As discussed in Exhibit 1.1, in the hotel industry, computers are widely used, from hotel reservations to billing and inventory of raw materials. Through the Internet, customers can

---

EXHIBIT 1.1   Why Study Computers?

Today, computers have become an essential part of everybody's life. Computers are being used in almost every field, in all walks of life, and also in new areas of advancement. The development of improved computer technology and efficient software has made a dramatic impact on science, technology, management, education, entertainment, and all other spheres of life.

As hotel management students, why should we learn about computers? The answer is, if we have good knowledge about computers, we can manage hotels in a better and more professional manner.

In hotels, computers are essentially used by managers and subordinates to keep a record of reservation of guests, allocation of rooms, settlement of guests' bills, and arrangement of events and banquets for the guests. In addition, computers are used for ordering food and beverages and requesting for special amenities such as movies and Internet facility. They are also used for preparing various sales, management information system (MIS), and inventory reports for the management and the proprietors.

The best approach to learning computers is to first be aware of the fundamentals of computers and then learn their various applications, such as network handling and computer languages.

easily reserve hotel rooms, book a hotel's banquet facility, and also obtain detailed information on various hotels in a particular area, places of tourist interest, weather, directions, etc.

EVOLUTION OF COMPUTERS

In ancient times, people used various devices for computing. The devices and methods used for calculation were not perfect and lacked speed. Manual computing devices, primarily used for simple calculations, were later replaced by automated computing devices. Charles Babbage, a nineteenth century professor at Cambridge University, is regarded as the father of modern digital computers. In 1842, Babbage came out with a new idea of a completely automated analytical engine for performing basic mathematical functions at an average speed of 60 additions per minute. Babbage's efforts established a number of principles, which are fundamental to designing any digital computer.

DATA AND INFORMATION

A computer usually accepts input in the form of data. Data is the plural form of the Latin word *datum*, which means 'something given'. In computer science, the term data refers to numerical, alphabetical, and special characters (e.g., $, %, +, -, !, @, #), which are represented in ways that can be processed by a computer. The Oxford English Dictionary defines 'fact' as something that is known to have happened, something that is true, or something that exists. Thus, data is a representation of facts. A set of numbers, 9102225551212, could be an example of data.

Data is the raw material and information is the finished product (manipulation of raw facts). Information refers to data in a particular context, which helps us understand facts. The set of numbers mentioned earlier, as an example of data, would become information when one knows that +91(022)2555-1212 is the telephone number of a directory service in India, specifically Mumbai. It includes a country code 91, an area code 022, a telephone exchange 2555, and a number within the exchange 1212.

NUMBERING SYSTEM IN A COMPUTER

A computer is a digital system that stores and processes data in the form of binary digits—0s and 1s. A computer handles various types of data, which include numbers, alphabets, and even special characters. Different types of codes have been developed to represent data entered by users in a binary format. Computers use decimal and binary systems.

Decimal System

The decimal system uses 10 as a base to represent different values. In this system, 10 symbols are available for representing digits 0–9. The most common operations performed by decimal systems are addition, subtraction, multiplication, and division.

Binary System

The binary system uses 2 as a base to represent different values. In this system, two symbols, 0 and 1, are used. In computer terminology, 0 and 1 are known as bits. A bit is the smallest unit of information that is used in a computer system. The primary memory is usually specified in kB, MB, or GB as follows:

1 Nibble = 4 bits

1 Byte = 8 bits

1 Kilobyte (kB) = 1024 bytes (2^{10})

1 Megabyte (MB) = 1024 kB

1 Gigabyte (GB) = 1024 MB

1 Terabyte (TB) = 1024 GB

1 Petabyte (PB) = 1024 TB

1 Exabyte (EB) = 1024 PB

1 Zettabyte (ZB) = 1024 EB

Let us go through a few conversions:

To convert megabyte to kilobyte, multiply by 1024.

To convert kilobyte to megabyte, divide by 1024.

The aforementioned principle applies to other conversions as well.

To convert terabyte to megabyte, multiply twice by 1024.

To convert megabyte to terabyte, divide twice by 1024.

For example, if we wish to convert 12,664.35 kB to megabytes, we would divide it by 1024 to get 12.37 MB. If we wish to convert 5.23 GB to kilobytes, we would multiply 5.23 with 1024 to get 5355.52 MB. When this value is multiplied with 1024, we get 54,84052.48 kB.

SYSTEM

A system is a group of integrated parts that have a common purpose of achieving an objective. A computer comprises integrated components—input unit, output unit, storage unit, and central processing unit (CPU)—which work together to perform the steps necessary in a program. The input and output units cannot function if they do not receive signals from the CPU. In addition, the storage unit of the CPU alone does not have any utility. The usefulness of each unit is dependent on other units and is realized when all units are put together to form a system.

In general terms, processing is a series of actions or operations that converts input into useful output. In data processing, data is the input and information is the output. Hence, data processing is a series of actions or operations that converts data into information.

A data processing cycle consists of three basic steps (see Fig. 1.1).

Fig. 1.1   Basic block diagram of a data processing cycle

Input cycle   In this phase, data is prepared in a convenient form and on a medium most suitable for data entry into a processing machine. We input the data through various input devices.

Processing cycle   In this phase, the primary memory of the computer system can manipulate or combine the input data as per instructions. Processing is done automatically in accordance with a series of instructions called a program, which is stored in the primary memory of the computer system (discussed in the section on control unit).

Output cycle   Once data is processed, the results need to be produced in the most suitable form for the user. A user can view the result of the processed data with the help of a computer printer.

COMPONENTS OF A COMPUTER SYSTEM

A computer is an electronic device used for performing calculations and controlling operations that can be expressed in either logical or arithmetical form. A computer receives input in the

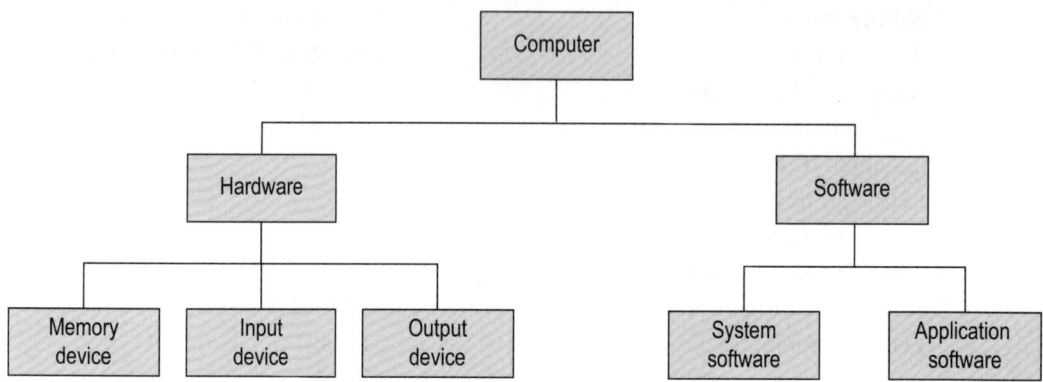

Fig. 1.2   Components of a computer

form of data, programs, and user replies. A program is a set of instructions that can be processed by a computer in either a serial or non-serial manner. A user reply is an input provided by the user in response to a question from the computer. A computer has a large information storage capacity, operates at high speed, and has 100 per cent accuracy. The applications of a computer are innumerable. Computers are seen as instruments for progress and their applications are visible in numerous areas such as education, hotels, tourism, industries, governments, entertainment, medicine, and scientific research work. Indeed, computers have left such an indelible impression on modern civilization that we cannot live without them in this information age.

Computers now come in a variety of sizes, shapes, and capabilities. However, all computers primarily have two components, as mentioned in Fig. 1.2:

- Hardware   • Software

Hardware   The tangible physical components of a computer system, which can be touched, seen, and felt, are known as hardware. Hardware units are used to input, store, and process data. In addition, they also help in displaying and storing the output for users. The basic hardware units are keyboard, mouse, CPU, and printer.

Software   It is a collection of data and information, using which a computer operates. A software is intangible, invisible, and of two types—application software and system software. In Chapter 2, we will discuss software in detail.

BASIC COMPUTER ORGANIZATION

The internal architecture of a computer usually differs from model to model though the basic organization remains similar for all computer systems. Figure 1.3 shows a block diagram of the organization of a basic computer.

The four units of a computer are input unit, storage unit, control unit, and output unit. Let us now discuss the functions of each of these in detail.

Input Unit

An input unit is an electromechanical device that accepts data from the user and translates it into a computer or machine-understandable language. Some input devices that are presently available are shown in Fig. 1.4 and are listed here:

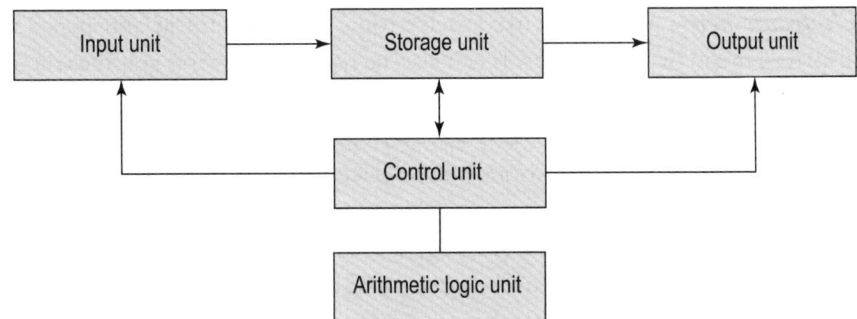

Fig. 1.3   Block diagram of a computer system

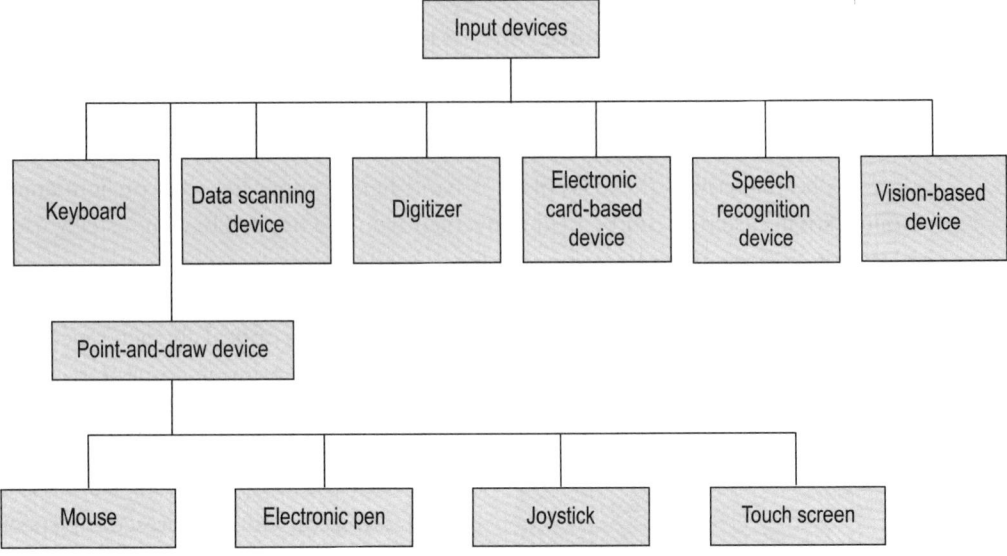

Fig. 1.4   Input devices of a computer

- Keyboard
- Data scanning device
- Electronic card-based device
- Vision-based device
- Point-and-draw device
- Digitizer
- Speech recognition device

Keyboard   The most commonly used keyboard layout is QWERTY (so named because the letters on the keyboard are arranged in that sequence), which allows data entry into a computer system by pressing a set of keys.

Point-and-draw device   The mouse, electronic pen, joystick, and touch screen are examples of point-and-draw devices.

Mouse   It is a small hand-held point-and-draw device and the most popular. The movements of the mouse are reproduced by a graphic cursor on the screen.

Joystick   In most joysticks, a button is provided on the top to select the option pointed to by the cursor. The button is clicked to make a selection. Joysticks are mainly used in video games, flight simulations, training simulators, industrial cranes, and remote controls.

Electronic pen   A user can use an electronic pen to point at and make a selection from a displayed menu or set of icons. It can also be used for drawing graphics directly on the screen.

Touch screen   A touch screen is the simplest user-friendly input device that helps a user choose from available options or icons displayed on a computer screen by using a finger. It is often used as an information kiosk at various public places for people to access stored information. For example, at airports and railway stations, kiosks provide information to passengers on lodging, distance to reach the place of lodging, tourist spots, and ticket-booking, besides automatic check-in of hotel guests and ordering of food using electronic menu cards.

Data scanning devices   Data scanning devices are used for directly entering data from source documents into a computer system. They are of many types. The most commonly used ones are explained here:

Image scanner   An image scanner converts paper documents (typed text, pictures, graphics, or handwritten material) into an electronic form that can be stored in a computer. Image scanners are available in various shapes and sizes. These include flatbed scanners and hand-held scanners. A flatbed scanner is just like a copier machine and consists of a box with a glass plate on the top and a lid that covers the glass plate. A hand-held scanner has a set of light emitting diodes within a small case that can be held conveniently in the hand. They are usually used when the volume of documents to be scanned is low.

Optical character recognition device   It is an optical scanner that is capable of detecting alphanumeric characters typed or printed on paper using an optical character recognition (OCR) font. OCR devices are used for voluminous work such as printing computerized bills, reading ZIP codes or PIN codes in postal services, and reading passenger tickets.

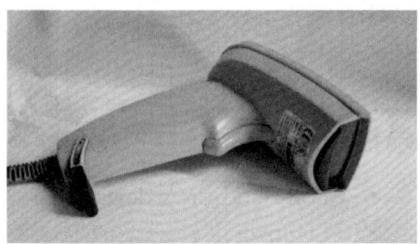

Fig. 1.5   Optical barcode reader

Optical barcode reader   An optical barcode reader (OBR, shown in Fig. 1.5) can scan a set of vertical bars of different widths representing specific data. It is used for reading tags in merchandise goods stores, medical stores, bookstores, and libraries.

Optical mark reader   An optical mark reader (OMR) is a special scanner used to recognize a type of mark made by a pen or pencil, and is specially used for checking answer sheets in examinations having multiple choice questions.

Magnetic ink character recognition device   A magnetic ink character recognition (MICR) device is generally used by the banking industry to directly read account numbers on cheques. A sample MICR cheque is shown in Fig. 1.6.

Digitizer   A digitizer is used for converting (digitizing) graphics, maps, pictures, and drawings into digital data form, which can be stored in random access

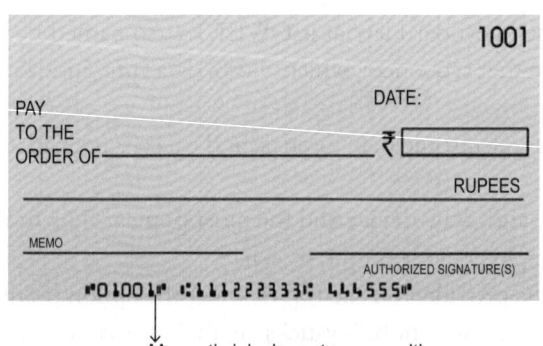

Fig. 1.6   MICR cheque

memory (RAM), and then displayed on the monitor. It can store data as x and y coordinates of a picture or drawing. It is used in computer-aided design (CAD) and geographical information systems (GIS).

Electronic card-based device   Electronic cards are small plastic cards having encoded data appropriate for a particular application. These are often issued by banks to customers to be used in automatic teller machines (ATMs).

Speech recognition device   Speech recognition devices are input devices that allow an individual to input data into a computer system by speaking to it.

Vision-based device   A vision-based device allows a computer to accept input by viewing an object. In this case, the input data is an object's shape and features in the form of an image. An example of a vision-based device is a computer with a digital camera.

Storage Unit

The main function of the storage unit, also known as memory, is to store data, either temporarily or permanently, in the computer system. The storage unit helps us retrieve the data for future use. The memory can be divided into parts as shown in Fig. 1.7.

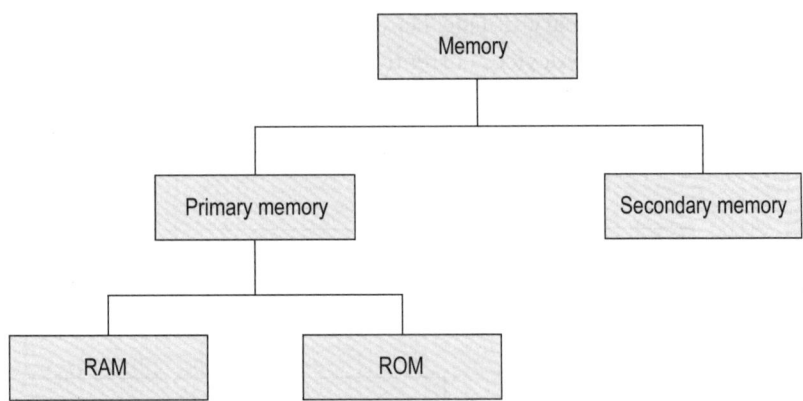

Fig. 1.7   Memory of a computer

Primary storage device   The primary storage of a computer is known as main memory. The main memory is used for holding parts of programs, instructions, data, intermediate results of processing, and results of the work presently processed by the computer. The storage capacity is limited and expensive. The primary storage of a computer system is presently made of semiconductor devices.

The main memory or the primary storage device is where we can store programs (instructions) and data to be used by the CPU. It usually consists of integrated circuits that are either in the motherboard or a circuit board adjacent to the motherboard. The motherboard is the main circuit board of a microcomputer and controls all other devices such as the display screen, keyboard, and disk drives. The main memory holds three types of content:

- Data for processing
- Programs for processing the data
- Information waiting to be sent to an output or a secondary storage device

The primary storage memory is of two types: random access memory (RAM) and read only memory (ROM).

RAM   The primary memory is built with volatile RAM chips. Volatile, here, means that it loses its contents as and when the power is switched off. It is also the main memory of the computer system, which stores the data temporarily, and can be accessed in any sequence.

ROM   This type of memory is permanently stored in the PC circuit and is non-volatile (i.e., it retains data even after the computer system has been switched off). All machine-level instructions are stored in the ROM and it is provided by the manufacturer. Data can be read from this memory but cannot be changed. In addition, the data has to be accessed sequentially, unlike RAM.

Read–write memory   Read–write memory is a computer memory that may be relatively easily written into as well as read from, unlike ROM. The term RAM is also used to describe writable memory. RAM refers to memory in which any location can be accessed in a constant amount of time.

Secondary storage device   Secondary storage devices have programs, instructions, data, and information pertaining to tasks that the computer system is currently not working on but may require for processing later. The secondary storage of a computer, also known as an auxiliary storage or non-volatile memory, has very high storage capacity. It is less expensive when compared to primary storage. Hard disks and DVD/CD-ROMs are the most commonly used secondary storage devices currently.

Figure 1.8 shows the classification of secondary storage devices.

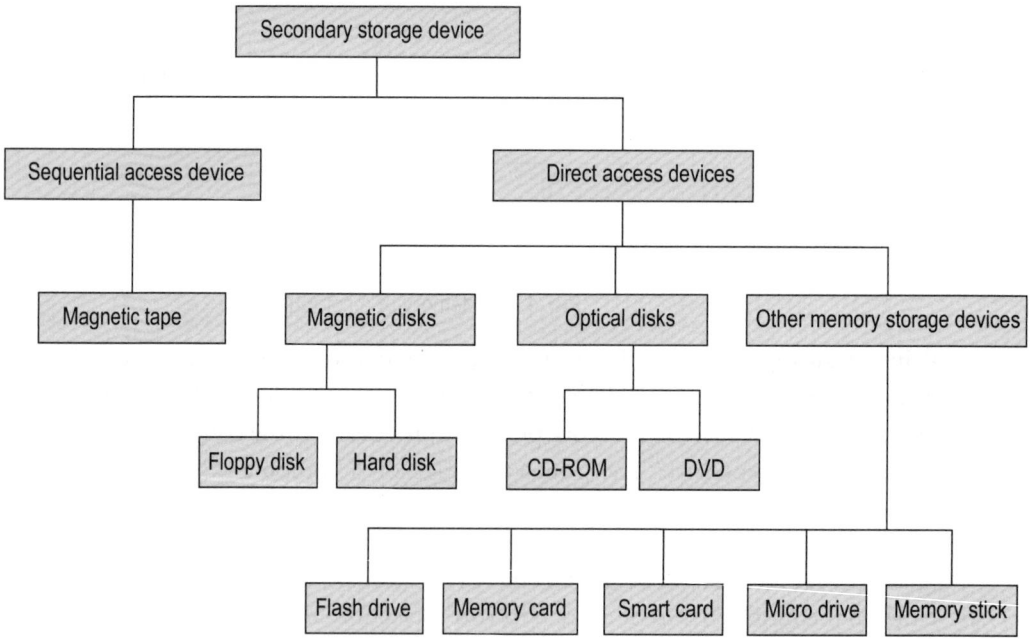

Fig. 1.8   Classification of secondary storage devices

Some of the secondary devices that are currently and frequently used are discussed here:

Magnetic disk   It is a type of storage device that has a layer of magnetic substance coated on a rigid or flexible surface. The drive is usually equipped with a read–write head assembly that can transfer data, represent it in the form of 0s and 1s, and convert it into a magnetic signal, which can later be stored on a medium.

Storage organization of a magnetic disk: The surface of a disk is divided into a number of invisible concentric circles called tracks. Each track is further divided into eight parts, known as sectors. This is shown in Fig. 1.9.

Hard disk: A hard disk is able to store a large amount of information within a small space (usually 40 GB). It is the primary online secondary storage device, made of rigid metal (frequently aluminium) platters, and comes in sizes ranging from 1–14 inches in diameter.

Optical disk   It is one of the most popular secondary storage devices, that can store a large amount of data within a restricted space. In this type of storage, signals are stored in the form of light. Binary digits 0 and 1 are converted into light information and stored in the read–write head assembly of the driver. While reading the disk, the bit patterns of 0s and 1s stored on the disk are retrieved.

Storage organization of optical disk: An optical disk has a long track starting at the outer edge and spiralling inwards to the centre. This spiral track is ideal for reading large blocks of sequential data such as audio and video (see Fig. 1.10). CD-ROM, CD-RW, and DVD are the commonly used types of optical disks. We will discuss CD-ROM and DVD disks now.

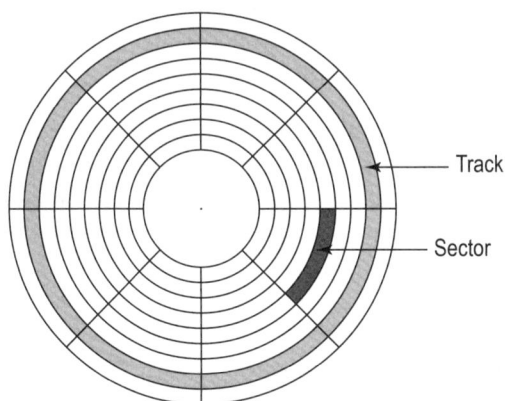

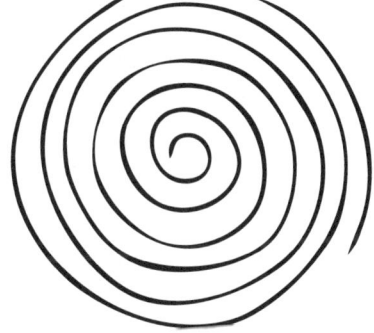

Fig. 1.9   Sector and track on a magnetic disk   **Fig. 1.10**   Track pattern on an optical disk

CD-ROM: CD-ROM stands for compact disk-read only memory. It is an optical disk impressed with a series of spiral pits on a flat surface. A standard CD-ROM can store between 650–700 MB of data.

DVD: DVD stands for digital versatile disk. A DVD can store 4.7–17 GB of data. It is used for storing high quality video and audio applications.

Other memory storage devices   We will now discuss other popular memory storage devices.

Flash drive (pen drive): A flash drive is a compact device the size of a pen (from which the name has been derived) that comes in various shapes and stylish designs. Nowadays, pen drives are smaller than pens. A pen drive is the most common way of transferring data from one computer to another. It is a plug-and-play device that can simply be plugged into a universal serial bus (USB) port of a computer. When the device is plugged into a computer, the computer automatically detects it as a removable drive. It does not require any battery, cable, or software and is easily compatible with most personal computers, desktops, and laptops.

Memory card: Like flash drives, flash memory cards are removable storage devices that are used in various kinds of electronic equipment. These cards are used in digital devices such as digital cameras and cell phones. Memory cards facilitate easy transfer of data from such devices to a computer, for storage in the hard disk or for further processing.

Smart card: A smart card, chip card, or integrated circuit card (ICC) is a pocket-sized card with embedded integrated circuits.

Micro drive: Micro drive is a brand name for a miniature, one-inch hard disk designed to fit into a compact flash slot.

Memory stick: A memory stick is a removable flash memory card. Memory stick digital data storage is designed to have a standard storage and is used to transfer media.

Control Unit

The CPU is the brain of a computer system. It is responsible for activating and controlling operations of other units of the computer system. In addition, it is responsible for processing data inside the computer system. The main functions of the CPU are as follows:

- Obtaining information from the memory
- Interpreting the code and then deciding what operations are to be performed
- Executing the given instructions
- Obtaining the result and storing it in the memory

These processing activities are readily carried out by a computer. A computer can accept input data and communicate the processed output to a large number of devices. A computer system is designed to facilitate, calculate, classify, sort, and summarize simple comparisons and then depending on the result, take a predetermined course of action.

The two basic components of a CPU are as follows:

- The control unit (CU) of a computer's CPU manages and coordinates with the entire computer system, including the input and output units. It obtains instructions from the program stored in the main memory, interprets these instructions, and issues signals with the help of which other units of the system run.
- The arithmetic logic unit (ALU) of a computer's CPU is the place where the actual execution of instructions takes place during data processing.

Output Unit

An output device is an electromechanical device that accepts data from the computer and translates it into a form that is fit to be used by users. Output devices produce output that are of two types:

Soft copy output   A soft copy output cannot be retrieved if not saved. Contents displayed on a monitor, words spoken by a voice response system, word documents, excel sheets, and PPTs are examples of soft copy output.

Hard copy output   A hard copy output is permanent as it is reproduced on paper or a tangible material that can be easily stored. It can thus be moved around. The output produced on paper by printers or plotters is an example of hard copy output.

The output devices, available now, are shown in Fig. 1.11.

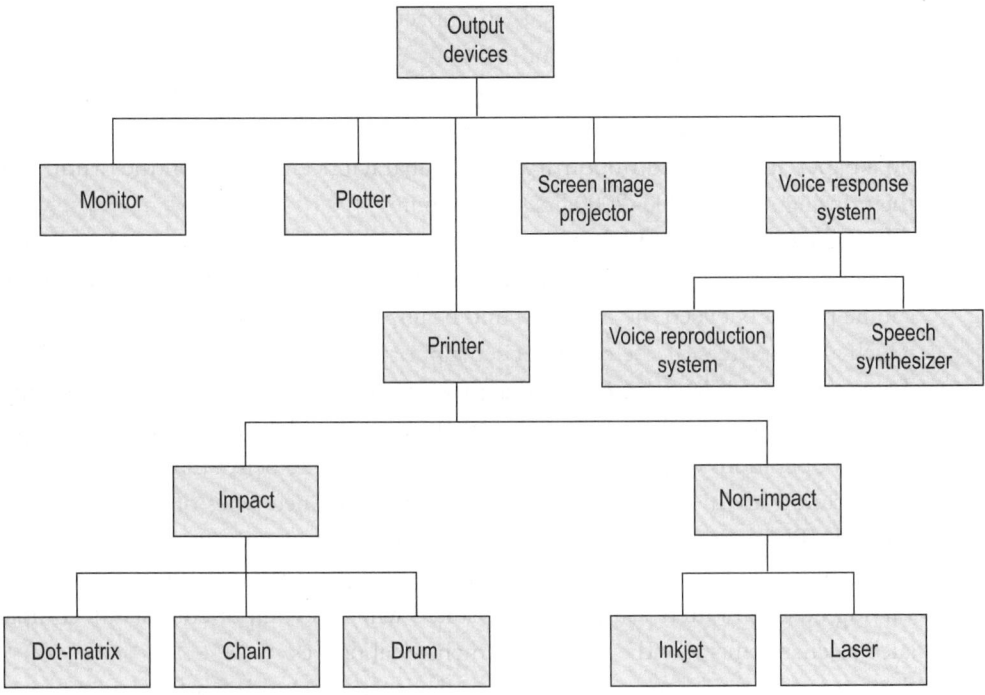

Fig. 1.11 Output devices of computer

Let us discuss the function of each of these in detail:

- Monitor
- Printer
- Plotter
- Screen image projector
- Voice response system

Monitor   It is the most popular output device for soft copy output. The output is displayed on a television-like screen. A monitor and a keyboard are usually linked with each other to form a video display terminal (VDT). The two basic types of monitors are cathode ray tube (CRT) and liquid crystal display (LCD) flat-panel. CRT monitors operate like television screens and are usually used in non-portable computer systems. LCD flat-panel monitors are thinner, lighter, and are used in desktops as well as portable computer systems.

Printer   This is a popular device for producing hard copy output from a computer system. Based on their working principles, printers are classified as impact and non-impact printers. Impact printers have mechanical contact with the printer head and paper, while non-impact printers print by spraying very small droplets of ink on the paper with the help of a nozzle. The classification of printers is shown in Fig. 1.11.

Dot-matrix printer   A dot-matrix printer is a character printer that forms characters and images as patterns of dots. It is an impact printer as it prints by hammering pins on an inked ribbon to leave an impression on the paper. It is considerably noisy and has slow printing speed (about 30–600 characters per second).

Inkjet printer   Inkjet printers are character printers that form characters and images by spraying drops of ink on the paper. They are non-impact printers as they print by spraying ink on the paper. They are less noisy than impact printers. The printing speed is comparatively slow (40–300 characters per second).

Drum printer A drum printer consists of a solid cylindrical drum with characters embossed on its surface in the form of circular bands. It is an impact printer as it prints by hammering on a paper and an inked ribbon against the characters embossed on the drum. It is noisy and the printing speed is 300–2000 lines per minute.

Chain/Band printer A line printer prints one line at a time. The chain/band printer has a set of hammers embossed in front of a chain/band in such a way that an inked ribbon and paper can be placed between the hammers and chain/band. It consists of a metallic chain/band on which all characters of the character set supported by the printer are embossed.

Laser printer It is a page printer that produces very high quality prints by forming characters and images with very tiny ink particles (as shown in Fig. 1.12). It is a non-impact printer that is quiet in operation. Its printing speed and print quality are better than other printers. However, laser printers are more expensive than the aforementioned types of printers.

Plotter Plotters are output devices especially used by professionals such as architects, engineers, city planners, and those who need to generate high precision, hard copy graphic output widely varying in size. Plotters are particularly suited when the output consists of complex graphics such as charts, maps, and even three-dimensional figures.

Screen image projector It is an output device used to project information from a computer onto a large screen so that it can be viewed by a group of people. It provides a temporary, soft copy output similar to a monitor (shown in Fig. 1.13).

Fig. 1.12 Laser printer

Fig. 1.13 Screen image projector

Voice response system In the section on input devices, we talked about a speech recognition system that allows a user to talk to a computer. On the other hand, a voice response system enables the computer to talk to a user. A voice response system has an audio-response device that produces audio output. The audio output is temporary and is a soft copy output. They are of two types:

- Voice reproduction system
- Speech synthesizer

CLASSIFICATION OF COMPUTERS

The classification of a computer is usually based on the following four categories:

- Purpose
- Technology
- Size and storage capacity
- Historical advancement

Based on the aforementioned criteria, the classification of computers is illustrated in Fig. 1.14.

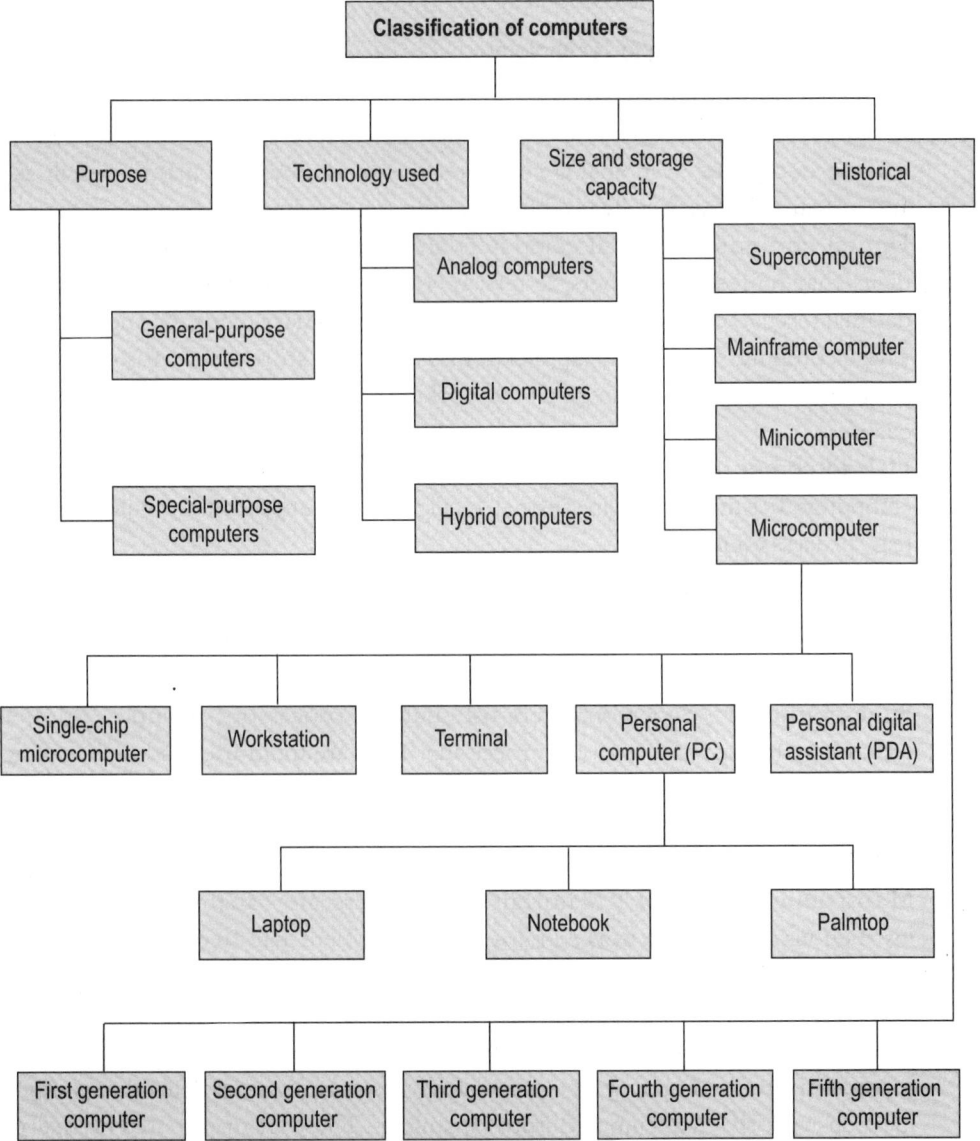

Fig. 1.14 Classification of computers

Classification According to Purpose

Computers are classified and manufactured as per the requirements and needs of users. Based on this, the classification is as follows:

General-purpose computer   Computers that follow instructions to meet general requirements such as analysis of sales and accounting, invoicing, inventory listing, management handling, and other tasks are called general-purpose computers.

Special-purpose computer   Computers that are designed from the start to perform special tasks such as weather forecasting and R&D, and have applications in science and space studies are called special-purpose computers.

Classification According to Technology

Computers can also be classified based on the technology involved in processing information, which could either be very simple or complex. Based on this, the classification is as follows:

Analog computer   An analog computer measures physical quantities such as temperature and pressure. It is especially used for scientific and engineering purposes.

Digital computer   Digital computers are general-purpose computers as they count the digits that represent numbers, letters, or other special symbols. A user may enter data in decimal or character form, which is converted into binary digits (0s and 1s) to store the data. These computers have many applications as they can store different sets of instructions and programs. We will discuss in detail the working and components of computers in subsequent chapters.

Hybrid computer   A hybrid computer system is a combination of both analog and digital computers. It is mainly used in artificial intelligence (robotics) and computer aided manufacturing (e.g., process control).

Classification According to Size

Sometimes, computers may have to process huge amounts of data at high speed. Thus, their sizes may vary as per the need and function. Based on this, their classification is as follows:

Supercomputer   Supercomputers use multiple processors, multiple ALUs, and superior technology for fast and complex processing. They are multi-user and multitasking computers, mainly used for complex scientific applications such as weather forecasting, engineering calculations, and data handling for large organizations or government agencies. Examples include CRAY, NEC 500, PARAM.

Mainframe computer   Mainframe computers are capable of executing billions of instructions and are used in a centralized location where many terminals (input/output devices) are connected with a single CPU as they allow thousands of users to work at a particular time. They are mainly used for railway and airline reservations, banking purposes, and in hotels, industries, and companies. IBM and DEC are the two major vendors of mainframe computers.

Minicomputer   Minicomputers are relatively faster and smaller than other computers. They are inexpensive with limited input–output capabilities. These are mainly used in small organizations. With respect to price, performance, and applications, minicomputers are more or less similar to microcomputers.

Microcomputer Microcomputers are the smallest category of computers and are designed with microprocessor (16-bit, 32-bit, and 64-bit) integrated circuits. They are used as personal computers (PCs). Microcomputers are further divided into five sub-groups.

Personal computer Personal computers are used for individual work. They can be found in hotels, homes, and small organizations. Personal computers (also known as desktops) are configured to have 16- or 32-bit processors (e.g., Intel Pentium IV), 32 MB or more RAM, a monitor (flat screen), a printer, 1–10 GB hard disk, a CD/DVD drive, special add-on card facilities (like a network interface card), and ports for connecting peripheral devices (such as mouse, pen drive, etc.).

PCs are further classified according to size, weight, and portability. They can be laptops, notebooks, and palmtops. Laptops are portable computers that are small, light weight, can be carried around easily, and are designed to operate with chargeable batteries that enable them to be used at places where there is no availability of an external power source. They can be easily used by placing them on an individual's lap and are hence called laptops. Palmtops are smaller in size and usually fit into our hands. Most laptops now have Wi-fi and Bluetooth connection facilities.

Fig. 1.15 Personal digital assistant

Courtesy: IDS

Personal digital assistant Personal digital assistants (PDAs) were initially introduced as a personal information manager (PIM). It is a pen-based hand-held mobile computer. PDAs now have *Wi-fi* and *Bluetooth* network capabilities. Figure 1.15 shows a personal digital assistant.

Workstation High-performance personal computers, called workstations, are applied in engineering and science (such as computer-aided engineering and application-specific integrated circuits). Workstations are mostly designed on reduced instructions set computing (RISC)-based processors.

Terminal Terminals are used to display information or accept input from a number of users. They are used in hotels to show the list of rooms available and to determine whether a guest can be checked in or not.

Single-chip microcomputer These computers are prepared on single chips that have microprocessors, a 64-byte read and write memory, 1–4Kbytes ROM and several single lines to connect to input/output.

Classification According to Generation

Computers have been in use since the 20th century. Over time, dramatic changes have been made to computer systems. Based on their advancement, computers are classified into five computer generations as follows:

First generation computer systems First generation computers used vacuum tubes to store data and programming instructions. Vacuum tubes consumed huge amounts of electricity, produced large quantities of heat, were relatively unreliable and bulky in size, and were prone to frequent hardware failure. Examples of first generation computers are electronic numerical integrator and calculator (ENIAC), electronic discrete variable automatic computer (EDVAC), electronic delay storage automatic calculator (EDSAC), UNIVersal Automatic Computer (UNIVAC, the first digital computer), and International Business Machines' IBM-701.

Second generation computer systems   John Bardeen, William Shockley, and Walter Brattain invented a new electronic switching device called a transistor. By using the transistor, computers became more reliable, powerful, smaller, and even cooler than first generation computers. In this era, high-level programming languages such as COBOL (COmmon Business Oriented Language), ALGOL (ALGOrithmic Language), and SNOBOL (StriNg Oriented symBOlic Language) were popularized.

Fig. 1.16   IC chip

Third generation computer systems   Jack St Clair Kilby and Robert Noyce invented integrated circuits (ICs), which had greater storing and instruction processing capacity compared to transistors. The IC technology, also known as *microelectronics* technology, integrates a large number of circuit components into a very small surface of silicon known as *chip* (see Fig. 1.16).

Fourth generation computer systems   In fourth generation computers, with large scale integration (LSI), it was possible to integrate thousands of electronic components into a single chip, and with very large scale integration (VLSI), a million components could be integrated into a chip. This progress led to the development of a *microprocessor*. During this period, high-speed computer networking (local area network, LAN and wide area network, WAN) and C programming language became popular. The revolution of the personal computer began from this generation.

Fifth generation computer systems   Ultra large scale integration (ULSI) superseded VLSI technology in the fifth generation, resulting in the production of microprocessor chips that had around ten million electronic components. Fifth generation computers have knowledge information processing systems and incorporate artificial intelligence. Computers based on human intelligence are self-learning systems that can store experiences and take decisions based on the information and logic stored in them. They can also process non-numerical information such as graphical representations and pictures. In the fifth generation, there has been a tremendous growth in computer networks.

CHARACTERISTICS OF COMPUTERS

Computers are unique when compared to other machines due to their multitasking capabilities.

The characteristics of a computer are as follows:

Speed   Many computers these days can perform hundreds of millions of processing operations in one second. Electric pulses travel at incredible speeds and since computers are electronic, their internal speeds are virtually instantaneous. The speed of a computer is generally measured in terms of million instructions per second (MIPS).

Accuracy   A computer performs all calculations with the same level of accuracy. We often refer to computer errors caused due to incorrect input data or unreliable programs as garbage-in-garbage-out (GIGO).

Storage   A computer can store voluminous amounts of data and information. The storage capacity of a computer is generally expressed in terms of megabyte (MB), gigabyte (GB), and terabyte (TB).

Versatility   Computers can handle a variety of applications and jobs, and are used in various fields. They are capable of performing tasks that can be reduced to a finite series of logical steps.

Diligence   Human beings can get tired and bored from doing calculations repeatedly. However, computers can efficiently perform tasks with high accuracy without feeling tired or bored.

Automation   Only minimum human intervention is required after the instructions and data are fed into the system. The computer follows all the instructions up to the last one.

Programmability   A computer is programmable. A program is a list of instructions informing the computer what it must do. Thus, computers are general-purpose machines.

Cost-effectiveness   Computers help in the reduction of manual work and hence, lower labour costs. The cost involved is a one-time investment of buying the system while the running costs are negligible.

LIMITATIONS OF COMPUTERS

Even machines with multi-function capabilities are not perfect. Though computers are a boon to mankind, they have some limitations. These include the following:

No IQ   A computer does not possess intelligence of its own. Its intelligence quotient (IQ) level is zero, at least till date. It has to be instructed on what it must do and in what sequence. Hence, only a user determines what tasks a computer will perform; the computer cannot take decisions on its own in this regard.

Lack of feeling   A computer is devoid of emotion. Since it is a machine, it neither has feelings nor instincts. Although human beings have succeeded in building a memory for a computer, it does not possess a human heart. We human beings can make certain judgements in our day-to-day life based on our feelings, taste, knowledge, and experiences. A computer, on the other hand, cannot make such judgements on its own. It can only make judgements based on instructions given to it in the form of programs that are written by us, human beings.

TYPES OF INFORMATION

The information need of an individual or a group is the desire to locate or obtain information to satisfy an individual or group's desires. They are of different types and can be classified as follows.

International information   An individual may be interested in knowing what is happening across the globe or what has happened in a neighbouring country in the past. For example, fluctuations in currency exchange rates differ among countries. Sometimes, a change of government in a country could result in a change in currency rates. This is an example of international information that may affect the interest of a very large number of people.

National information   Information pertaining to a particular country that might be of importance to its citizens is called national information. Newspapers, radio, and television provide data that is important from a national point of view. The share market rates of a country and prices of manufactured products are types of national information. It consists of important data required by different organizations in the country.

Corporate information   The management of a corporation may like to keep its employees informed on the various activities of the organization. The minutes of a meeting sent to an employee who has been on an official trip abroad is an example of corporate information.

Department information   Departments work by conveying progress and other information to their headquarters and sister departments for the successful running of their organization.

Departments benefit by sharing important information, which is referred to as department information. The total revenue generated by a hotel and the break-up of revenue generated by each department is an example of department information.

Individual information Information about an individual, detailing his/her personal achievements and qualifications, is called individual information. An organization stores employee information such as salary details and residential addresses. This data is of use to the person as well as the organization. This is an example of individual information.

SUMMARY

The chapter begins with a brief discussion on computers, data, information, and processing. The evolution of computers over the years has been explained. The chapter focuses on the various components of a computer system, input and output devices, their functions, as well as storage devices. The chapter also deals with the various types of computers, based on their classification, characteristics, and the technology used.

KEY TERMS

Analog computer This is a computer that measures physical quantities such as temperature and pressure.

Arithmetic logic unit (ALU) This is a unit of a computer that performs all the mathematical and logical operations. The ALU is part of the central processing unit.

Bluetooth It is an industrial specification for short-range wireless connectivity using globally unlicenced short-range radio frequency. It provides a way of establishing wireless connectivity between devices such as PCs, laptops, printers, mobile phones, and digital cameras for exchange of information.

Cathode ray tube (CRT) It is an electronic tube with a TV-like screen on which information is displayed.

CD-ROM It stands for compact disk-read only memory. A CD-ROM disk is an enormous optical storage device. It is used for archiving read-only data. It has a storage capacity of 650 MB. Newer disks have a storage of 700 MB.

Central processing unit (CPU) The control unit and arithmetic logic unit of a computer system together constitute the CPU of a system. It carries out all the calculations and comparisons performed by the computer. It is also responsible for activating and controlling the operations of other units of the computer.

Chain/Band printer These printers print by striking a set of hammers on an inked ribbon and paper placed against a metallic chain. All characters of the character set supported by the printer are embossed on the metallic chain.

Computer A computer is an electronic device that can accept and store input data, process it, and produce output by interpreting and executing programmed instructions.

Computer-aided design (CAD) It refers to the use of computers in automatic design and drafting operations.

Computer-aided manufacturing (CAM) It is the use of computers to automate manufacturing operations.

Computer system The various components (input–output, storage devices, CPU) of a computer that are integrated together to perform the steps in a program are together called a computer system.

Control unit (CU) The unit of a computer system that manages and coordinates the operations of all the other components is called the control unit.

Data processing system It is a system that accomplishes data processing and includes the necessary resources (procedures and devices) needed to process data.

Digital computer This is a computer that works with discrete quantities and uses numbers to stimulate real-time processes.

Digitizer It is an input device used to convert (digitize) pictures, maps, and drawings into digital form. It is mainly used in geographical information systems (GIS).

Dot-matrix printer It is a character printer that prints characters and images as a pattern of dots.

Drum printer This is a line printer that prints characters by striking a set of hammers on an inked ribbon and paper placed against a solid cylindrical drum with characters embossed on its surface in the form of circular bands.

DVD The digital video or versatile disk is primarily designed to store and distribute movies and videos.

Electronic card-based reader It is an input device normally connected to a computer and is used to read data encoded on an electronic card to transfer it to the computer for further processing.

Electronic delay storage automatic calculator (EDSAC)
Professor Maurice Wilkes, a British scientist at Cambridge University, developed the EDSAC. In this machine, addition and multiplication took 1500 microseconds and 4000 microseconds, respectively.

Electronic discrete variable automatic computer (EDVAC)   It was used to store programs in binary concepts, as a string of binary numbers in the memory.

Electronic numerical integrator and calculator (ENIAC)   It was the first all-electronic computer developed for military purposes. It was mainly used for solving ballistics-related problems.

Fifth generation computer system   These computers have knowledgeable information processing systems, network facilities, and can process non-numerical information such as graphs and pictures.

First generation computer system   Computers that used vacuum tubes and were programmed in an assembly language are referred to as first generation computers.

Flash drive   A flash drive is a compact device that comes in various shapes and enables easy transport of data from one computer to another.

Fourth generation computer system   This system is used in large scale integrated circuits, semiconductor memories, high-level languages, and operating systems.

Garbage-in-garbage-out (GIGO)   It refers to computer errors caused due to incorrect input data or unreliable programs.

Gigabyte (GB)   One GB is equal to 1,07,37,41,824 (2^{30}) bytes in a computer.

Hard disk   A magnetic disk's storage is made of rigid metal (mostly aluminium). The hard disk plotter comes in diameter sizes ranging from 1 to 14 inches.

Hybrid computer   It is a combination of analog and digital computers.

Impact printer   It is a printer where there is mechanical contact between the print head and paper. The print head strikes the individual character on the ink ribbon and transfers the same to the paper

Information   Processed data that is obtained as the output of data processing is called information. Users use it to achieve specific purposes.

Inkjet printer   It is a character printer that forms characters and images by spraying small drops of ink on the paper.

Input device   It is used to enter data and instructions into a computer. Examples include a keyboard and mouse.

Integrated circuit (IC)   It consists of several electronic components such as transistors, resistors, and capacitors on a single chip of silicon. They eliminate the need for wired interconnections between components.

Joystick   It is an input device that serves as an effective pointing device for application in video games, flight simulations, and industrial robots.

Keyboard   It is an input device that enables data entry into a computer by pressing a set of keys, which are neatly mounted on a board connected to the computer system.

Kilobyte (kB)   One kB is equal to 1024 (2^{10}) bytes in a computer.

Laptop   It is a personal computer that is portable, light weight, works on chargeable batteries, and can be comfortably used by placing it on an individual's lap.

Laser printer   It is a page printer that produces very high quality output of characters and images using very tiny ink particles. It uses a combination of laser beams and electrophotographic techniques to create printed outputs.

Liquid crystal display (LCD)   LCD flat-panel monitors are thin, light, and are commonly used with portable computer systems like notebook computers.

Local area network (LAN)   It connects a computer with other peripheral devices within a limited geographical area of a few kilometres.

Magnetic disk   It is the most popular direct-access secondary storage device. A magnetic disk is a thin circular plate/platter made of metal or plastic and coated on both sides with a recording material, which is magnetized.

Magnetic ink character recognition (MICR)   This technology is used in the banking industry for faster processing of large volume of cheques. This technology also ensures accuracy of data entry as most of the information on the cheque is pre-printed and directly fed into the computer.

Mainframe computer   It is a computer system that is mainly used for processing of bulk data. It is used in environments where a large number of users need to share a common computing facility such as in research groups, educational institutions, and engineering firms.

Megabyte (MB)   One MB is equal to 10,48,576 (2^{20}) bytes in a computer.

Memory card   Flash memory-based cards, usually used in digital cameras and cell phones, are removable storage devices.

Microcomputer   It is the smallest category of computers with a microprocessor and integrated circuits such as RAM, ROM, and other input–output interface chips.

Microprocessor   It is an IC chip that contains all the circuits needed to perform arithmetic, logic, and control functions (core activity of all computers).

Minicomputer   It is small, inexpensive, and uses limited input–output devices.

MIPS   MIPS stands for million instructions per second.

Monitor   It is a device for producing soft copy output. It displays the output on a screen.

Mouse   It is a small, hand-held input device that serves as an effective point-and-draw device.

Optical barcode reader   It is an input device that is able to interpret combinations of marks (bars) that represent data.

Optical character recognition (OCR) device   It is an input device equipped with a character-recognition software and is used to input text documents and store them in a form suitable for word processing.

Optical mark reader (OMR)   It is an input device that is capable of recognizing a pre-specified type of mark made by pencil or pen.

Output device   In a computer system, an output device is used to supply information and results of computation to the outside world. Monitors and printers are examples of such devices.

Palmtop computer   It is a battery-operated computer with personal assistants tools. It can easily fit into a pocket and a user can operate it by keeping it on the palm.

Personal computer (PC)   It is a small and inexpensive computer used for personal work, entertainment, education, and hobbies.

Personal digital assistant (PDA)   Also known as pocket PC, the PDA is a pocket-sized computer having facilities such as calendar, calculator, notepad, Internet access, spreadsheets, word processing, and other application packages.

Plotter   A plotter is an output device frequently used by architects, engineers, and other professionals who need to routinely generate high-precision, hard-copy, graphic output of varying sizes.

Point-and-draw device   Such a device can be used to rapidly point to and select a particular graphic icon or menu item from multiple options displayed on the screen.

Primary storage   It is a temporary storage area built into the computer hardware where instructions and data of a program reside, mainly when the program is being executed.

Printer   It is an output device used to produce hard copy output.

Random access memory (RAM)   It is the memory in which the time to retrieve stored information is independent of the address where it is stored.

Read only memory (ROM)   It is a non-volatile memory chip where data is stored permanently and cannot be altered by the programmer.

Scanner   It is an input device used for direct data entry into a computer from source documents.

Secondary storage device   It is a storage device that supplements the main memory of a computer. It is a part of the computer's memory that is non-volatile and has low cost per bit stored, though its operating speed is generally far lower than that of the primary memory.

Second generation computer system   This system used transistors in the CPU and high-level languages such as FORTAN and COBOL.

Single-chip microcomputer   This computer system is designed on a single-chip microprocessor and connected to various input and output devices through a single line.

Software   It is a set of computer programs, procedures, and associated documents (flowcharts, manuals, etc.) related to the effective operation of a computer system.

Speech synthesizer   It is a voice response system that converts text information into a spoken sentence.

Terabyte (TB)   One TB is equal to about one trillion (10^{12}) bytes.

Terminal   Computer systems that are used for displaying or accepting information from a number of users are called terminals.

Third generation computer system   This system used integrated circuits in the CPU, high-speed magnetic core main memories, powerful high-level languages, and time-sharing operating systems.

Touch screen   It is a simple, intuitive, and easy-to-learn input device. It enables users to choose from available options by simply touching the desired icon or menu item displayed on a touch-sensitive computer screen, with their finger.

Transistor   It is a controlled electronic switch fabricated using a semiconductor. It is extensively used for designing various electronic devices.

Universal automatic computer (UNIVAC)   It was the first digital computer. It marked the arrival of commercially available digital computers for business and scientific applications.

Vacuum tube   It is a fragile glass device that uses filaments as a source to control and amplify electronic signals.

Voice reproduction system   It reproduces audio output by selecting a suitable audio output from various sets of pre-recorded audio responses.

Wide area network (WAN)   It is a computer network that interconnects computers spread over a large geographical area. It may also enable LANs to communicate with each other. This type of network operates nationwide or worldwide. The transmission mediums used are normally public systems such as telephone lines and satellite links.

Workstation   It is a powerful desktop computer designed to meet the needs of engineers and architects. It has greater processing power, larger storage, and better graphic display facility when compared to a normal PC.

REFERENCES

Arora, A. and S. Bansal, *Computer Fundamentals*, First edition, Excel Books, New Delhi, 2000.

Balagurusamy, E., *Fundamentals of Computers*, Tata McGraw Hill, New Delhi, 2009.

Gupta, S. and S. Gupta, *Computer Aided Management*, Excel Books, New Delhi, 2004.

Jain, S., *'O' Level Made Simple*, BPB Publications, New Delhi, 1999.

Mukherjee, P. and S. Bandhopadhay, *Introduction to Computer Science*, Vol. I, Deep Prakashan, Kolkata, 2001.

Sinha, P.K. and P. Sinha, *Computer Fundamentals*, Fourth edition, BPB Publications, New Delhi, 2009.

Web References

http://1.bp.blogspot.com/_D3TOTHZeE9Q/TDa2RXlbm2I/AAAAAAAAABs/yPQX-T7VsjI/s1600/computer-block-diagram.jpg&w=419&h=289&ei=BeywTpu5CleGrA ecraRo&zoom=1&iact=rc&dur=230&sig=117970912863622786612&page=1&tbnh=99&tbnw=143&start=0&ndsp=9&ved=1t:429,r:4,s:0&tx=68&ty=48, last accessed on 2 November 2011.

http://3.bp.blogspot.com/-MxnFZY0IZm0/Tf25MRr9TbI/AAAAAAAAAE4/ulcL-kXTNdY/s1600/inkjet-printer.jpg&w=347&h=374&ei=JgOxTt20Hc_QrQefr6lm&zoom=1&iact=rc&dur=551&sig=117455315472194794439&page=1&tbnh=105&tbnw=91&start=0&ndsp=10&ved=1t:429,r:1,s:0&tx=27&ty=40, last accessed on 2 November 2011.

http://codeidol.com/hardware/a-certification/Storing-Data/Working-with-Disk-Storage/&docid=se2mctEnyE0P9M&imgurlt, last accessed on 9 November 2011.

http://freebarcodefonts.dobsonsw.com/images/code128bar.jpg&w=916&h=316&ei=IgGxTuGPHsbsrAf1rYw2&zoom=1&iact=rc&dur=541&sig=117455315472194794439&page=1&tbnh=51&tbnw=148&start=0&ndsp=10&ved=1t:429,r:2,s:0&tx=112&ty=32, last accessed on 2 November 2011.

http://images.itreviews.com/beta.itreviews.com/photos/hardware/h1376.jpg&w=358&h=400&ei=k_-wTrKOAYjYrQf_qcxk&zoom=1&iact=hc&vpx=290&vpy=88&dur=2203&hovh=237&hovw=212&tx=146&ty=210&sig=117455315472194794439&page=1&tbnh=115&tbnw=109&start=0&ndsp=5&ved=1t:429,r:2,s:0, last accessed on 2 November 2011.

http://images.techtree.com/ttimages/story/wheel_mouse_optical.jpg&w=450&h=393&ei=DP2wTp_oAoLyrQfL7-BL&zoom=1&iact=rc&dur=190&sig=117455315472194794439&page=1&tbnh=101&tbnw=116&start=0&ndsp=10&ved=1t:429,r:4,s:0&tx=41&ty=57, last accessed on 2 November 2011.

http://ityard.blogspot.com/2010/11/what-is-plotter.html&docid=Vm9IauQrsAVLKM&imgurl9, last accessed on 27 June 2012.

http://microsoftergonomickeyboard.info/wp-content/uploads/2011/03/microsoft-wireless-ergonomic-keyboard.jpg&w=1000&h=484&ei=E-2wTsOtGom0rAfDvbBp&zoom=1&iact=rc&dur=591&sig=117970912863622786612&page=1&tbnh=89&tbnw=204&start=0&ndsp=6&ved=1t:429,r:2,s:0&tx=50&ty=63, last accessed on 2 November 2011.

http://nuclearstreet.com/nuclear_power_industry_news/b/nuclear_power_news/archive/2009/04/03/cray.aspx&docid=9iDFxDCU8s3wvM&imgurl, last accessed on 2 November 2011.

http://ocrscanners.net/ocr-handwriting-software-read-this-before-you-buy/, last accessed on 27 June 2012.

http://photo-dict.faqs.org/phrase/4285/digitron-vaccum-tube.html, last accessed on 2 November 2011.

http://rocky.digikey.com/weblib/Zilog/Web%252520Photos/Z-18-DIP.jpg&w=640&h=640&ei=oOuwTpXWGIXWrQejsqhx&zoom=1&iact=rc&dur=0&sig=117970912863622786612&page=1&tbnh=106&tbnw=169&start=0&ndsp=10&ved=1t:429,r:2,s:0&tx=92&ty=58, last accessed on 2 November 2011.

http://static.howstuffworks.com/gif/laser-printer4.jpg&w=400&h=290&ei=wQOxTuChN4nTrQfw7bE8&zoom=1&iact=rc&dur=240&sig=117455315472194794439&page=1&tbnh=103&tbnw=167&start=0&ndsp=10&ved=1t:429,r:3,s:0&tx=92&ty=52, last accessed on 2 November 2011.

http://steelbridgemedia.com/desktop-vs-mobile/&docid=Id-eUMICiaj1JM&imgurl, last accessed on 28 June 2012.

http://the-gadgeteer.com/assets/ipen-r10.jpg&w=498&h=375&ei=Fv6wTriYBY-HrAeD2_k4&zoom=1&iact=rc&dur=140&sig=117455315472194794439&page=1&tbnh=100&tbnw=136&start=0&ndsp=10&ved=1t:429,r:0,s:0&tx=70&ty=72, last accessed on 2 November 2011.

http://www.bbc.co.uk/scotland/learning/bitesize/standard/computing/images/magnetic_ink_cheque.gif&w=546&h=280&ei=J7g3T6u8CIm0rAePI5nWBQ&zoom=1&iact=rc&dur=220&sig=108666040211155976721&page=2&tbnh=102&tbnw=198&start=18&ndsp=22&ved=1t:429,r:15,s:18&tx=41&ty=57, last accessed on 12 February 2012.

http://www.bridgat.com/files/Supply_Optical_Mark_Reader_Scanner_2.jpg&imgrefurl%3D584%26tbm%3Disch&um=1&itbs, last accessed on 2 November 2011.

http://www.businesscomputers.com/touch_screen_monitor.htm&docid=suAk_WGE2oYsHM&imgurl, last accessed on 21 November 2011.

http://www.cyberindian.net/wp-content/uploads/xerox-phaser-3435-laserprinter.jpg&w=450&h=396&ei=LwaxTqOYFYrlrQehx5RT&zoom=1&iact=rc&dur=591&sig=117455315472194794439&page=1&tbnh=101&tbnw=115&start=0&ndsp=5&ved=1t:

429,r:0,s:0&tx=21&ty=60, last accessed on 2 November 2011.

http://www.givoe.com/en/service/software-solution/Restaurant-Digital-Menu-System-Solution-on-Tablet-PC-Menu-X.html, last accessed on 12 February 2012.

http://www.greatcanadianopportunity.ca/2011/02/how-smartphones-stack-up-to-the-pcs-of-yesteryear/&docid, last accessed on 2 November 2011.

http://www.insidemind.net/wp-content/uploads/2010/03/ipaq-pda-pocket-pc.jpg&w=300&h=300&ei=L-KwTqOCLYTlrAe9roU3&zoom=1&iact=rc&dur=480&sig=117970912863622786612&page=1&tbnh=120&tbnw=120&start=0&ndsp=9&ved=1t:429,r:0,s:0&tx=79&ty=52, last accessed on 2 November 2011.

http://www.ixbt.com/dvd/infocus/x1/projector.jpg&w=600&h=304&ei=hwaxTs-LE8uxrAeHtY1p&zoom=1&iact=rc&dur=190&sig=117455315472194794439&page=2&tbnh=68&tbnw=135&start=5&ndsp=10&ved=1t:429,r:7,s:5&tx=70&ty=34, last accessed on 2 November 2011.

http://www.karbosguide.com/images/4c_045.gif&w=287&h=272&ei=Hnl6T4bME4nJrQebplTHCA&zoom=1&iact=rc&dur=252&sig=108666040211155976721&page=4&tbnh=152&tbnw=160&start=68&ndsp=24&ved=1t:429,r:10,s:68&tx=98&ty=115, last accessed on 14 February 2012.

http://www.netguruonline.com/wp-content/uploads/2011/07/Portable-Handheld-Scanner.jpg&w=450&h=318&ei=GgCxTpz7GJGGrAeY0Olm&zoom=1&iact=rc&dur=190&sig=11745

5315472194794439&page=1&tbnh=93&tbnw=131&start=0&ndsp=10&ved=1t:429,r:3,s:0&tx=36&ty=63, last accessed on 2 November 2011.

http://www.npcil.nic.in/hrdbarc/htmldocs/exam_instructions.htm&docid=_yBvpZC2pjGnAM&imgurl, last accessed on 2 November 2011.

http://www.onlinemca.com/mca_course/kurukshetra_university/semester5/computergraphics/img_computer_graphics/joystick.jpg&w=402&h=591&ei=uf2wTvHOMYeyrAel4olk&zoom=1&iact=hc&vpx=493&vpy=51&dur=1743&hovh=272&hovw=185&tx=120&ty=225&sig=117455315472194794439&page=1&tbnh=131&tbnw=89&start=0&ndsp=9&ved=1t:429,r:3,s:0, last accessed on 2 November 2011.

http://www.smartconsumerz.com/images/Panasonic-dot-matrix-printer.jpg&w=300&h=244&ei=yAKxTtuEJsPlrQfU2ZxE&zoom=1&iact=rc&dur=761&sig=117455315472194794439&page=1&tbnh=103&tbnw=127&start=0&ndsp=10&ved=1t:429,r:2,s:0&tx=83&ty=71, last accessed on 2 November 2011.

http://www.tollyandhra.com/2011/08/features-of-pen-drive.html&docid=C6G2ybtD04PUqM&imgurl9&start, last accessed on 14 February 2012

http://www.vcu.edu/vcu/ucc/tour.htm, last accessed on 2 November 2011.

www.infoimpact.com/IQBook/Ch2-Book-Defining_Info_Quality.pdf, last accessed on 21 November 2011.

EXERCISES

Concept Review Questions

1. What do the terms 'data' and 'information' mean?
2. List and explain some important characteristics of a computer.
3. In computer terminology, what does the term 'generation' mean? How many generations have been there till date?
4. Expand the following abbreviations used in computer terminology: ENIAC, EDVAC, EDSAC, UNIVAC, IBM
5. What is an IC? What are the advantages of IC over transistor technology?
6. What is a laptop?
7. What are the five basic operations performed by any computer system?
8. Draw a block diagram to illustrate the basic organization of a computer system and explain the functions of the various units.
9. What is an input device? Name some commonly used input devices.
10. Differentiate between impact and non-impact printers. Write the features of each printer.
11. Differentiate between the characteristics of primary and secondary storage systems of a computer.
12. Expand the following abbreviations:
    (a) GIGO   (d) LCD   (f) OCR
    (b) OMR   (e) MICR   (g) DVD
    (c) CPU
13. Write down some similarities between a human being and a computer.
14. What operating systems do workstations normally use?
15. Why is an electronic pen sometimes preferred over a mouse as an input device?

Multiple Choice Questions

1. The father of the modern digital computer is
   (a) Howar A. Aiken   (c) Charles Babbage
   (b) Rober Noyce   (d) John Bardeen
2. A very small surface of silicon is known as a
   (a) transistor   (c) circuit
   (b) chip   (d) vacuum tube
3. The brain of a computer system is
   (a) the CPU   (c) a magnetic disc

(b) an input device
(d) information

4. An example of a point-and-draw device is
   (a) printer
   (b) keyboard
   (c) scanner
   (d) mouse

5. Which of the following is a non-impact printer?
   (a) Laser printer
   (b) Drum printer
   (c) Dot-matrix printer
   (d) All of these

6. CD-ROM stands for
   (a) casual disk for read and write
   (b) compact disk-read only memory
   (c) common drive-ROM
   (d) none of these

7. Data is the raw material for
   (a) a computer
   (b) a user
   (c) information
   (d) storage devices

Project Work

1. You are setting up a kiosk to facilitate five desktop publishing (DTP) operators. What kinds of hardware would you consider purchasing and why?

2. A book consisting of 500 pages, each page having roughly 250 characters (including spaces), is to be stored on a secondary storage device. What should the capacity of such a device be?

3. With the rapid advancements in handwriting-recognition software, do you think the keyboard will continue to be a preferred input device for generation next? Which alternative—speech-recognition or handwriting-recognition devices—do you think has a better chance of replacing the keyboard as a primary device for the input of text?

4. Convert the following:
   (a) 12,64,735 kB to GB
   (b) 4096 TB to MB
   (c) 24,93,668 PB to GB
   (d) 12,268 TB to GB

CASE STUDY

Nancy Singh visited Melting Pot, a restaurant at Prince Residency hotel. She was happy to find a digital menu there, and noticed at each table there was a tablet PC to order food using the digital menu system. In addition to the price, Nancy was able to get information such as the ingredients used to prepare the dish, size of a portion, calorific value, and a photograph of the plated food item, with the help of the digital menu. Though there was no waiter in the restaurant, Nancy felt comfortable as she did not have to wait for somebody to take her order. The tablet does not force customers to order food without giving them time to think. Since the order is routed directly from the customer to the main server, billing errors are also avoided. The kitchen staff can start preparing the dish as soon as they receive the order, rather than depending on the waiter to give the kitchen order ticket (KOT).

The *smart search* option on a digital menu, mostly offered as an application on the tablet PC, gives the diner a visual tour of the dishes with pictures, wine pairings, and brief descriptions on the ingredients used. This enhances the customers' experience and tempts them to try out new dishes.

The digital menu, called *Titbit*, has been developed by Prince Residency hotel. Order customization by guests and zero waiter intervention in ordering have become a reality now.

Guests are greeted with a welcome message when they come in. They can place orders, request for bills, and even call for a waiter if necessary.

In addition, guests can also read online newspapers, check their e-mails, browse the Internet, chat online, play short games, and try out many other features that would make the services in a restaurant more appealing. This innovative product has been developed specifically for restaurants and coffee shops, thus ensuring that customers are served better and faster.

(a) As a new generation hotel manager, will you go in for a digital tablet menu? Justify your answer.
(b) Will conventional menu cards become extinct?
(c) Will a fine-dining restaurant convert to a waiter-less restaurant and use tablet PCs? Justify your answer.

Answers to Multiple Choice Questions

1. (c)   2. (b)   3. (a)   4. (d)   5. (a)   6. (b)   7. (c)

CHAPTER 2

Software Concepts and Operating Systems

LEARNING OBJECTIVES

After reading this chapter, you will be able to understand the following:

- Concept and classification of software
- Concept of an operating system
- Functions and classification of operating systems
- Introduction to DOS, Windows, and Linux
- Programming languages

The terms *hardware* and *software* are frequently used in connection with computers. Hardware refers to the physical devices of a computer system. Input, storage, processing, control, and output devices are hardware. In Chapter 1, we learnt about the hardware of computer systems. The term *software* will be introduced in this chapter and discussed at length in the chapters to follow.

CONCEPT OF SOFTWARE

Although capable of performing many useful functions, computer hardware does not work by itself. Instructions need to be given to it. The kind of instructions that the hardware understands are primitive tasks such as moving data in or out of the memory and adding numbers. For more complex operations, the instructions are grouped together. A sequence of a finite number of instructions is called a computer program (discussed in the next section).

EXHIBIT 2.1 Basic Computer Operation

Kapil joined the front office department of a hotel at a very young age. He had a pleasing personality and impressive communication skills. However, his strengths were overlooked as he had very little knowledge of computers. He did not know how to carry out simple tasks, such as preparing a directory and searching for files, on a computer. For this, he had to ask his colleagues, who used to avoid him as they were jealous of the other skills he possessed.

He realized that if he had basic knowledge of computers, he would not have to request his colleagues for help. Keeping this in mind, Kapil joined a basic computer course to learn MS-DOS and Windows. This helped him understand the functioning of a computer system, the storage and retrieval of data, the creation of files and folders, and the drive in which data is stored.

When instructions or programs are placed in the primary memory of the computer, the central processing unit (CPU) starts executing them one by one. A software is a collection of computer programs and data that tells the computer what to do and how it must go about it. By creating appropriate software, a computer may be able to do perform almost any task within its capacity.

CLASSIFICATION OF SOFTWARE

Software can be classified into the categories shown in Fig. 2.1.

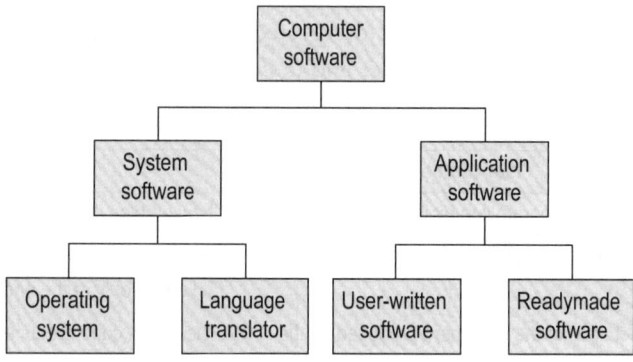

Fig. 2.1 Categories of software

System Software

System software or system packages are collections of programs that directly control the functioning of all hardware resources. They are general programs that are written to help users use a computer system, by performing tasks such as controlling operations, moving data in and out of the computer, and carrying out the steps in executing an application program. System software supports the running of packages by communicating with secondary devices such as printers, card readers, disks, and tapes. They are also used in developing software and monitoring the use of various hardware resources such as memory, peripherals, and CPU. System software help in making the operation of a computer system more effective and efficient.

Examples of system software: The most important system software are operating systems such as disk operating system (DOS), Windows, OS/2, Macintosh, UNIX, and Linux. We will also discuss operating systems currently available in the market, later in this chapter. Other system software include compilers, assemblers, and interpreters that are a part of the language translator.

Language Translator

We will now discuss the different kinds of language translators, such as compilers, assemblers, and interpreters.

Compiler   A compiler is a program that can translate a high-level language program into a machine language program. A compiler checks all kinds of limits, ranges, and errors. It occupies a large part of the memory as it takes longer than an interpreter to execute programs. It thus has low speed. A compiler produces an object code that cannot be read by people. A *self* or *resident compiler* is a sequence of bytes that encodes specific machine instructions that would be executed by a microprocessor when it runs (executes) the program.

Assembler   An assembler is a program that translates an assembly language program into a machine language program. An assembler that runs on a computer for which it prepares object codes (machine codes) is called a *self assembler* (*resident assembler*).

Interpreter   An interpreter is a program that translates one statement of a high-level language program into machine code and then executes it. The process is continued till all the statements of the program are translated and executed. The object program produced by the interpreter is not saved. If an instruction has to be repeated, it must be interpreted once again, and translated into the machine code. For example, during repetitive processing of the steps in a loop, each instruction in the loop must be reinterpreted every time the loop is executed. Table 2.1 shows the differences between a compiler and interpreter.

Table 2.1   Differences between compiler and interpreter

Interpreter	Compiler
It translates the program line by line.	It assembles the whole program.
The debugging process is easy.	The debugging process is complex as it generates errors only at the end of compilation.
The object code of the statement produced by the interpreter is not saved.	The object code produced by the compiler is permanently saved for future reference.
It is a smaller program compared to a compiler. Thus, it occupies lesser memory space and has a lower execution time.	It is a complicated process compared to an interpreter. Thus, it has a higher execution time and occupies larger memory space.
It is a slow process.	It is 5–25 times faster than an interpreter.
It is used in FORTRAN program.	It is used in C language program.

Application Software

Application software is software that instructs the computer to perform a specific set of tasks. There are application software to meet the needs of users. To make a computer operational, application software has to be installed along with the operating system.

Examples of application software: The most frequently used application software is of two types:

- General-purpose application software (database management packages, word processors, spreadsheets, etc.)
- Special-purpose application software (accounting, inventory, production management, property management system, etc.)

User-written software   An organization may develop customized software, depending on its needs.

Readymade software   Different kinds of software that can be used for accounting or management of hotels are available in the market.

Difference between System Software and Application Software

Good system software allow application packages to be run on the computer with very little effort. If there is no system software, application packages cannot be run. The making of system software is a complex task that requires extensive knowledge and specialized training in computer science. System software consists of programs that are designed to control different operations of a computer. It controls the activities of the computer hardware and interacts with the application software to perform a particular task.

Application software consists of programs that are designed to perform specific tasks for the user. They are said to be end-user programs as they are used to obtain the required results.

OPERATING SYSTEM

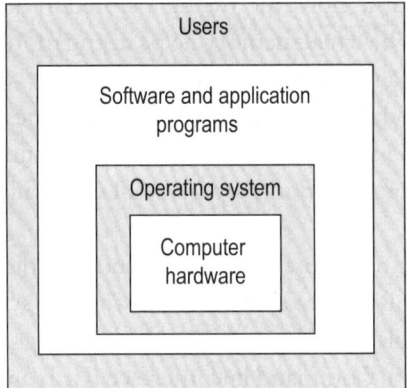

Fig. 2.2 Logical architecture of a computer system

An operating system is the most essential system software to manage the operation of a computer. It regulates the flow of signals from the CPU to various parts of the computer.

It is the first program to be loaded (copied) on to the computer's memory after the computer has been switched on. The operating system is an important component of the computer as it sets the standards for the application programs that can run on it. The logical architecture of a computer system is given in Fig. 2.2.

MS-DOS, OS/2, Windows, UNIX, and Linux are popular operating systems.

Functions

Windows and Linux are the commonly used operating systems. The function of an operating system is to manage the resources (files) of the computer system. These are shown in Fig. 2.3.

Data management The major function of an operating system is to keep track of the data on the disk. The application program does not know where the data is actually stored or how to retrieve it. When a program is ready to accept the data, it signals the operating system with a coded message. The operating system finds the data and delivers it to the program. On the other hand, when the program is ready to output, the operating system transfers the data from the program onto the next available space on the disk.

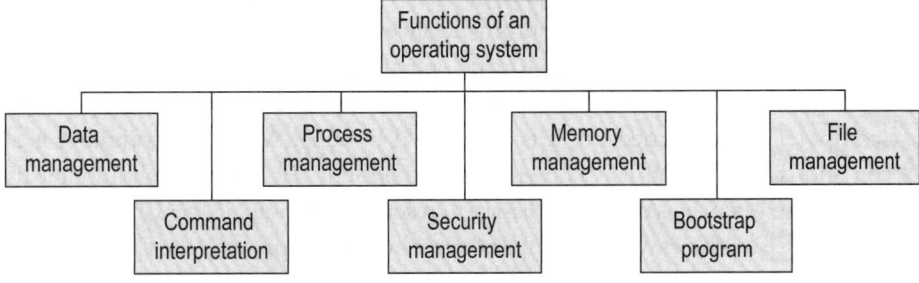

Fig. 2.3 Functions of an operating system

Process management   It is a module that deals with the creation and deletion of processes, scheduling of system resources to the different processes requesting them, and providing mechanism for synchronization and communication among the processes.

Memory management   The memory management module deals with the allotment and de-allotment of memory space to programs in need of this resource. Either primary memory or secondary memory can be used. *Virtual memory* is a feature of an operating system that enables a process to use a memory (random access memory, RAM) address space that is independent of processes running in the same system, and use a space that is larger than the actual amount of RAM present. *Virtual memory* helps in increasing the available memory of the computer by substituting the memory available in the hard disk; however, the data stored in the hard disk has to be returned to the RAM for usage. Though virtual memory helps to run more programs, it slows down the computer because of the data transfer processes.

File management   The file management module performs file-related activities. A file is a collection of information or data that is stored in a computer system and has a unique name. The main functions of this module include the organization, storage, retrieval, naming, sharing, and protection of files.

Command interpretation   An operating system interprets user commands and directs system resources to process those commands. With this mode of interaction with the system, users are not very concerned about the hardware details of the system. Two broad categories of user interfaces are supported by this operating system. Command line interface (CLI) and graphical user interface (GUI) act as an intermediary between the application and user by facilitating communication between them.

GUI helps users to interact with the operating system by point-and-click operations. It contains various pictorial representations of objects such as files, directories, and devices. CLI enables a user to interact with the operating system by issuing specific commands that the user types on the command line.

Security management   Multi-user operating systems maintain a list of authorised users and provide password protection against unauthorized users who might try to gain access to the computer system. The commonly used approaches to security management are user authentication, access control, and cryptography.

Bootstrap program   Boot means to start or make the computer system ready to take instructions. The word 'boot' comes from *bootstrap*. Booting the computer helps in getting the read only memory (ROM) loaded in the main memory. In a personal computer, there is a small bootstrap routine in a ROM chip that is automatically executed when the computer is turned on or reset. The bootstrap routine searches for the operating system, loads it, and then passes control to it.

We can boot our computer system in two ways: cold booting and warm booting. When the computer is first turned on, it is called cold booting. When the computer is already switched on and is being reset, it is known as warm booting. If a personal computer uses a single task operating system, it is usually necessary to reset the computer if it crashes. For example, in IBM PC compatible machines, we can perform warm booting by pressing the Ctrl + Alt + Del keys together.

Measuring System Performance

Efficiency of an operating system and overall performance of a computer system are usually measured by its *throughput, turnaround time*, and *response time*. Throughput is the quantity of work/job that the system processes per unit time. *Turnaround time* is the time gap between the submission of a job to the system and the completion of the job. *Response time* is the time gap between the time of submission of job to the system, to the first response obtained from the system.

CLASSIFICATION

All operating systems contain the same components, whose functionalities are also almost similar. For instance, all operating systems perform the functions of storage management, process management, and security. The procedures and methods used to perform these functions might be different and may have some distinguishing features. The following is the classification of operating systems:

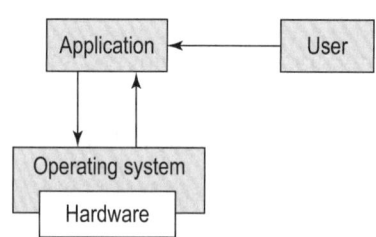

Fig. 2.4 Single user–single processor system

Single User–Single Processor System

It is the simplest of all computer systems. It has a single processor, runs a single program, and interacts with a single user at a time. A representative example is MS-DOS. Figure 2.4 shows a single user–single processor system.

Multi-user System

A multi-user operating system allows a number of users to work on a single computer. All users are provided with a terminal connected to a single computer. Examples include Linux, Windows NT, and Novell. Windows NT and Novell will be discussed in Chapter 4.

Batch Processing System

A computer remains idle while an operator loads and unloads jobs and prepares the system for a new job. This causes enormous wastage of valuable computer time. A method of automatic job-to-job transitions, known as batch processing, was devised to reduce this idle time. When one job is finished, the system control is automatically transferred back to the operating system that performs housekeeping jobs (like clearing data remaining in the memory from a previous job). Batch processing system is used in COBOL programming.

Another term that is commonly used in a batch processing system is job scheduling, which is the process of sequencing jobs so that they can be executed on the processor. It recognizes different jobs on the basis of first-come-first-served (FCFS) basis. Figure 2.5 shows the batch processing system.

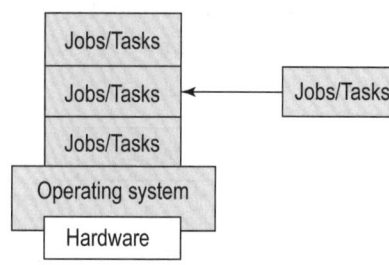

Fig. 2.5 Batch processing system

Multiprogramming System

Multiprogramming is the simultaneous execution of two or more different programs (jobs) by a computer. Some of the most popular multiprogramming operating systems are UNIX, VMS, and Windows NT.

In some multiprogramming operating systems, only a fixed number of jobs can be processed at once (*multiprogramming with fixed tasks, MFT*), while in others, the number of jobs can vary (*multiprogramming with variable tasks, MVT*). The area occupied by each job residing simultaneously in the main memory is known as *memory partition*. In multiprogramming systems, an entire program is loaded into its own block of memory, called memory partition. The early systems implemented multiprogramming with fixed partitions. As the size of the programs grew, it became difficult to find partitions that were large enough to accommodate these programs and also have a multiprogramming level that was high enough to produce good CPU utilization.

In order to achieve high CPU utilization with larger programs, multiprogrammed systems were combined with the virtual memory. This combination allows the selection of a partition that is smaller than the address space of the program. It relies on a paging policy to manage the contents of the partition.

Multitasking System

Often, a single-user system has multiple tasks to process. For example, while editing a file in the foreground, a sorting job could also be given in the background. In such a situation, the position of each of the tasks is viewed on the computer's screen by portioning the screen into multiple windows. Thus, the progress of different tasks is viewed on different windows in a multitasking system.

Difference between Multiprogramming and Multitasking Systems

Multiprogramming is the execution of multiple jobs between layers (of the same or different users) in a multi-user system, whereas multitasking is the execution of multiple jobs (often referred to as tasks of the same user) in a single-user system.

Multiprocessing System

The use of an input/output (I/O) processor improves the efficiency of a computer system by making simultaneous input, processing, and output operations possible. The CPU performs arithmetic and logical operations, while the I/O processor carries out I/O operations simultaneously. Figure 2.6 describes a single CPU and Fig. 2.7 shows the multi-CPU architecture of a computer system.

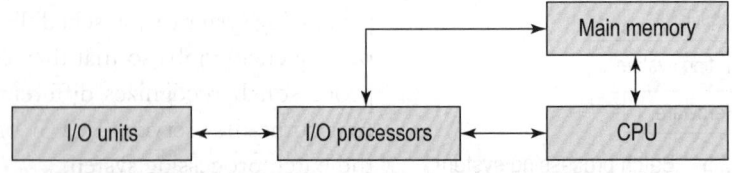

Fig. 2.6   Architecture of a computer system with a single CPU

The idea of using I/O processors to improve system performance has been carried a step further by designing systems with multiple CPUs known as *multiprocessing systems* (also popularly known as *parallel processing*). In these systems, multiple CPUs are used to either process instructions from different and independent programs or different instructions from the same program, at the same time.

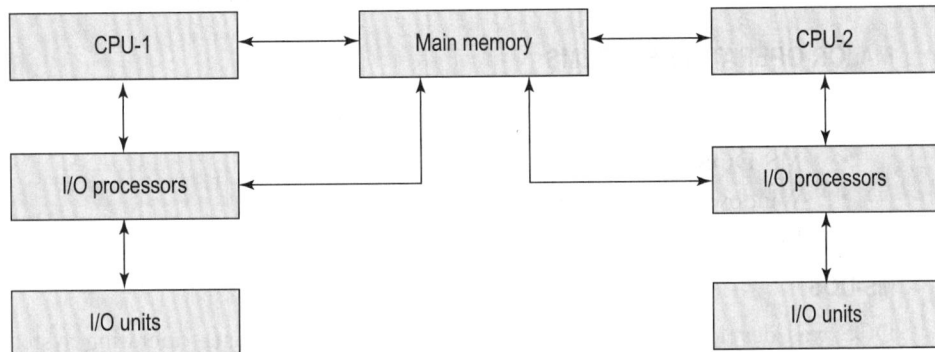

Fig. 2.7   Basic organization of a typical multiprocessing system

An example of a multiprocessing system is Encore's version of UNIX for the Multimax computer. This computer can be configured to employ dozens of processors, all running a copy of UNIX. The differences between multiprogramming and multiprocessing are given in Table 2.2.

Table 2.2   Differences between multiprogramming and multiprocessing

Multiprocessing	Multiprogramming
Simultaneous execution of two or more processes by a computer system having more than one CPU	It is the execution of multiple jobs of two or more processes by a single CPU system
Simultaneous execution of several program segments of the same or different programs	Completion of a portion of one program, then a portion of another, and so forth in brief successive periods

Time-sharing System

Time-sharing is a method of simultaneously providing interactive use of a computer system to many users in such a way that each user is the sole user of the system. A time-sharing operating system uses CPU scheduling and multiprogramming to provide each user with a small portion of a time-shared computer. Its advantages include minimizing CPU idle time, providing quick response time, and offering good computing facility to small users.

Distributed System

The current trend in computer systems is to distribute computation among several processors. The processors communicate with each other through various communication lines like high-speed telephone lines. These systems include small microprocessors, work stations, and minicomputer systems. These processors are referred to by a number of different names, such as sites, nodes, and computers, depending on the context in which they are mentioned.

Real-time Systems

A real-time system is used when there are strict time requirements on the operation of processors or the flow of data, and thus is often used as a control device in a dedicated application.

MAJOR OPERATING SYSTEMS

In this section, we will discuss three different operating systems:

- MS-DOS
- Windows
- Linux

MS-DOS

DOS was developed by Microsoft, primarily as a single-user operating system working on personal computers. This operating system was simple and small enough to be stored on a floppy disk, hence the name disk operating system. It provides a command line interface with which a user types a command on the command line for performing a particular task. The command line interface is also known as DOS prompt. DOS is a character-based operating system, that is, we type the command with the help of a keyboard. From 1981 till date, many versions starting from Ver 1.0 to 6.2 have been released. The successive versions have been enhanced for improving the management of computer resources.

The *command prompt* of MS-DOS allows us to execute files with the following extensions:

- .com (command files)
- .bat (batch files)
- .exe (executable files)

The structure of MS-DOS comprises the following programs:

Io.sys   It is a hidden, read-only system file of MS-DOS that is used for starting the computer system and is responsible for the management and allocation of hardware.

Msdos.sys   It is a hidden, read-only file that is executed after the execution of the io.sys file. It is responsible for managing the memory, processors, and input–output devices of a computer system.

Config.sys   It is a system file that is used to configure the various hardware components of a computer system.

Command.com   It is a command interpreter that is used to read and interpret the commands issued by users.

Autoexec.bat   These are batch files consisting of a list of commands that are executed as the computer system starts up.

Accessing MS-DOS

To work with MS-DOS, install it in the computer system. After installing, start MS-DOS using one of the following two interfaces:

- Start menu
- Run command

Using Start menu   We can start MS-DOS by performing the following steps using the Start menu:

- Select Start ⟶ Programs ⟶ Accessories, as shown in Fig. 2.8.

Fig. 2.8   Selection of a program in Windows XP

- Click the Command Prompt option to display the command prompt window of MS-DOS, as shown in Fig. 2.9.

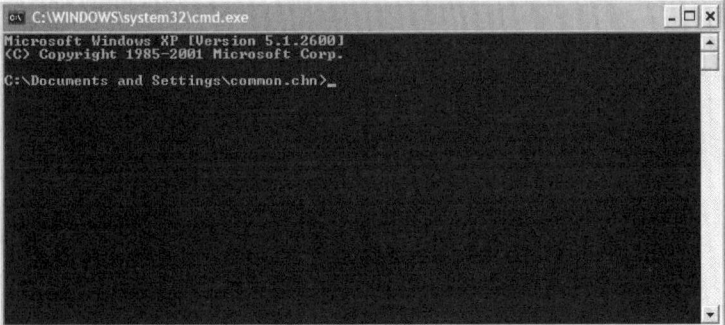

Fig. 2.9   Command prompt window in Windows XP

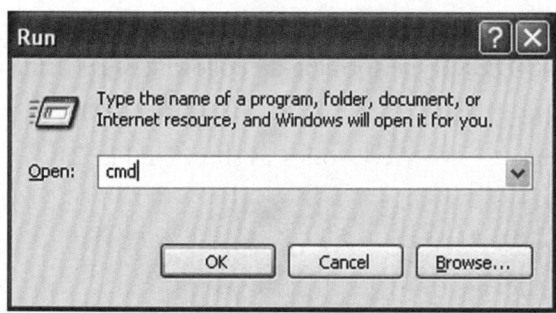

Fig. 2.10   Run dialog box in Windows XP

Using Run command   We can also start MS-DOS by performing the following steps using the Run command:

- Select Start ⟶ Run to display the Run dialog box, as shown in Fig. 2.10.

- Type 'cmd' in the Open text box and click OK to display the command prompt window of MS-DOS, as shown in Fig. 2.10.

MS-DOS Commands

The two kinds of MS-DOS commands are as follows:

- Internal commands
- External commands

Internal commands   These commands are automatically loaded on to the memory at the time of loading the operating system. Internal commands can be understood by MS-DOS itself. The DOS system files (io.sys, msdos.sys, command.com) are sufficient. Hence, no other files are required to execute internal commands.

To explain the DOS commands, let us consider a directory structure as given in Fig. 2.11. To execute the DOS commands from the proper prompt (c:\>, a:\>, etc.), we have to first type the command and then press the Enter key.

With the directory structure shown in Fig. 2.11, we can execute various commands.

Some of the key internal commands of MS-DOS are DATE, TIME, DIR, MD, CD, RD, COPY CON, TYPE, COPY, DEL, and CLS.

- DATE command: The DATE command is used to display the current date of the computer system. To check the current date, type the following command and press Enter.

C:\> Date

The execution of the aforementioned command will display the current date of the computer system. It also enables us to enter a new date, if we want to change the current date of the computer system.

- TIME command: The TIME command is used to display the current time of the computer system. To check the current time, type the following command and press Enter.

C:\> Time

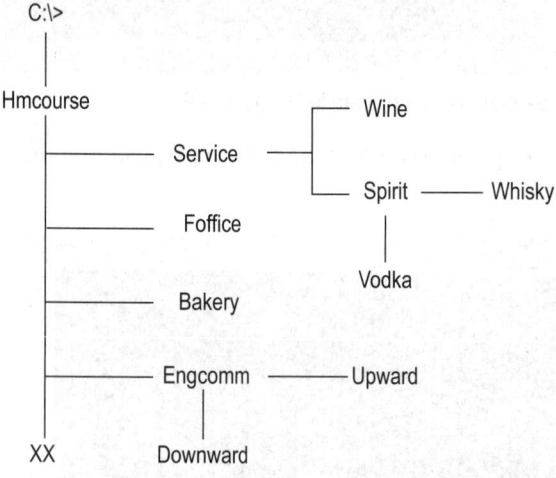

Fig. 2.11   Directory structure

The execution of this command will display the current time of the computer system. It also enables us to enter a new time, if we want to change the current time of the computer system.

- DIR command: The DIR command is used to display the names of files and subdirectories present in a directory, along with their size in bytes, and the date and time of the last modification performed. To run the DIR command, we need to type the following commands at the command prompt and then press Enter. Table 2.3 lists the options available with the DIR command,

Table 2.3   Options with DIR command

C:\> DIR commands	Purpose
DIR/P	Displays list in the form of pages on the screen
DIR/W	Displays list in the wide format on the screen
DIR/O	Displays list in a sorted order
DIR/S	Displays files in specified directories and subdirectories
DIR/*.*	Displays all the variable files and subdirectories in the current working directory

A set of files having the same extension can be displayed by using *wildcard* characters such as asterisk (*) and question mark (?).

- MD command: The MD command is used to create a new directory at the specified location in the computer system. MD stands for make directory and is also known as the MKDIR command. To create a new directory, type the following command and press Enter.

   C:\>MD Hmcourse\Service

- CD command: The CD command is used to move from one directory to another specified directory on the computer system. CD stands for change directory. It is also known as CHDIR command. To change the directory, type the following command and press Enter.

   C:\>CD Hmcourse
   The execution of this command will make Hmcourse the current working directory. This will change the command prompt in the command prompt window as C:\ Hmcourse>.

The CD command can also be used to move from the current working directory to the root directory of the computer system. To change the current working directory to the root directory, type the following command and press Enter.

   C:\Hmcourse>CD\

The execution of the aforementioned command will change the prompt from the current working directory Hmcourse to the root directory, that is, c:\>.

- RD command: The RD command is used to remove an empty directory from the computer system. It is also known as RMDIR command. To remove an empty directory, type the following command and press Enter.

```
C:\Hmcourse\service>RD Wine
```

The execution of the aforementioned command will remove the empty directory `Wine` from the specified location, that is, `C:\Hmcourse\Service>`.

- `COPY CON` command: This command is used to create a file. Let us create a file named `xx` under the `Hmcourse` directory. The following command will open up a file named `xx` in the `Hmcourse` directory.

```
C:\Hmcourse>Copy Con xx
```

This command enables a user to type some content and create a file. After completing the typing, we have to close the file. The command for closing the file after saving is Ctrl+Z. Alternatively, F6 key can be pressed. Here `copy con` means copy to the console. Conversely, `con copy` command helps us to view the contents of a file.

```
C:\Hmcourse>Con Copy xx
```

- `TYPE` command: To view the contents of a file in its current location, type the following command and press Enter.

```
C:\Hmcourse> type xx
```

The execution of the aforementioned command will display the contents of the file `xx` present in the `Hmcourse` directory.

- `COPY` command: The `COPY` command is used to copy a file from one location to another on the computer system. To copy a file from one directory to another, we type the following command and then press Enter.

```
C:\>Copy Hmcourse\xx D:
```

The execution of this command will copy the file `xx` available in the C drive, to the D drive. We can also copy a file in the current directory with the same content but with a different name. To copy a file with a new name in the current directory, type the following command and press Enter.

```
C:\> Copy Hmcourse\xx aa
```

The execution of this command will copy the file `xx` with a new name `aa` in the same location, that is, C drive.

- `DEL` command: The `DEL` command is used to delete one or more files from the computer system. `DEL` stands for delete and is similar to the `ERASE` command. To delete a file from a specified directory, type the following command and press Enter.

```
C:\Hmcourse>del xx
```

The execution of this command will delete the file `xx` from the `Hmcourse` directory.

- `CLS` command: The `CLS` command is used to clear the screen of the command prompt window. To clear the screen, type the following command and press Enter.

```
C:\>CLS
```

The execution of the aforementioned command clears the screen and places the cursor at the top of the left-hand side of the command prompt window.

External commands   These commands are auxiliary program files that may or may not be present on the disk. We can use these commands only if the corresponding program files are available. They are utility programs that are normally supplied with the DOS diskettes. These programs can only be used if they are available on the disk. There are a number of external commands and each requires a file support. These commands have the extension.com, .exe, or .bat and are not used as frequently as the internal commands. Some of the most important external commands of MS-DOS are CHKDSK, COMP, EDIT, TREE, DELTREE, FIND, and HELP.

- CHKDSK command: The CHKDSK command is used to check the status of hard disk drives on the computer system for errors. We should run this command frequently in order to detect errors on the disk. To check the status of the current disk, type the following command and press Enter.

    C:\>CHKDSK C

The execution of the aforementioned command will display the status of the current disk and to execute the command, the required file is CHKDSK.EXE.

- COMP command: The COMP command is used to compare two existing files. To obtain the difference between the two files, type the following command and press Enter.

    C:\>COMP Hmcourse/Service/Spirit/Whisky Vodka

On execution, the command will compare the two files, Whisky and Vodka.

- EDIT command: The EDIT command is used to create, open, or edit a particular file. It helps in opening and editing the contents of the existing file. To edit a file, we type the following command and press Enter.

    C:\>EDIT Hmcourse/xx

- TREE command: The TREE command is used to view the files and subdirectories of the current directory in the form of a tree. The purpose of this command is to view the structure of the current directory in an easier manner. To view files and subdirectories, we type the following command and press Enter.

    C:\>TREE

As the command is executed, it will display the structure of the C drive and the available files and subdirectories in the C drive, in tree format.

- DELTREE command: The DELTREE command is used to permanently delete a directory with all the files in it, from the computer system. To delete the files and directories, we type the following command and press Enter.

    C:\>DELTREE Hmcourse

With this command, all the files and subdirectories present in the directory Hmcourse will be deleted.

- FIND command: The FIND command is used to search for and display a chain of characters in the specified field. To find the chain, type the following command and press Enter.

```
C:\>FIND "Spice" Hmcourse\xx
```
On execution of the command, the chain **spice** in file **xx** will be displayed.

- **HELP** command: The **HELP** command is used to display information related to various commands in MS-DOS. For this, type the following command and press Enter.

```
C:\>HELP
```
On execution of the command, the command and its description are displayed on the command prompt.

Quitting MS-DOS

Exit command: This command quits MS-DOS operating system.

Type **Exit** and press Enter.
```
C:\>Exit
```
We have discussed the DOS operating system in detail. We will now discuss a popular and more interesting user-friendly software, Microsoft Windows.

Microsoft Windows

This operating system was developed by the Microsoft Corporation to overcome the limitations of its own MS-DOS operating system. The first successful version of this operating system was Windows 3.0, released in 1990. Subsequently, Windows 95, Windows 98, Windows 2000, Windows XP, Windows XP Professional, Windows Vista, Windows 2007, Windows 7, and Windows 8 were released.

The main features of Microsoft Windows XP are as follows:

- Its native interface is a graphical user interface. GUI allows users to select files, programs, or commands by pointing to the pictorial representation on the screen, rather than typing long, complex commands on the prompt. Hence, for a new user, it becomes easier to learn and use a computer system.
- Object linking and embedding (OLE) is a way by which we transfer and share information among Windows-based applications. When an object (e.g., an image file created with Paint program) is linked to a compound document (e.g., a presentation document created with an image), the document contains only a reference to the object; any changes made to the contents of a linked object will also be seen in the compound document. When an object is embedded in a compound document, the document contains a copy of the object; any alterations made to the contents of the original object will not be seen in the compound document unless the embedded object has been updated.
- It is a single-user multitasking operating system, that is, a user may run more than one program at a time. The monitor's screen can be portioned into multiple windows and the progress of different programs can be viewed on different windows.

Starting Windows

To start Windows, we perform the following steps:

- Switch on the computer system and wait for a few seconds. The computer system loads Windows.

- After successfully logging into Windows, the Windows desktop (shown in Fig. 2.12) is displayed.

Let us now discuss the working of various devices, icons, programs, and support functions.

Desktop   The desktop is the opening screen of the Windows operating system that displays the taskbar and various icons such as the recycle bin and folders. The Windows desktop can be easily customized.

A user interface helps us interact with Windows to perform various tasks. We will now discuss the usefulness of a mouse in Windows.

Mouse   A mouse is an input-cum-pointing device in a GUI. Generally, a mouse can be used in the following four ways:

- Single clicking with the left button
- Double clicking with the left button
- Single clicking with the right button
- Drag and drop

Folder   In Windows, the directory is also called a folder. We can store many files in a folder. A folder helps us organize our files efficiently. Figure 2.12 shows a folder icon. To create a new folder, right-click the mouse on the desktop. Choose the New option. Then select the Folder option. In this way, a new folder that can be named as per the user's need is created.

Icon   We can quickly access a file, program, or any other resource by selecting the icon. Thus, an icon is also called a shortcut.

My Computer   This folder contains icons of the different drives in the system and also the three main system folders, namely the control panel, printer, and scheduled tasks.

My Documents   It is a system folder (shown in Fig. 2.12) where we can store all our documents that need to be accessed easily.

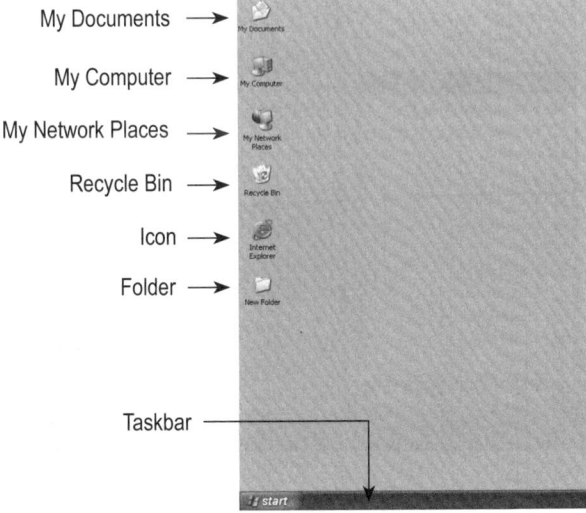

Fig. 2.12   Windows XP desktop

Recycle Bin   All deleted files and folders go to this special system folder (shown in Fig. 2.12). We cannot retrieve any file or folder from the recycle bin unless it is restored to its original location. Sometimes, we may delete a file by mistake. However, it is possible to recover the file from the recycle bin.

My Network Places   If the computer that we are working on is connected to a network, we can use this option (shown in Fig. 2.12) to browse the network resources in the same way we browse through the contents of our computer.

Taskbar   This is a horizontal bar (shown in Fig. 2.12), found at the bottom of the desktop. It contains a start button and a clock, and shows the name of the program that is currently running, thus helping the user work with multiple windows.

Start menu   This button displays a list of commands and shortcuts that we can point to, to accomplish a task. Sometimes, on the menu, there is an arrow that points to the right. This means that additional choices are available on a secondary menu, called a submenu. Parts of the Start menu button are shown in Fig. 2.13.

Help and Support   The help option aids us in getting information on procedures and instructions to execute any option or command in Windows XP. To work with the Help and Support option, do the following:

- Click the Start button.
- On the Start menu, click the Help and Support option (as shown in Fig. 2.14) and write down the requirement in the search option.

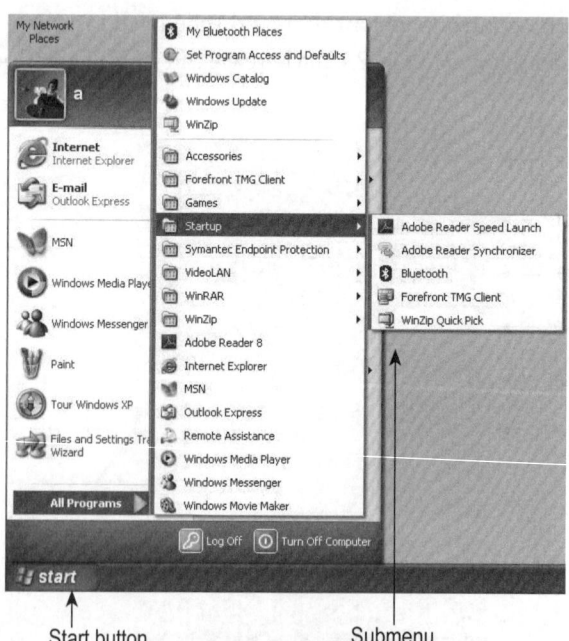

Fig. 2.13   Parts of Start button in Windows XP

Fig. 2.14   Help and Support option in Windows XP

Search option   This helps us search for files or folders in the system.

- Click the Start button.
- Choose the Search option. The search window will appear.
- Type the name of the file or folder that you wish to search for.
- Click on the Search button.
- The search results will be displayed.

Programs   Windows XP makes it easy for us to add, remove, start, and quit programs.

Adding programs   We can easily add or install any kind of software through the add program facility in the control panel in Windows XP. Software installation requires a source disk/CD of the particular software.

Removing programs   We can easily remove or erase any kind of software from our computer system with the help of this facility in the control panel. A user should be very cautious while removing a program as removal will lead to permanent deletion from the system. The procedure to add or remove programs is as follows:

- Insert the source disk.
- Click the Start button, point to Settings, and then click the Control Panel.
- Double click the Add/Remove Programs option.
- Click on Change or Remove Programs, or Add New Programs (Fig. 2.15), and then follow the instructions.

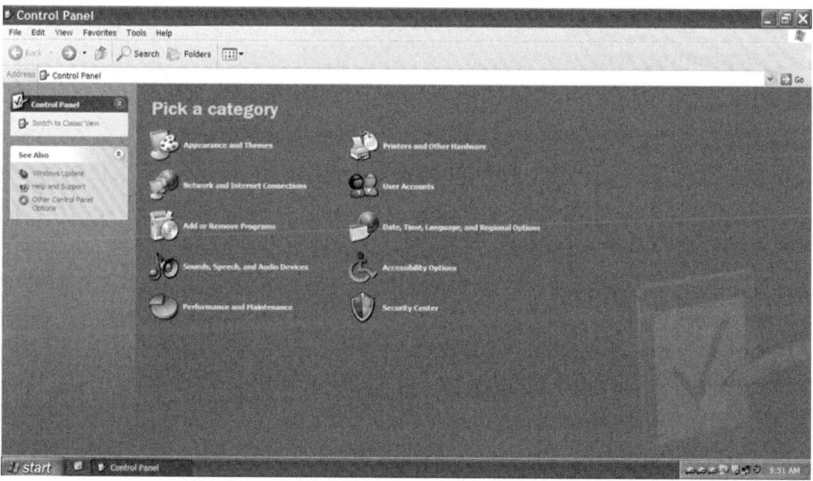

Fig. 2.15   Contents of the Control Panel in Windows XP

Starting and quitting programs   Most programs installed in a computer are located in the Programs section of the Start menu. The procedure to start or quit programs is as follows:

- Click the Start button, and then point to Programs. The Programs menu appears (Fig. 2.16). Choose from the available options.

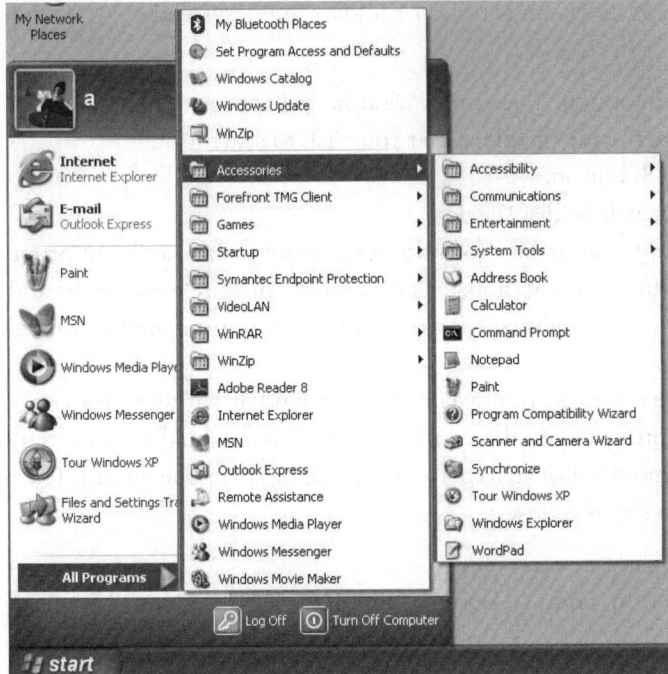

Fig. 2.16   Contents under Programs option in Windows XP

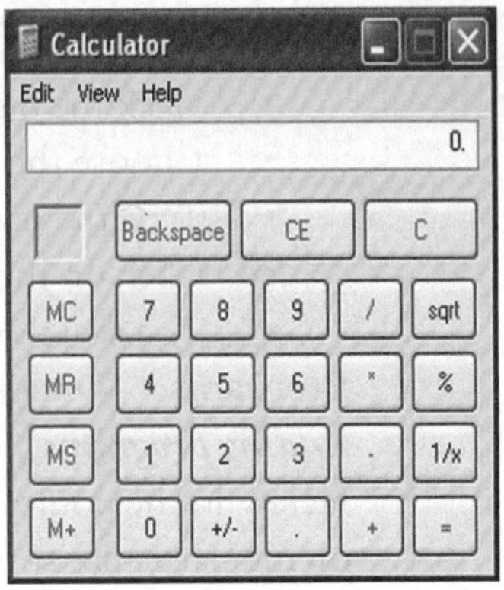

Fig. 2.17   Calculator in Windows XP

- Point to a group (e.g., Accessories) that contains the program we want to work with and then click the program name, for example, Calculator (shown in Fig. 2.17).

- To quit a program, click the Close button on the upper right-hand corner of the program window.

Managing files and folders   File names in Windows XP can be up to 255 characters long, including spaces, but cannot contain characters such as :, \, /, :, *, ?, ", <, >, and |.

Control Panel   It is a system folder that is configured by Windows InstallShield during installation of Windows. From here, one can customize the setting of the hardware and software.

Customizing desktop

- Click the Start button, point to Settings, and then Control Panel.
- Double click on the Appearance and Themes icon.
- The Appearance and Themes screen will give us the option of picking a task. Click on the Change the Desktop Background option (shown in Fig. 2.18).
- Then, the Display Properties screen will appear (shown in Fig. 2.19).
- Choose from the many options under the Background drop-down menu (as an e.g., Tulip has been shown in Fig. 2.19). Select Apply, and then click the OK button. On doing this, the desktop background will change to Tulip.

Windows Explorer   Like *Internet Explorer*, this is also a browsing tool. To select this option, click the Start button, point to Programs, Accessories and then click Windows Explorer. It helps us view the folders and files in our computer in a hierarchical order. There are two ways of opening Windows Explorer:

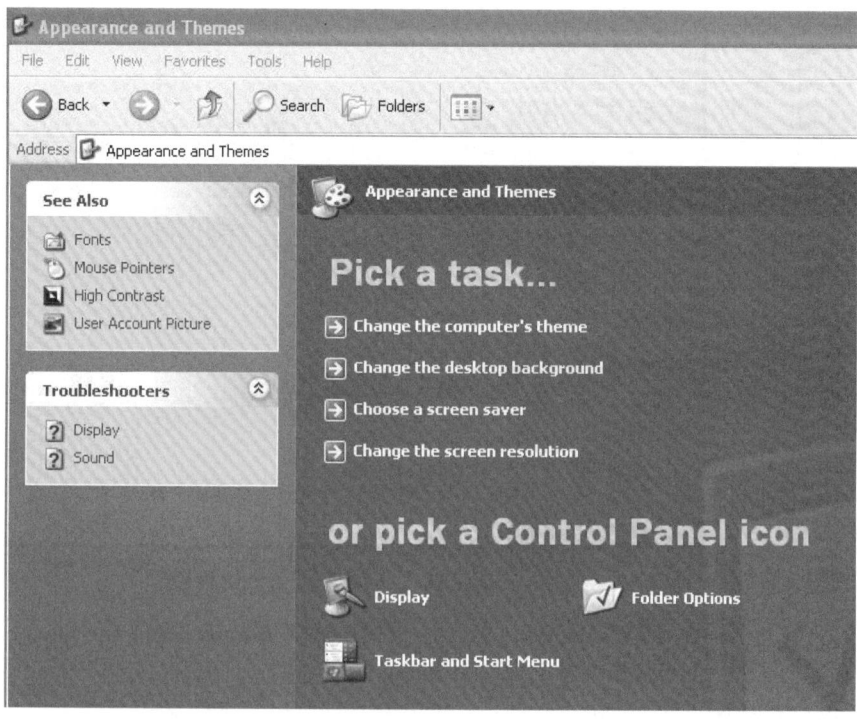

Fig. 2.18   Appearance and Themes option in Windows XP

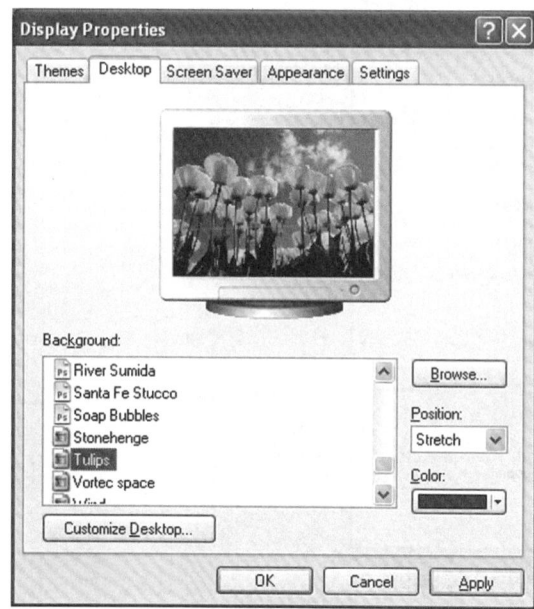

Fig. 2.19   Display Properties showing background options in Windows XP

- If Windows Explorer exists in the default location, it can be opened by selecting Programs and then Accessories from the Start menu (shown in Fig. 2.20)
- Windows Explorer can also be opened by right-clicking on the Start button and then choosing the Explorer command.

The *Windows Explorer* window is shown in Fig. 2.20. It has two panes—on the left and right. As shown in Fig. 2.20, the left pane of the window displays the desktop and all files and folders including My Computer. The right pane displays the contents of the selected drive or folder. Windows Explorer displays a '+' or '−' sign to the left of the folder or drive. The '−' sign indicates that Windows Explorer is displaying all subfolders of the selected drive/folder. On the other hand, the '+' sign indicates that there are some subfolders under the selected drive/folder that have not been displayed.

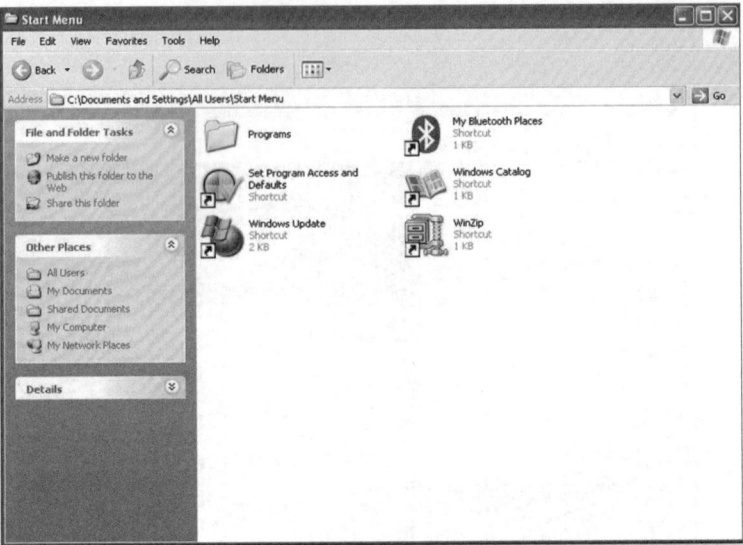

Fig. 2.20   Windows Explorer in XP

Logging off from Windows   To log off from Windows, go to the Start button and choose Log Off (as shown in Fig. 2.21). A dialog box opens and asks if we wish to log off. We can either choose Yes or No.

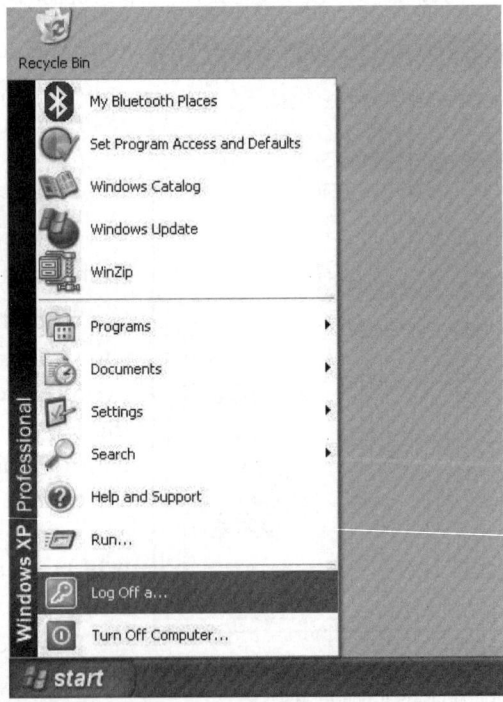

Fig. 2.21   Log Off option in Windows XP

The next operating system that we're going to discuss is Linux.

Linux

Linux is an operating system used in personal computers. It is an open-source, multitasking, multiprocessing operating system, enhanced and backed by programmers worldwide. The name Linux is derived from its inventor Linus Torvalds. Linux is different from other operating systems as it is freely available from different vendors and can be modified.

Features

The following are some significant features of the Linux operating system:

- Allows multiple users to work simultaneously
- Allows execution of several programs and processes at the same time to ensure efficient utilization of the processor
- Implements the concept of virtual memory, which enables the Linux operating system to execute a program whose size is larger than the main memory of the computer system.

The following are the core components of the Linux operating system:

- *Kernel* is the central part of the Linux operating system, which controls and manages communication between the different hardware and software components of the computer system.
- *Shell* is a user interface of the Linux operating system, which acts as an intermediary between the user and the kernel (shown in Fig. 2.22). Shell is a program that takes the commands issued by the user and interprets them in an efficient manner to obtain the results.

A user who works in a Linux operating system sends a request to the *shell* using a command. Then, the *shell* transfers the information of the user to the *kernel*, which in turn passes it to the hardware. A similar route of transmission is followed when information has to move from the hardware to the user.

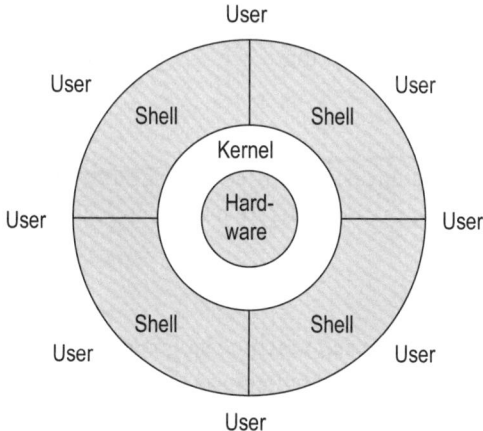

Fig. 2.22   User interaction between kernel and shell

Logging into Linux   If a computer has only Linux operating system, then on booting it would run Linux. However, many computers may have two operating systems such as Windows and Linux. In such a situation, we have to choose one of the operating systems when the computer boots. The option to choose is time-bound and if the user does not select one of the options, the system will switch to the default option.

File permission   Since Linux is a multi-user operating system, it provides a feature known as file permissions, which protects a user's files from being tampered with by other users. A file/directory has three types of permissions as mentioned here.

- Read permission: This allows a user to read files/directories.
- Write permission: This lets a user modify or write the data in the file.
- Execute permission: This allows a user to run the file as a program.

File structure in Linux   Linux treats even folder, documents or presentation, as files. Even a directory that contains entries for several other files, is treated as a file. All hardware devices, such as input–output devices, storage devices, and others, are treated as files. The Linux file system is organized in a hierarchical manner. The system administrator starts with the root directory. Working at the root is dangerous as any mistake can seriously damage or destroy the system. Don't work at the root unless absolutely necessary. The root is represented by a forward slash (/). Under the root directory, there are several system directories and the home directory.

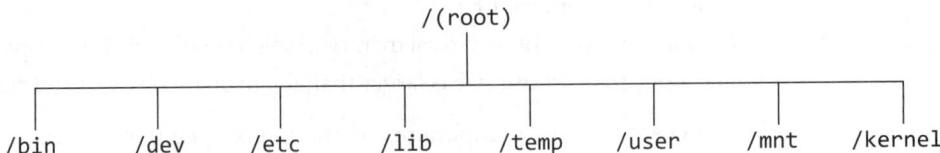

`/bin`: contains executable program files

`/dev`: contains special device files

`/etc`: contains all system-wide configuration information

`/lib`: contains library files

`/temp`: contains all temporary files

`/user`: contains the home directory of the user

`/mnt`: contain sub-directories 'floppy' and 'CD-ROM', which when mounted, show the contents of the floppy disk and CD-ROM disk respectively

`/kernel`: contains all kernel-specific code

Logging off from Linux   If you are working in text mode screen, press <Ctrl+Alt+Del>, wait for the system to reboot, and then switch off the PC. If you are working under Windows XP system, press <Ctrl+Alt+Backspace> first, then <Ctrl+Alt+Del>. Never switch off or reset the PC directly as this could damage the file system.

We will now discuss some common programming languages.

PROGRAMMING LANGUAGE

A language that is accepted by a computer system is known as a programming language or computer language. The method of writing the instructions in such a language for a planned program is called programming or coding.

Classification

Computer languages can be classified into four categories: machine language, assembly language, high-level language, and object-oriented programming language, as given in Fig. 2.23.

Machine Language

A programming language that a computer understands without using a translation program is called machine language. The language is normally written as strings of binary digits 1 and 0.

Advantage Programs written in machine language can be executed very quickly, as machine instructions are understood by the computer without the need for translation.

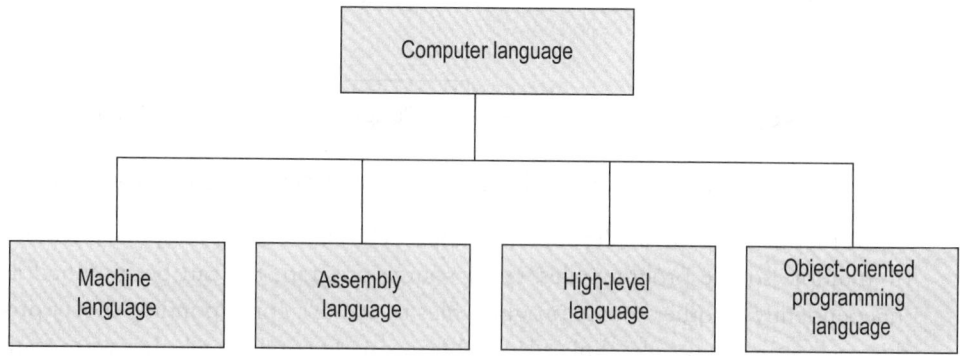

Fig. 2.23 Classification of computer languages

Limitations A machine language programmer must be familiar with the operation code number and hardware structure of the computer as this language is totally machine dependent. It is difficult and time-consuming for individuals to check instructions to locate errors. Machine language programs do not execute without compilation.

Assembly Language

A language in which the instructions and storage locations are represented by letters and symbols instead of numbers is called assembly language or symbolic language.

An assembler is a translator that translates an assembly language program (source program) into its equivalent machine language program (object program), as shown in Fig. 2.24.

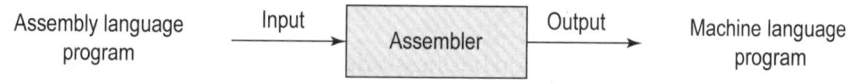

Fig. 2.24 Translation process of assembler

Advantages This language is easier to understand and use when compared to a machine language. In addition, the programmer does not need to memorize the storage locations of data and instructions while writing an assembly language program.

Limitations Assembly languages are often referred to as low-level programming languages as they are machine dependent. A programmer must have a good knowledge of the characteristics and logical structure of a computer system. As assembly language programs are still written at the machine-code level, writing them is very time consuming. Insertion or removal of some instructions of a program is difficult. Assembly language programs do not execute without compilation.

High-level Language

High-level programming languages have been designed to overcome the limitations of machine and assembly languages. COBOL, BASIC, PASCAL, Python, C#, C, and C++ (discussed later) are examples of high-level programming languages.

A high-level programming language must be converted (translated) into its equivalent machine language program before it can be executed by a computer. This translation is done with the help of a translator program called a compiler. Figure 2.25 shows the process of translating a high-level programming language to a machine language program using a compiler. For example, a BASIC compiler translates high-level language into machine language BASIC program.

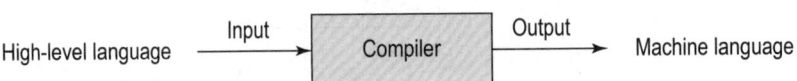

High-level language — Input → Compiler — Output → Machine language

Fig. 2.25 Translation of high-level language to machine language

A modular (use of small programs) approach is used for developing software and consists of multiple source program files. Each source program file can be modified and compiled independently of other source program files to create a corresponding object program file. In this case, a program called linker is used to properly combine all object program files of the software and convert them into a final executable program file.

An interpreter is another type of translator that is used to translate a high-level language program into its equivalent machine language program.

Advantages High-level programming languages are machine independent. Hence, they are easy to learn and use. Writing programs in these languages requires lesser time and effort. Insertion or removal of certain instructions from a program is also possible without any compilation and with very little effort.

Limitations These languages require more main memory space as execution takes a longer time. High-level languages are less flexible than assembly languages because they do not normally have instructions or mechanisms to control a computer's CPU, memory, and registers.

Popular high-level languages

COBOL It stands for common business-oriented language. It was mainly designed for business purposes but can also be used for other applications. The language is simple, portable, and easily maintainable.

BASIC It stands for beginner's all-purpose symbolic instruction code. It is a general-purpose language and easy to learn as it uses common English words.

PASCAL Pascal helps the programmer develop well-structured programs that are easy to maintain and modify, and have applications in research projects and computer games.

C It is a general-purpose high-level programming language. It is a highly efficient programming language as the programs developed in the language are executed very rapidly by the CPU of the computer system.

C++ It is an extended version of C language. It has several object-oriented programming features such as classes, virtual functions, inheritance, and templates.

Python It is an object-oriented high-level programming language that is used for developing software for a variety of applications. Programs developed with Python can run on different platforms and operating systems. Thus, it is said to be a platform-independent language.

C# It is pronounced 'C-sharp' and was developed by Microsoft to work with Microsoft's .net platform.

Object-oriented Programming Language

Object-oriented programming languages deal with solving a problem by identifying the real-world objects of the problem, processing requirements of those objects, creating simulation of those objects, their processes, and the required communication between the objects. This concept has now become a major programming language and is widely used in software development.

Advantages   Object-oriented programming languages are easy to maintain and modify, as new objects can be created with small differences to existing ones. They also provide a good framework for code libraries where supplied software components can be easily adapted and modified by the programmer.

Limitations   Object-oriented programming languages take more time to execute. They also offer lesser number of functions when compared to low-level programming languages, which interact directly with the hardware.

Programming languages helps us perform specific tasks by providing a set of instructions to the computer system. A proper language should be chosen after considering factors such as purpose of the program, programmer's experience, and performance and efficiency of the language.

SUMMARY

The chapter discusses software, their classification, as well as their need and uses. The concept of operating system and its categorization are also dealt with. We learnt how to work with DOS, Windows, and Linux operating systems. Under DOS, we discussed the creation of directories and files, and internal and external commands. In addition, we learnt how to rename, delete, copy, and move files and directories. External commands in DOS are those that are used to format disk, check disk facility, and copy disk contents. In the later part of the chapter, Windows, a user-friendly single-user multitasking operating system based on GUI system, was explained. We were thus able to understand the functions of My Computer and Control Panel, and the creation of user groups and password settings. The last part of the chapter examines Linux, a multitasking and multiprocessing operating system that uses both, CLI and GUI. We were also able to familiarize ourselves with kernels, shells, and directory structure commands. The classification, advantages, and disadvantages of various programming languages were also discussed.

KEY TERMS

Assembler   It is a translator program that translates (converts) an assembly language program (source program) into its equivalent machine language program (object program).

Assembly language   It is a low-level programming language that uses principles and strategies of memory for writing the instructions for a program.

Batch processing system   This operating system processes jobs by organizing them into groups known as batches, though it processes only one job at a time.

Compiler   It is a translator program that translates (converts) a high-level language program (source program) into its equivalent machine language program (object program).

Cold booting   When the computer system is turned on, it is known as cold booting.

Cryptography   The study and practice of a technique in which data and messages are kept secured by the authorized user to prevent, unauthorized access to the data.

Command line interface (CLI)   CLI refers to the fact that you have to take the help of a keyboard to type commands to interact with the computer. We can only type commands in the form of text, as in MS-DOS or a command prompt. There are no images or graphics on the screen and it is a primitive type of interface.

Data management   The storing and retrieving of data by the operating system is known as data management.

File management   It is the process of managing files in a computer system.

Graphical user interface (GUI)   Most modern computers make use of GUI, an interface that makes use of graphics, images, and other visual clues like icons.

High-level language   It is a machine-independent language that uses words similar to the English language, for developing software applications.

Interpreter   It is a program that translates one statement of a high-level language program into machine code and executes it.

Kernel   It is the central part of a Linux operating system that manages and controls the communication between various hardware and software components.

Linux   It is an open source operating system enhanced and backed by thousands of programmers worldwide. It has two interfaces: CLI and GUI.

Machine language   It is a low-level programming language where instructions are specified in terms of 0s and 1s.

MS-DOS   This stands for Microsoft disk operating system.

Microsoft Windows   It is a single-user multitasking operating system whose native interface is a GUI.

Memory management   This function of the operating system helps in the allocation of memory space to process data at the time of execution.

Multi-user operating system   This system allows multiple users to work simultaneously on the same computer system.

Multitasking operating system   This operating system allows a user to carry out multiple tasks at the same time on a single computer system.

Multiprocessor operating system   This operating system allows the use of multiple central processing units in a computer system, for executing multiple processes at the same time.

Operating system   It is an integrated set of programs that controls the resources (CPU, memory, input–output devices) of a computer system and provides its users with an interface.

Object-oriented language   An object-oriented programming language deals with solving a problem by identifying the real-world objects of the problem, processing requirements of those objects, creating simulations of those objects, their processes, and the required communication between the objects.

Programming language   It is used to write computer programs in order to instruct the computer system to perform specific tasks.

Process management   It involves the execution of tasks such as creation and scheduling of processes, and the management and termination of process, as needed.

Real-time operating system   It is used for processing critical applications that are to be executed within a specified time.

Response time   It is the interval between the time of submission of a job to the system for processing, to the time the system produces the first response to the job.

Single user–single processing system   A single processor that runs a single program and interacts with a single user at a particular time is called single user–single processing system.

Security management   This function of the operating system helps in securing and protecting the computer system, both internally and externally.

Shell   It is a user interface of the Linux operating system that acts as an intermediary between the user and the kernel of the operating system.

Software   It is a collection of computer programs that allow the user to use the computer for a specific purpose.

System software   It is a set of one or more programs designed to control the operation and extend the processing capability of a computer system.

Throughput   It is the amount of work that a system is able to do per unit time. It is measured as the number of jobs (processes) completed by the system per unit time.

Turnaround time   It is the interval between the time of submission of a job to the system for processing, to the time of completion of the job.

Virtual memory   A computer has limited memory (as per the RAM installed in it). Due to this limitation, a computer can run short of memory if many programs are running at once. Virtual memory helps to increase the available memory of the computer by substituting the memory available in the hard disk; however, the data stored in the hard disk has to be returned to the RAM for usage. Though virtual memory helps in running more programs, it slows down the computer because of the data transfer processes.

Warm booting   When the computer is already switched on and is being reset, it is known as warm booting.

REFERENCES

Arora, A. and A. Bhatia, *Information Systems for Managers*, Excel Books, New Delhi, 1999.

Balagurusamy, E., *Fundamentals of Computers*, Tata McGraw Hill, New Delhi, 2009.

Gupta, S. and S. Gupta, *Computer Aided Management*, Excel Books, New Delhi, 2004.

Jain, S., 'O' Level Made Simple, BPB Publications, New Delhi, 1999.

Mukherjee, P. and S. Bandhopadhay, *Introduction to Computer Science*, Vol. I, Deep Prakashan, Kolkata, 2001.

Sinha, P.K. and P. Sinha, *Computer Fundamentals*, Fourth edition, BPB Publications, New Delhi, 2007.

Web References

http://searchstorage.techtarget.com/definition/virtual-memory, last accessed on 1 August 2012.

http://www.cs.cmu.edu/~dst/DeCSS/object-code.txt, last accessed on 27 July 2012.

http://www.differencebetween.com/difference-between-cui-and-vs-gui/#ixzz3Nt9HONXb, last accessed on 10 September 2011.

http://www.extropia.com/tutorials/unix/shells.html&docid, last accessed on 29 July 2012.

EXERCISES

Concept Review Questions

1. What is an operating system? Why is it necessary for a computer system?
2. Write down the difference between system software and application software.
3. What are the differences between a compiler and interpreter?
4. Explain the classification of an operating system.
5. Write down the differences between multiprogramming and multiprocessing.
6. Differentiate between DEL and DELTREE DOS commands.
7. Write the main features of Microsoft Windows XP operating system.
8. What do we understand by the term 'programming languages'?
9. Write down the advantages and limitations of object-oriented programming languages.
10. Describe the file structure of Linux in brief.
11. 'An operating system works as the resource manager of a computer system.' Give suitable reasons supporting the statement.
12. 'An operating system is an example of system software.' Explain this sentence giving suitable points.
13. When do we call a computer language machine dependent? What is the main disadvantage of such a language?

Multiple Choice Questions

1. An operating system is
   (a) a collection of hardware components
   (b) a collection of input–output devices
   (c) a collection of software routines
   (d) none of these
2. An operating system
   (a) links a program with the subroutines
   (b) provides a layered, user-friendly interface
   (c) enables a programmer to draw a flow chart
   (d) none of these
3. While using the DIR command in DOS, you can use certain switches with it. Which one of the following cannot be used with DIR?
   (a) /p   (b) /w   (c) /s   (d) /T
4. The execution of two or more programs by a single CPU is known as
   (a) multiprocessing   (c) multiprogramming
   (b) time-sharing   (d) none of these
5. What is a machine language?
   (a) It is an object-oriented language.
   (b) It is a language of 0s and 1s.
   (c) It is a high-level language.
   (d) None of these
6. Which of the following programs is used to convert assembly programs into machine instructions?
   (a) Compiler   (c) Assembler
   (b) Interpreter   (d) None of these

Project Work

1. In DOS, create the directory structure shown in the following figure in the C drive:
2. In DOS, remove the aforementioned directory structure from the C drive using external commands.
3. Customize the desktop background in Windows operating system.

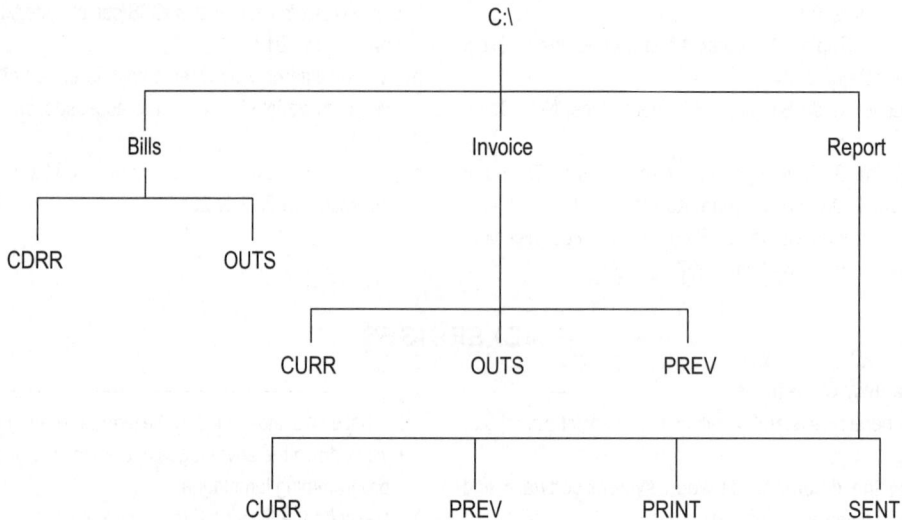

CHAPTER 3

Microsoft Office 2003

LEARNING OBJECTIVES

After reading this chapter, you will be able to understand the following:

- Features of MS Word
- Working of the Edit menu and Mail Merge
- Spreadsheet and its features
- Working of formulae and functions
- Preparing sample worksheets and graphical representations
- Features of PowerPoint and preparing a presentation
- Preparing organization charts in a presentation
- Working of MS Access
- Preparing databases, tables, and forms

EXHIBIT 3.1   Working with MS Office

Debraj Adhikari has been working in the Food and Beverage (F&B) department in Templeton Club for quite some time. He was recently promoted as the F&B manager. Neeraj Bisht, a new resident manager who had joined the club, had requested all department heads to give a presentation followed by a report on their departmental budget, within 24 hours.

Debraj was puzzled. Even though he had been working for about eight years in the industry, he had not given any presentation. He had basic knowledge of MS Word and used to oversee the monthly calendar of events that was sent to all club members using Mail Merge.

Though Debraj was hardworking and had all the details for the report, he was worried about the presentation,

considering the time given by Neeraj. An anxious Debraj called his friend Kapil Sachdeva and shared the details of the presentation to be made.

Kapil came down to Debraj's residence and explained the use of MS Excel for easy calculations like summation, the uses of various formulae, and calculation of average of a set of data and its representation with the help of graphs. Kapil also gave pointers on MS PowerPoint and its features to help Debraj make an effective presentation. He explained the steps involved in choosing a slide design, background, colour scheme, and design template. He also described the use of charts, graphs, and media clips. Debraj was satisfied with Kapil's suggestions and made an excellent presentation before the resident manager the following day.

MICROSOFT OFFICE SUITE

Personal computers have become powerful and versatile, and are popular among individuals and organizations. There has been a rapid increase in the usage of computers due to the availability of systems and user-friendly software.

Microsoft has a suite of software packages that meet the standard requirements of most organizations. The software suite of Microsoft Office includes various application packages such as word processing, spreadsheets, presentations, and database management.

Word Processing

Word processing is perhaps the most popular activity that users carry out on their personal computers. It helps in the accomplishment of both simple (e.g., writing a letter or notice) and complex (e.g., designing a web page or newsletter) tasks.

A word processor processes words or text and then edits, modifies, deletes, or prints it. Thus, word processors help in moving and shuffling typed text as per the requirement of the user.

Word processing refers to activities pertaining to the presentation of text and graphics, usually used for the purpose of printing or viewing online (i.e., on the web). Several years ago, word processing used to be carried out by manual or semi-manual systems such as electronic typewriters and teletypewriters. Word processing has come a long way since then.

Spreadsheet

We can tabulate information by arranging numerical data in rows and columns in a spreadsheet. Though Excel can be used for processing text, it is mostly used for the analysis of numerical data. With the help of Excel, we can use features such as formula, function, and logic, as well as create tables, charts, and graphs for data analysis.

Presentation

Sometimes we need to create a presentation for a board meeting, departmental meeting, or to address a large audience. For this, a PowerPoint package is necessary to connect the system to a visual output device. Using PowerPoint, we can create slides, transparencies, and speakers' notes. Tables, graphs, charts, images, and audio and video clips can also be added to a slide to make a presentation more interesting and realistic.

Database Management

A hotel usually stores a lot of data pertaining to guests and employees such as guest IDs, names, addresses, phone numbers, fax numbers, and e-mail addresses. It is difficult to operate and retrieve guest data stored in manual registers. MS Access, the database management package, helps us create, store, retrieve, and query information easily.

To work with MS Word, MS Excel, MS PowerPoint, and MS Access, we have to load the MS Office 2003 suite in the system.

MS WORD 2003

Microsoft Word is a Windows-based word processing application. We can start MS Word from the Start menu or by using the Run command. Figure 3.1 shows MS Word being opened using the Start menu.

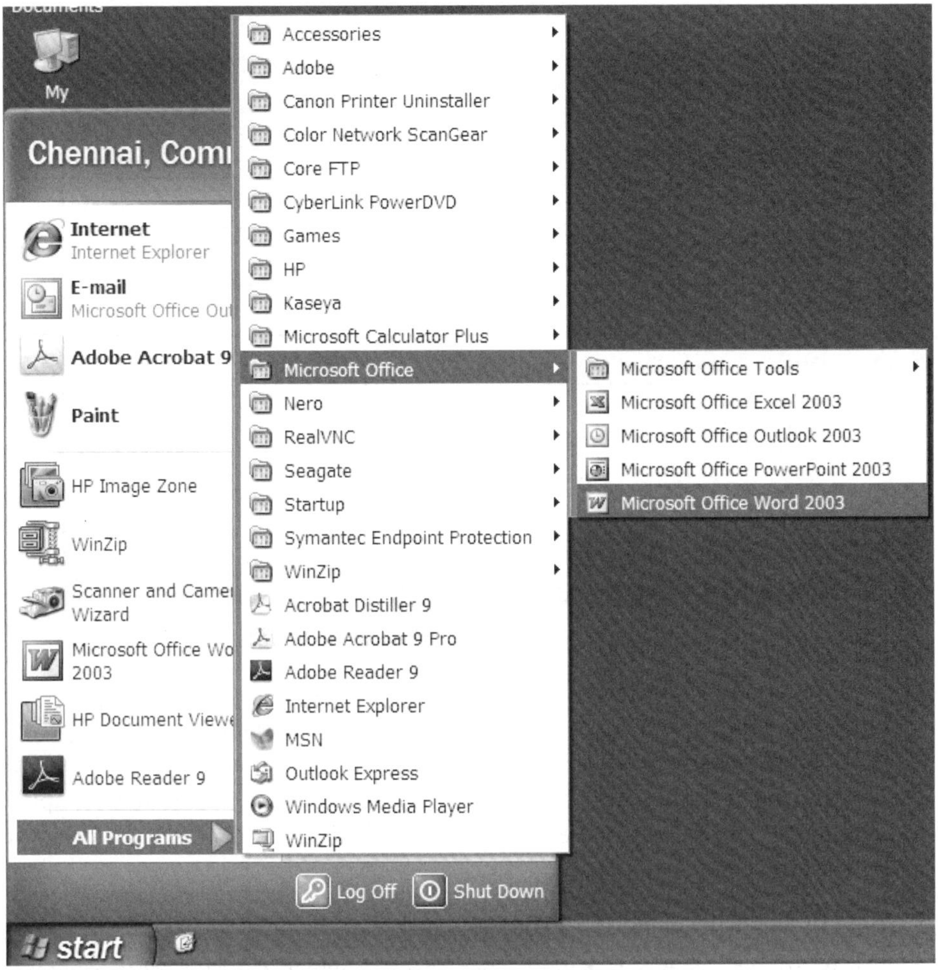

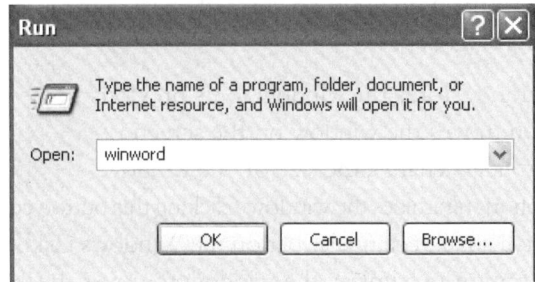

Fig. 3.1   Starting MS Word in Windows XP

- Using the Start menu

Step 1: Click the Start button on the taskbar.

Step 2: Choose Programs.

Step 3: Select Microsoft Office.

Step 4: Click Microsoft Word 2003.

- Using the Run command

Step 1: Click the Start button on the taskbar.

Step 2: Choose the Run option to display the Run dialog box.

Step 3: Type winword in the text box and then click OK (as shown in Fig. 3.2).

Fig. 3.2   Starting MS Word using Run command in Windows XP

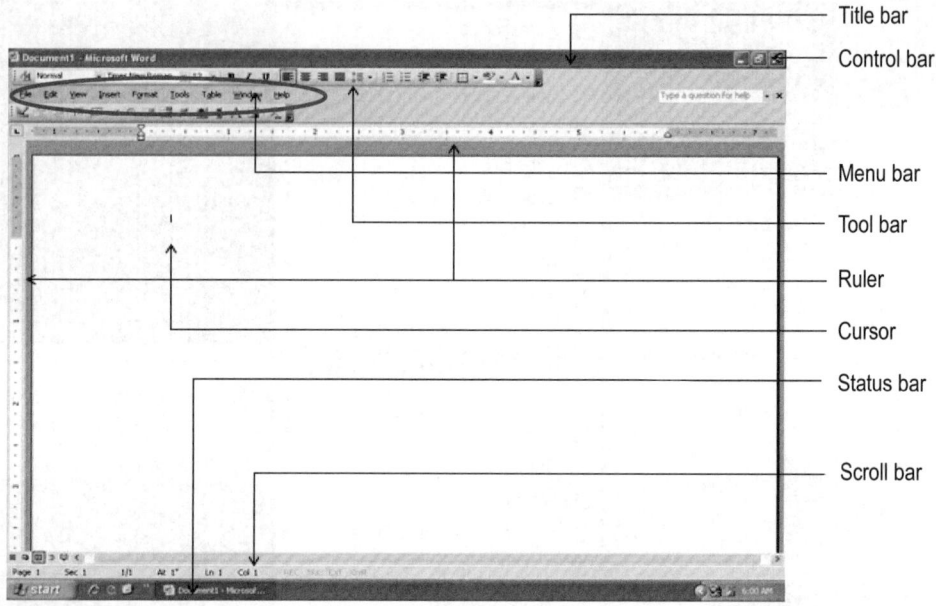

Title bar

Control bar

Menu bar

Tool bar

Ruler

Cursor

Status bar

Scroll bar

Fig. 3.3 Typical MS Word window

Execution of this process will start MS Word 2003 on the computer. As a result, the MS Word window will appear on the screen and Document1 will be displayed (as shown in Fig. 3.3) on the left-hand corner of the Title bar. By default, MS Word opens Document1. This can be renamed and saved. The look and size of this Windows application can be customized and hence it varies from one installation to the other. A typical Word window is shown in Fig. 3.3.

Major Components of MS Word Window

The Word window is a rectangular box through which the Microsoft program interacts with the users. Its components are explained here:

Title (Caption) bar The Title bar shows the name of the Windows application and the name of the currently opened document (e.g., Document1). The window automatically names the document numerically as Document1, Document2, and so on (as shown in Fig. 3.3).

Control box The Title bar also has a Control box and a few control buttons. When we right-click the mouse anywhere outside the Control box (but on the Title bar), a menu containing the following items appears (as shown in Fig. 3.4):

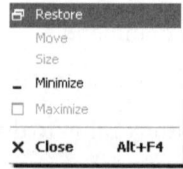

Fig. 3.4 Control box in MS Word

- The *Restore* option restores the window to its previous size.
- The *Move* option moves the window on the screen.
- The *Size* option helps change the size of the window.
- The *Minimize* option minimizes the window. Clicking this button compresses the window into a small rectangular bar on the Windows taskbar.
- The *Maximize* option maximizes the window to cover the entire area of the desktop.
- The *Close* option closes the window and terminates the program.

Menu bar   The Menu bar displays menus that are available to the users. A menu is a group of logically connected commands. This bar is located under the Title bar (as shown in Fig. 3.3). File, Edit, View, Insert, Format, Tools, Table, Window, and Help are some of the menus on the Menu bar.

File menu   The File menu has a set of commands that help in maintaining file operations such as opening an existing document, saving the current document, and also printing a document.

Edit menu   The Edit menu has a set of commands that perform operations related to editing and manipulation of the content.

View menu   The View menu contains a set of commands that can be used to display the document in different views such as Normal, Web layout, or Print layout.

Insert menu   The Insert menu contains a set of commands that can be used to insert various objects such as clip art, auto shapes, and text boxes in a document.

Format menu   The Format menu contains a set of commands that can be used to alter the look and layout of the content present in the document. These include changing the font type, font colour, font size, and font style.

Tools menu   The Tools menu contains a set of commands that can be used to check and correct spellings and grammatical mistakes, count the number of words and characters, etc. The Mail Merge function will be discussed later in this chapter.

Table menu   The Table menu contains a set of commands that help to create, modify, and delete tables in a document.

Window menu   The Window menu contains a set of commands that help us perform tasks related to the active window in which we are working such as opening a new window, splitting an active window into different panes, and arranging opened documents into separate windows.

Help menu   The Help menu provides information on MS Word from sources such as the Office Assistant tool and Microsoft Office Web.

Toolbar   A toolbar (as shown in Fig. 3.3) is a group of icons (referred to as tool buttons), each of which represents a menu command. On clicking these buttons, the respective commands can be executed. Thus, this is an easier and faster way of command execution. Toolbars generally appear below the Menu bar but can be placed anywhere by clicking and dragging them. Toolbars can be created with customized tool buttons as per one's requirement. When a mouse pointer is placed over a tool button, a brief message is displayed for a short while. This is known as a tool tip.

Ruler   The MS Word window is supplied with one horizontal and one vertical ruler, which are displayed along the top and left margins of the document, respectively (as shown in Fig. 3.3). Rulers can be used to set document margins and indents (a blank space between a margin and the beginning of a printed or written line) in an easy manner, providing measurements while formatting the page so that consistency is maintained throughout the document.

Cursor   The cursor (as shown in Fig. 3.3) is an MS Word pointer (insertion pointer) that helps us to type text, insert graphics (tables), and overwrite. A cursor can be moved and placed on the document using a pointing device (mouse) or keyboard arrow buttons.

Status bar   This bar displays the current position of the cursor, status of important keys on the keyboard (for e.g., if the Caps Lock key is on or off), and other relevant information. It is located at the bottom of the window (as shown in Fig. 3.3).

Scroll bar   Scroll bars are sliders that can be moved with the help of a mouse. As the scroll bar is moved, the window spans through the document exposing different portions of the document. The two types of scroll bars—horizontal and vertical—are placed at the bottom and right-hand side of the document (as shown in Fig. 3.3).

Working

When we open a word document, a blank word document (e.g., Document1 as shown in Fig. 3.3) appears. To start working on a word document, it is essential to know the various short cut keys available. Some of them are Home, End, Page up, Page down, Enter, and key combinations with Ctrl, Shift, and Alt. Keyboard short cuts help in easy and quick navigation through computer software. Table 3.1 lists out the keyboard short cuts for moving the cursor.

Table 3.1   Key combinations for cursor movements

Movement of the cursor	Keys
One character to the left or right	← or →
One word to the left or right	Ctrl + ← or Ctrl + →
One line up or down	↓ or ↑
One paragraph up or down	Ctrl + ↑ Ctrl + ↓
To the beginning or end of a line	Home or End
Up or down one screen	PgUp or PgDn
To the top or bottom of the current screen	Ctrl + PgUp or Ctrl + PgDn
To the beginning or end of the document	Ctrl + Home or Ctrl + End
To the top or end of the window	Alt + Ctrl + PgUp or Alt + Ctrl + PgDn
To the location of the insertion point when the document was last closed	Shift + F5
To the next line	Enter
To remove a character to the right of the cursor	Del
To remove a character to the left of the cursor	Backspace

Operations Performed

MS Word is generally used for creating professional as well as personal documents in an efficient manner. We will now discuss the basic operations performed in MS Word.

- Creating a document
- Opening a document
- Saving a document
- Printing a document
- Closing a document
- Editing a document
- Mail Merge

Creating Documents

We can create a document by clicking the File menu and choosing New; pressing Ctrl+N; or clicking the Blank document icon on the Standard toolbar.

After the execution of the command, the MS Word window will appear (as shown in Fig. 3.5) showing the task pane. It is a rectangular pane found on the right-hand side of the window and appears by default. We can consider the task pane to be an open menu that can be used to open a new or existing document.

Note: If the task pane is not visible, we can open it by selecting the View menu and then choosing the task pane option or by pressing Ctrl+F1 simultaneously.

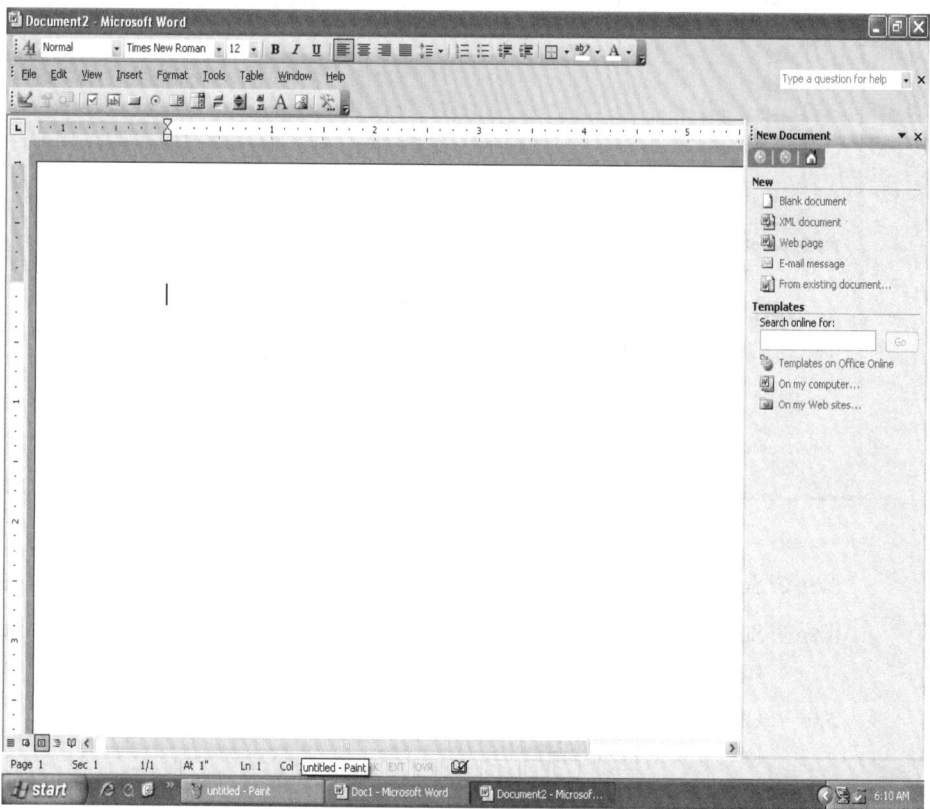

Fig. 3.5   Opening a blank document in MS Word

There are many types of pre-designed documents—blank document, XML document, web page, e-mail message, from existing document—available in the task pane (as shown in Fig. 3.5)

- Blank document: This option is used if the user wants to create a printed document.
- Blank web page: This option is selected if the user wants to display the document's content on an intranet or Internet through a web browser. A web page opens in a Web layout view. Web pages are saved in HTML format with a .html extension.
- Blank e-mail message: If we use Outlook 2000 or Outlook Express, we can use this option to compose and send a message or document to others, directly from Word.

Opening Documents

We can open an existing document by performing the following steps:

Step1: Select the Open option from the File menu; press Ctrl+O; or click the Open tool from the Standard toolbar. The Open dialog box will appear (as shown in Fig. 3.6).

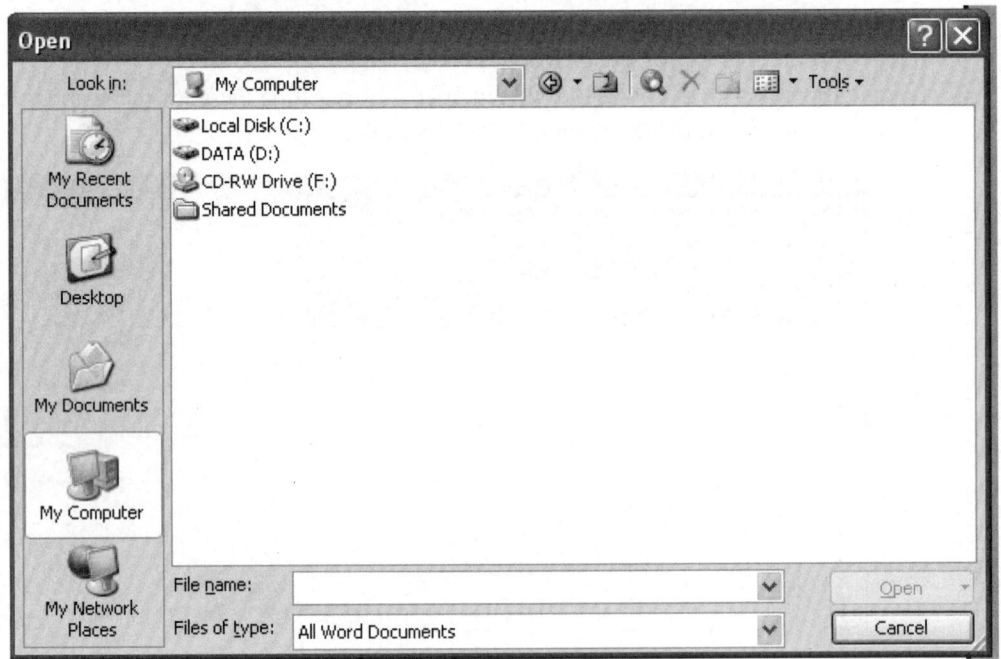

Fig. 3.6   Open dialog box in MS Word

Step 2: Select the appropriate folder.

Step 3: Select the required file from the file window or write the required file's name in the file name tab.

Step 4: Click the Open button on the right-hand side of the dialog box or press the Enter key on the keyboard.

Saving Documents

We can save the active or current document by performing the following steps:

Step 1: Select the Save command in the File menu; press Ctrl+S; or click the Save tool on the Standard toolbar. If you are saving the file for the first time, the Save As window will appear (as shown in Fig. 3.7).

Step 2: Choose the appropriate folder from the Save in box to save the document.

Step 3: Write a document name in the File name text box.

Step 4: Click the Save button.

If we close an open document without saving the latest content, MS Word prompts us to save the changes in the document.

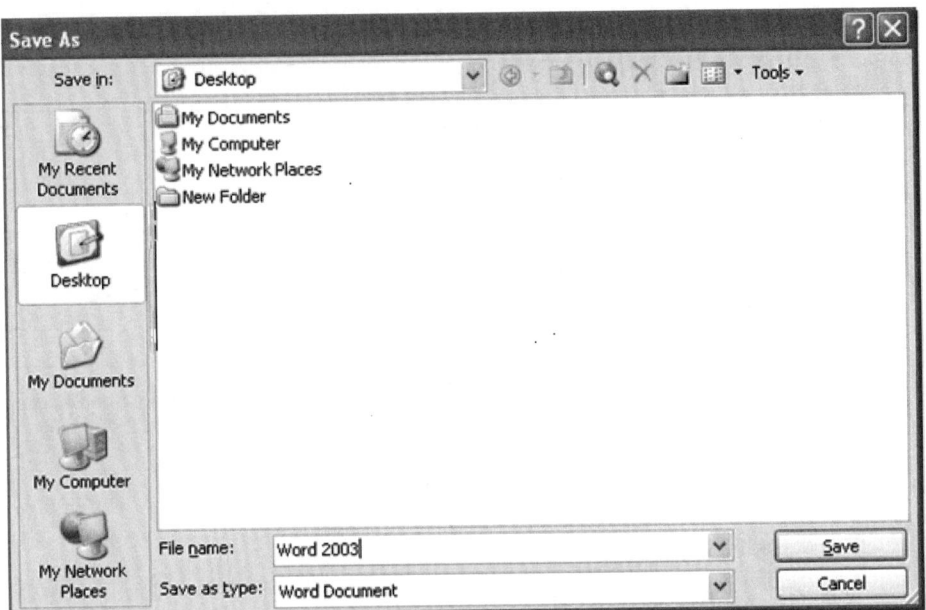

Fig. 3.7   Save As window in MS Word

Save As   This option is used for saving a copy of the document in the same or with a different name.

Step 1: Select the Save option from the File menu.

Step 2: Click the Save As option.

Step 3: Repeat the steps mentioned under Save option.

Both commands—Save and Save As—save a document. The difference between them is that the 'Save As' command allows the user to save a file with a different name and format, whereas, the Save option saves the document in the same name as it was saved for the first time.

Save as web page   This option will save the word document in the web page format. We can also view it using the web browser.

Printing Documents

The following steps describe the Print preview and Print options.

Print preview   This option would display the document as it would look when printed on a page. The steps involved are as follows:

Step 1: Select the File menu.

Step 2: Click the Print Preview option or click the Print Preview tool.

Print   This option prints the active document by default. The steps involved are as follows:

Step 1: Select the File menu.

Step 2: Click the Print option; click the Print tool; or press Ctrl+P. The print dialog box appears (as shown in Fig. 3.8).

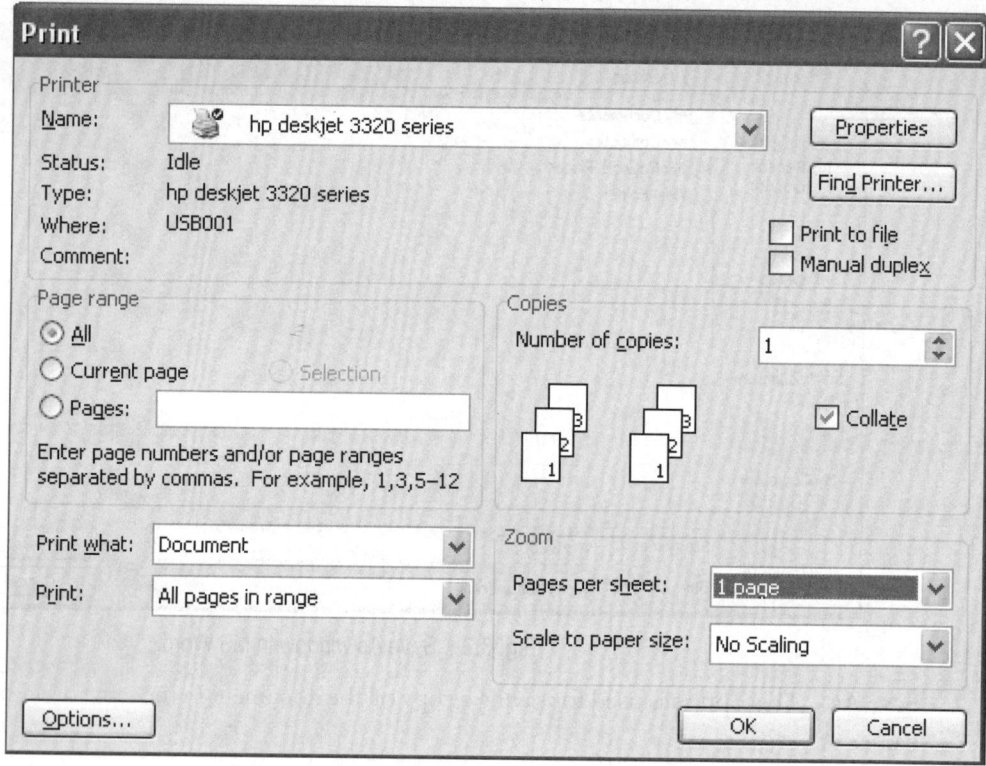

Fig. 3.8 Print dialog box in MS Word

Step 3: Select Page range 'All' (as shown in Fig. 3.8).

Step 4: Select OK to print.

Difference between Print Preview and Print Print Preview helps us view a full page layout of the active document and also helps us make changes before a printout of the document is taken. It is a soft copy output. On the other hand, by using the Print option, we can print the active document without any alteration. It is thus a hard copy output.

Closing Documents

We can close the active or current document and quit from the MS Word window.

Closing files To close the current document, select the File menu and click the Close command. **Exiting from word** To exit from the MS Word window, carry out one of the following steps: select the File menu and click the Exit command; press Alt+F4, or click the Exit tool on the toolbar.

Difference between Close command and Exit command While working with multiple documents in MS Word, if we choose the Close option, the active document will be closed. On the other hand, if we opt for the Exit option, we will quit from all the documents in MS Word.

Editing Documents

All necessary editing commands have been grouped under the Edit menu and on the toolbar. The Edit menu is shown in Fig. 3.9.

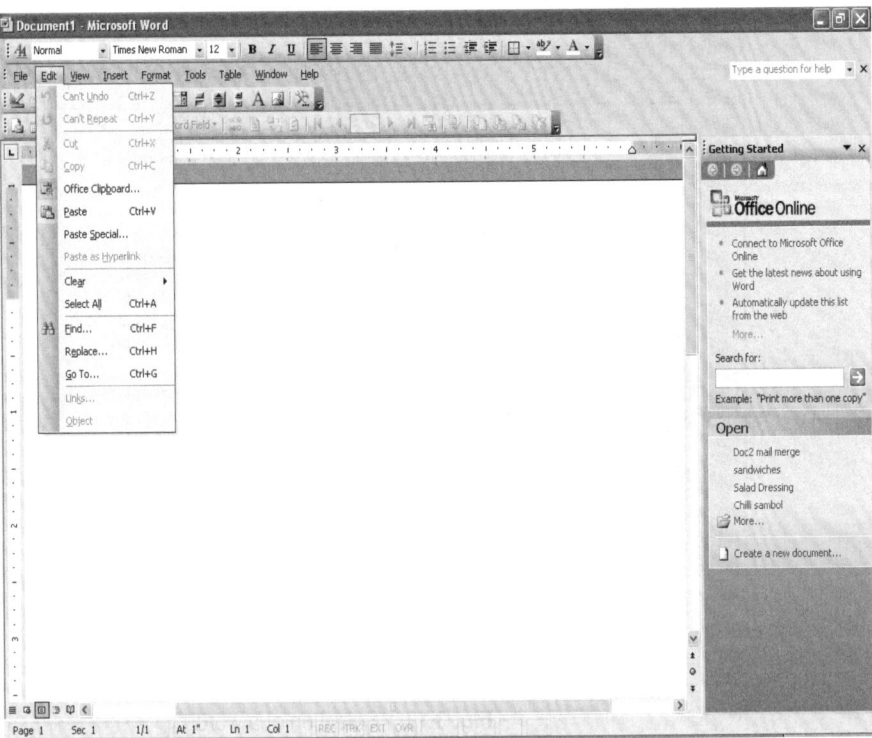

Fig. 3.9   Edit menu

Selection Option

In MS Word, we can select characters, words, lines, or paragraphs that are part of a document.

- To select a word, one of the following steps is carried out:
  - Place the mouse pointer on the word and double-click it;
  - Place the cursor before the word, press the left button of the mouse, and drag it till the end of the word; or
  - Place the cursor before the word and then press Shift + Ctrl + Right arrow key.

The selection of a word is shown in Fig. 3.10.

- To select a line in a text document, one of the following steps is carried out:
  - Point the mouse to the left margin of the document. The mouse pointer will change to an arrow pointing opposite to usual direction. Now click the left mouse button only once.
  - Place the cursor in front of the first character of the line and press Shift + End key (as shown in Fig. 3.11).

- To select a paragraph in a text document, one of the following steps is carried out:
  - Place the mouse pointer to the left margin of the document and double click. The entire paragraph will be selected.

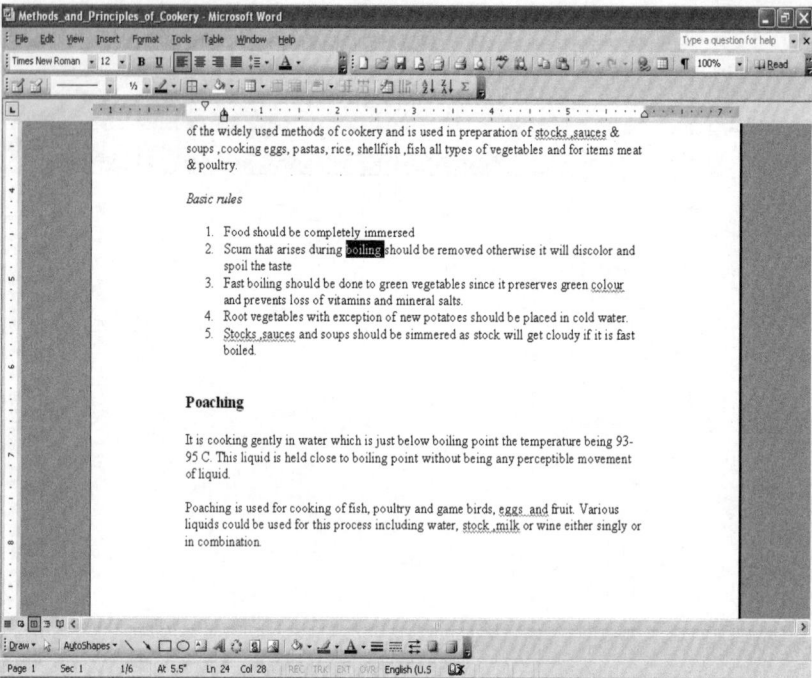

Fig. 3.10   Selection of a word

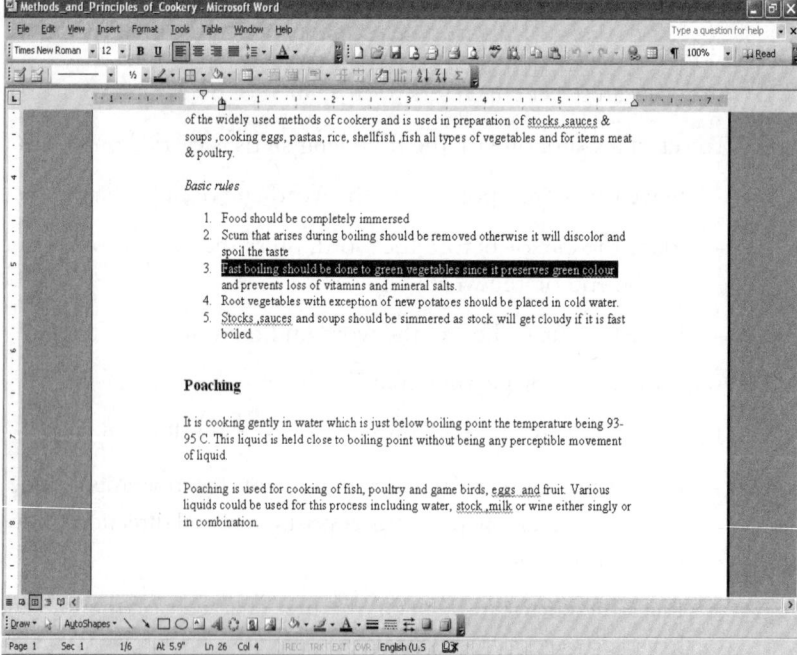

Fig. 3.11   Selection of a line

– Place the cursor on the first character of the paragraph and press Ctrl + Shift + Down arrow keys (as shown in Fig. 3.12).

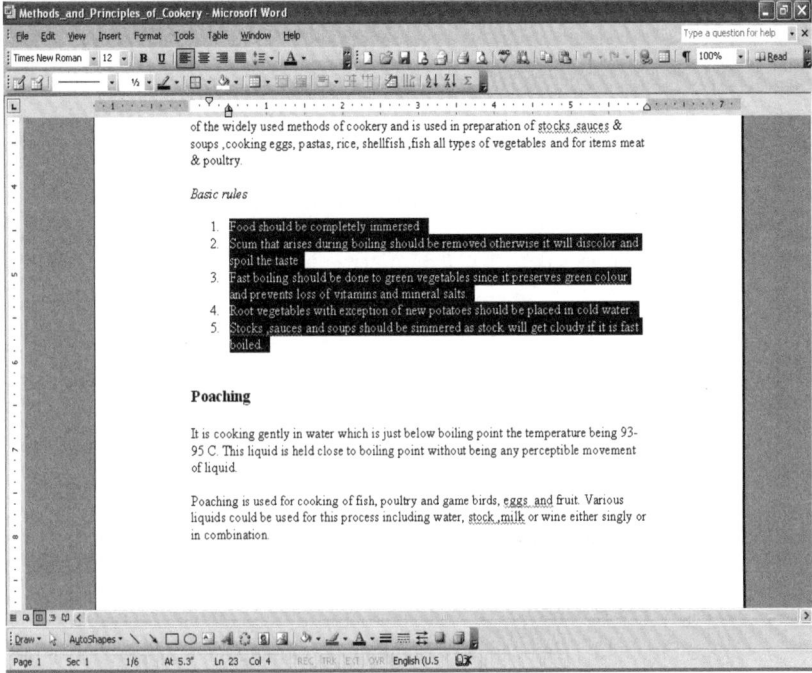

Fig. 3.12   Selection of a paragraph

Undo/Redo Command

Sometimes, after making certain changes, we might wish to annul those changes. MS Word provides a command for undoing whatever was done in the previous step.

Undo   The procedure to cancel the changes made is as follows:

Step 1: Select the Undo option under the Edit menu; click the Undo button on the Standard toolbar; or press Ctrl+Z.

Step 2: The undo option displays all recent actions that MS Word can undo. We can select the appropriate action to be undone from this list.

Redo   If an undo option has to be reversed, it is known as Redo. The procedure to execute this command is as follows:

Step 1: Select the Redo option under the Edit menu; click the Redo button on the Standard toolbar; or press Ctrl+Y.

Cut and Paste

We will now discuss the Cut and Paste options in MS Word.

The Cut operation removes the selected object or objects from the present document and puts them in a memory called the clipboard. The clipboard allows pieces of information to be temporarily stored so that it can be retrieved later by another application. The object may be a text, picture, table, or any component that is available in the document.

The Paste operation inserts content (that might be from any source) from the clipboard at the insertion point in the active document.

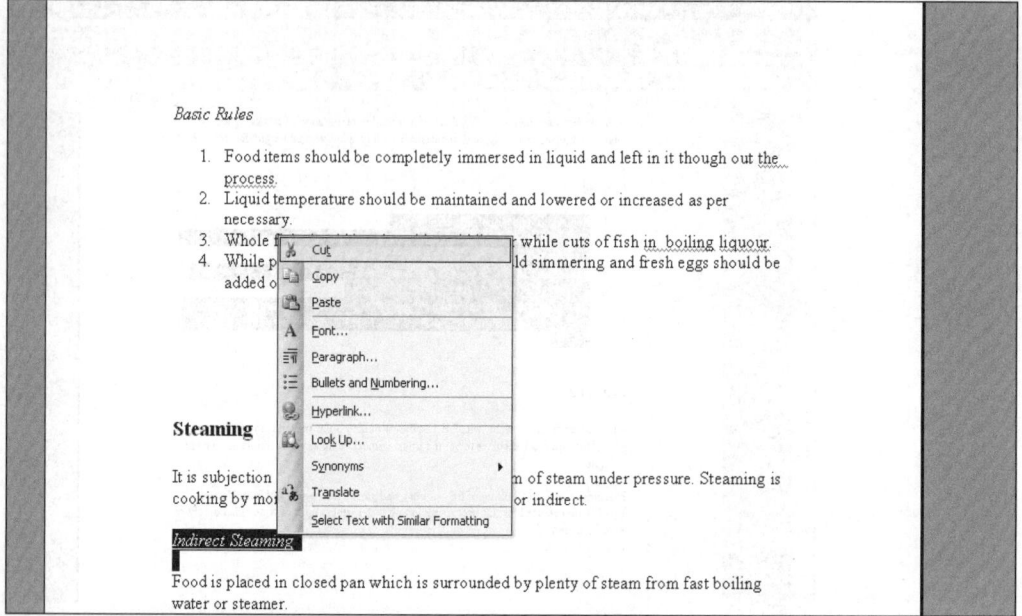

Fig. 3.13   Selecting text and then performing the cut operation in MS Word

To cut text or an object, perform one of the following steps after selecting the text(s) or object(s):

– Select Cut from the Edit menu; click the Cut button on the Standard toolbar; or press Ctrl+X.

– Move the mouse pointer on the selected text, right-click, and select Cut (as shown in Fig. 3.13).

To paste from the clipboard, perform the following steps:

Step 1: Move the cursor to the location where the text/object is to be pasted.

Step 2: Select Paste from the Edit menu; click the Paste button on the Standard toolbar; press Ctrl+V; click the right mouse button and select Paste from the context menu (as shown in Fig. 3.14).

Copy and Paste

Copying is the duplication of the selected text/object of a document or any other source, to a desired place in the same or different document. The procedure for copying text is similar to that of moving text. The only difference is that cutting removes the object(s) from the original place whereas copying leaves it as it is. To copy a particular text/object, perform the following steps:

Step 1: Select the text or a block of the text.

Step 2: Select Copy from the Edit menu; click the Copy button on the Standard toolbar; press Ctrl+C; or click the right mouse button and select Copy from the context menu.

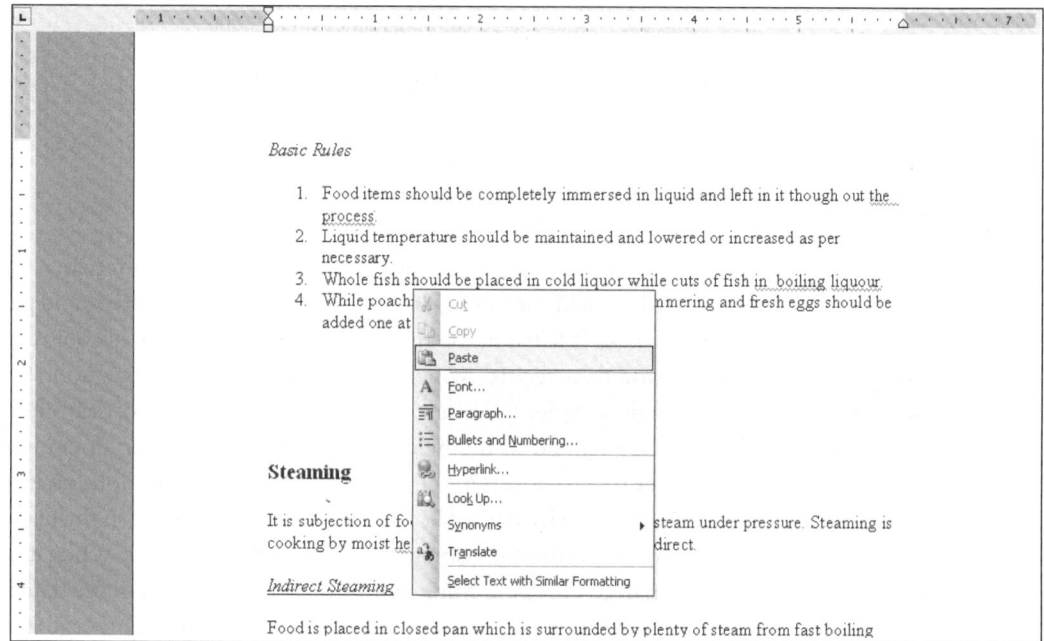

Fig. 3.14   Paste operation after cut operation has been carried out in MS Word

The Paste operation is similar to that described in the section on Cut and Paste.

MAIL MERGE

Mail Merge allows us to merge a document (e.g., a letter) and a set of information (e.g., the names and addresses of all clients) together. When we merge the letter and list of names, we can send mass mail. Each copy of the document we print will be personalized with a different name and address. Typing a letter once, combining it with the client list, and then printing it, helps us to conveniently mail many clients. It would have been time-consuming and cumbersome if mails had to be typed out separately for each mail address. Mail Merge can be used for sending invitations to guests for special events in hotels and for sending weekly or monthly bills to guests in clubs or residential hotels.

Getting Ready for Mail Merge

Before we start learning about Mail Merge, we need to understand the two documents that are required for this.

Main document: The main document is one that contains text and graphics to be included in all the copies of the merged document. For example, the body of an invitation is the main document. The main document can be in the form of a letter, an invoice, or any other type of document. The idea is that the document can go to many recipients and some unique information can be included to personalize each copy.

Data source: It is a document that contains text and graphics that would vary with each version of a mail merge document, for e.g., a list of names and address to whom the document should be sent to (as shown in Table 3.2).

Table 3.2 Data source showing names and organizations

First name	Last name	Organization
Pritam	Saxena	Reliance
Piyush	Sharma	Glaxo
Sandeep	Bhatnagar	Nestlé

We can consider a data source to be a table. Each column in the data source corresponds to a category of information or data field such as first name, last name, street address, and postal code.

The name of each data field is listed in the first row of cells, called the header record. Each subsequent row contains one data record, which is a set of related information (for e.g., the name and address of a single recipient). When we complete the merge operation, individual recipient information will be mapped to the fields we included in the main document.

Merge fields: It is a placeholder that we insert in the main document. For example, on inserting the merge field <city>, Word inserts a city name like 'New Delhi' that is stored in the city data field.

Merge fields are used to customize the content of individual documents. When inserted into the main document, they map to corresponding columns of information in the data source. If MS Word doesn't find the information it needs by automatically mapping merge fields to the headings in the data source, it gives us the opportunity to do so, when inserting addresses and greeting fields or previewing the merge.

Mail Merge Wizard

The Mail Merge Wizard helps us create form letters, mailing labels, envelopes, directories, mass e-mails, and fax distributions. To complete the basic process, perform the following steps:

- Open or create a main document.
- Open or create a data source with individual recipient information.
- Add or customize merge fields in the main document.
- Merge data from the data source into the main document to create a new, merged document.

The wizard guides us through the aforementioned steps. If we prefer to work without the help of the wizard, we can use the Mail Merge toolbar. The end result is that each row (or record) in the data source would produce individual form letters, mailing labels, envelopes, or directory items. The Mail Merge Wizard completes the process in a few steps. The steps involved are as follows:

Step 1: Go to Tools, choose Letters and Mailings, and then select Mail Merge (as shown in Fig. 3.15).

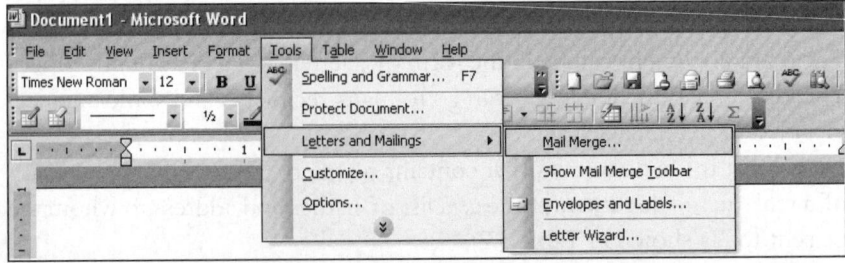

Fig. 3.15 Choosing Mail Merge from the toolbar

Step 2: Select the document type from the available options (as shown in Fig. 3.16).

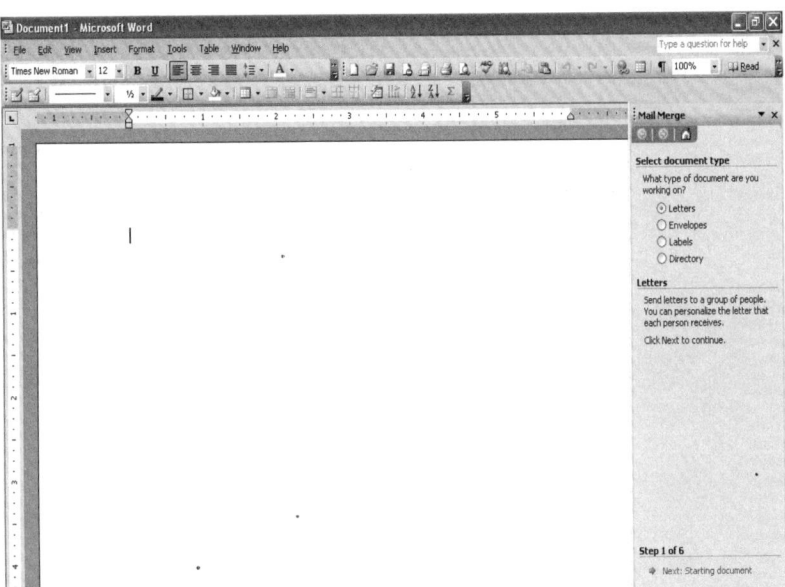

Fig. 3.16   Selection of document in Mail Merge

Step 3: Select recipients. Choose one of the given options to start with your letter or main document. We can type a new address list for the source document (as shown in Fig. 3.17).

Fig. 3.17   Selecting recipients

Step 4: Now we have to <u>S</u>ave the list (as shown in Fig. 3.18).

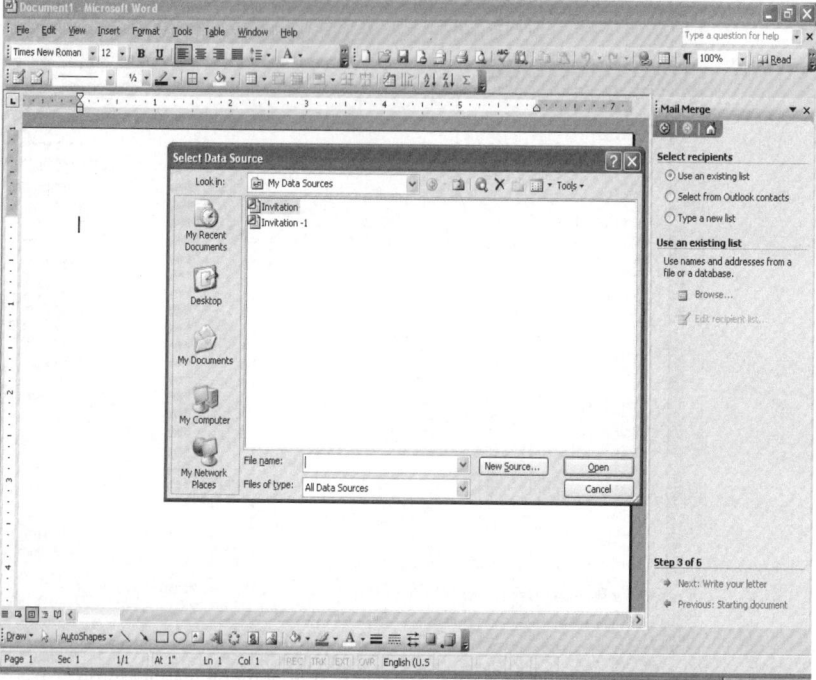

Fig. 3.18   Address list being saved

All the records in the Address list can be seen in the Mail Merge Recipients list box. We can also edit the list, and add or delete a record in the list (as shown in Fig. 3.19).

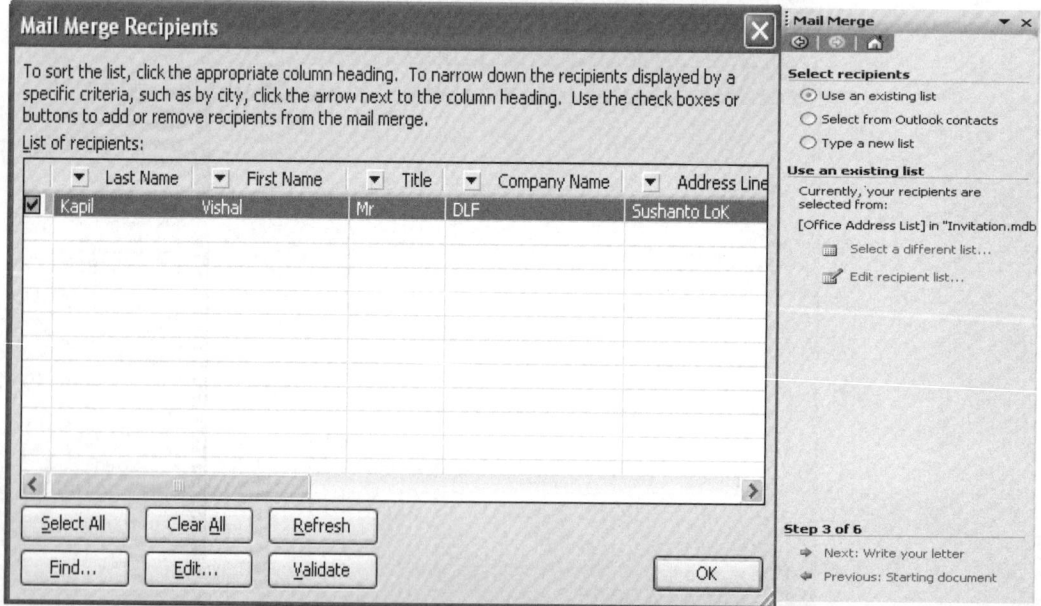

Fig. 3.19   Addresses in the Mail Merge Recipients list in MS Word

Step 5: Now we can write the letter and insert the address block (as shown in Fig. 3.20).

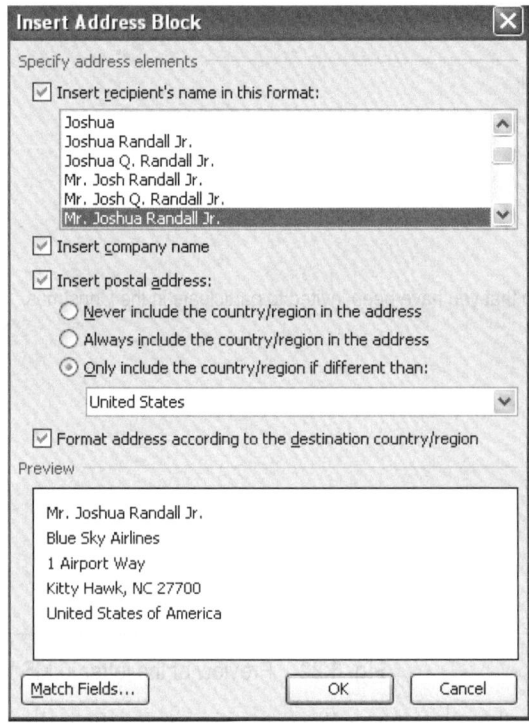

Fig. 3.20   Sample address list in MS Word

Step 6: The main document will look as shown in in Fig. 3.21.

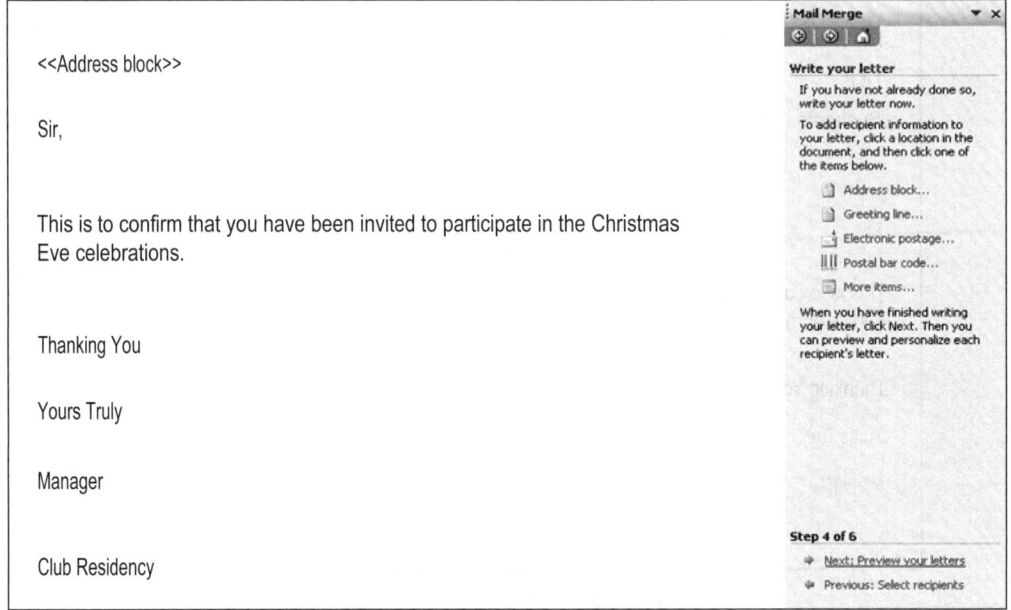

Fig. 3.21   Main document in MS Word

Then preview the letter (as shown in Fig. 3.22).

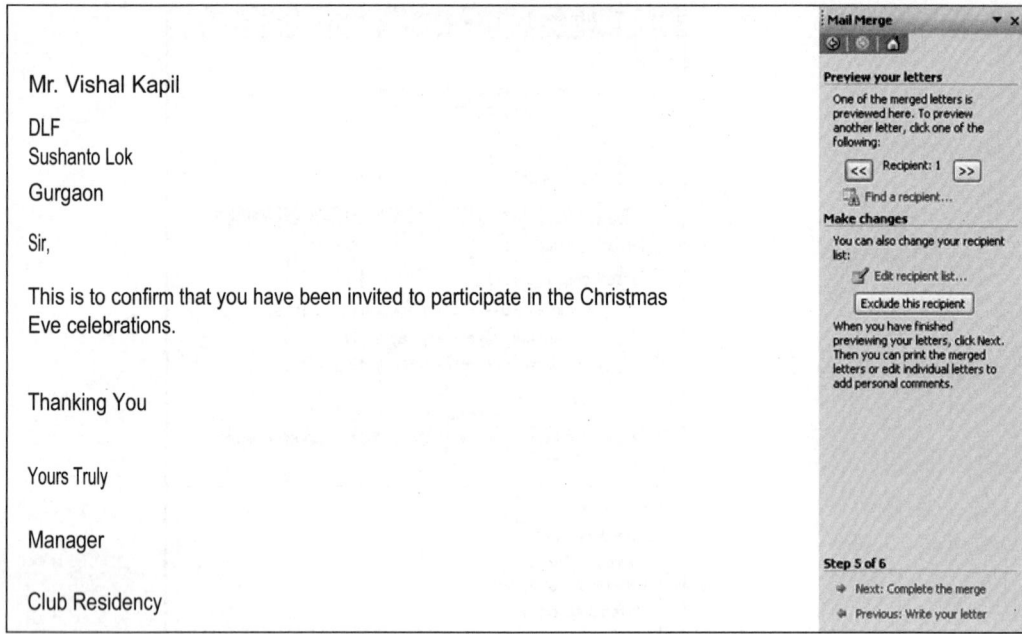

Fig. 3.22 Preview of the letter in MS Word

Step 7: Complete the merge. We can edit or print the individual letters if required (as shown in Fig. 3.23).

Fig. 3.23 Completed merge file in MS Word

Features

We will now discuss the features of MS Word. The features of this application can be used for searching, creating, editing, formatting, assembling, and managing documents. MS Word also offers other options that help us save time. They are as follows:

Converting text to table or table to text   In MS Word, we can easily convert a text to a table and also separate text with a comma, tab, or other separator characters to indicate where a new column should begin. We use a paragraph mark to begin a row. When we select a table and want to convert it to text, we have to remove all separators. By doing this, the original content will appear just like normal text.

Word count   This is very handy when we are typing a document (for e.g., an article or a note for a lecture) where limitation of words is mandatory. Using this facility, we can calculate the number of words in the entire document. MS Word displays the number of words on a miniature toolbar and Status bar.

Documents last worked on   If we need to revert to documents we have worked on, in order to make additional changes, we have the option of seeing up to nine of these documents on the File menu. By default, it lists the last four documents we have opened.

Templates   We can use templates to create legal forms, business forms, financial statements, address labels for CDs or DVDs, confidential labels, envelope seals, stickers, seasonal decorations, greeting cards, reward certificates, etc.

Spelling and grammar checker   When we type a document, MS Word automatically checks the document, helping users identify spelling and grammatical mistakes. MS Word offers suggestions of possible corrections from which users can choose the right one. For this, MS Word maintains internal dictionaries. One dictionary is active at a time. By default, spelling errors are underlined in red colour, whereas grammatical errors are underlined in green.

Formatting   Once the text of a document has been typed, users would like to give it a suitable look and layout based on their needs. This is referred to as formatting. Formatting affects the appearance of the document and not its textual/graphical contents. The Format menu helps in giving a meaningful format to the document so that a user can read and understand the contents easily.

Thesaurus   If we encounter any difficulty in understanding a word or if we want a synonym for a particular word, we can take the help of the thesaurus. It will list out synonyms for the selected word that are stored in the inbuilt dictionary.

Mail Merge   This feature allows us to quickly create personalized letters by merging information from two different files. Thus, it is used to insert variable data into a fixed format by combining two files (data source and main document). Using this feature, we can generate mass mails without any error.

Word wrapping   This feature of text editors and word processors allows the program to automatically insert line endings.

Page break   Using this option, the point of flow of text in a document moves to the top of a new page. Most word processing programs create an automatic page break when the material on the page reaches a specified maximum limit.

Auto correct   The auto correct feature automatically fixes common typographical errors and also replaces specific key strokes with special characters. It can be customized as per need. Auto correct can be used to replace abbreviations with the words they represent.

Auto text   The auto text feature stores frequently used text and graphics that can be chosen from a list of Auto text entries or inserted using short cut keys. We can create our own auto text entries for any amount of text and graphics.

Header and footer   Headers and footers refer to text or graphics that are printed at the top and bottom of every page in the document. A header is printed on the top margin, whereas a footer is printed on the bottom margin.

We have so far discussed the creation and editing of a document in MS Word. We also talked about Mail Merge, using which we can send individually addressed letters to a large number of recipients.

MS Word does not support numerical calculations, tabulations, and representations of data. For this we need MS Excel, which is discussed in detail in the following section.

MS EXCEL 2003

Excel is an integrated electronic worksheet (spreadsheet) program developed by Microsoft Corporation. It performs different kinds of calculations and displays the results on the screen in the form of figures or graphs. Excel helps in the preparation of data in an organized, orderly, and meaningful format.

Applications

An electronic spreadsheet program is used to perform calculations, store information in the memory of the computer, and display information or results in the required format on the computer screen. This application is used to create reports, use formulae, and perform complex calculations. It is best suited for scientific and statistical analyses. Excel is used to prepare profit and loss statements, annual reports, and budget of hotels. In addition, Excel is also used in the evaluation of hotel or business firms, banking and other financial firms, and in the compilation of tax statements. The applications of this package are also visible in the preparation of analytical reports including statistical analysis, forecasting, and regression analysis. Many different kinds of charts can be prepared and presented in an attractive way, thus helping in the depiction of data in a structured manner. MS Excel is also used to create a relationship between different types of data.

Structure

The smallest unit of an addressable data container in Excel is the *cell*. A cell is formed by the intersection of a *row* and a *column*. In a worksheet, each row has a unique number such as 1, 2, 3, ..., 65536; each column has a unique name such as A, B, C, ..., Z, AA, ..., IV. Therefore, each cell has a unique *cell address* in a worksheet. For example, in D13, D denotes the column name and the number 13 denotes the row number.

A *worksheet* is rectangular collection of many cells. A single Excel worksheet contains 65,536 rows and 256 columns and therefore 1,67,77,216 (65,536 × 256) cells. By default, each worksheet in a workbook has a unique name (Sheet1, Sheet2, etc.). The user can also change the sheet name.

A *workbook* is a collection of one or more worksheets. A workbook is saved as a file in the secondary storage device with .xls as its default file extension. Each workbook is stored as a file with a unique name, for example, hotelraj.xls. Each Excel workbook can have a maximum of 255 worksheets.

Starting MS Excel

MS Excel can be opened in any of the following ways:

- The Excel directory contains a main executable file. To start Excel from DOS, this file can be included in the path statement of the Autoexec.bat file in the root directory. Once the path is set, all Windows applications close. Now the user will have to type Win Excel and press the Enter button.
- By double-clicking an icon already created on the desktop
- From the quick launch box icon
- From the Windows taskbar in the following way:

Step 1: Click the Start button on the Windows taskbar.

Step 2: Select the Program option.

Step 3: Click the Microsoft Excel 2003 option (as shown in Fig. 3.24).

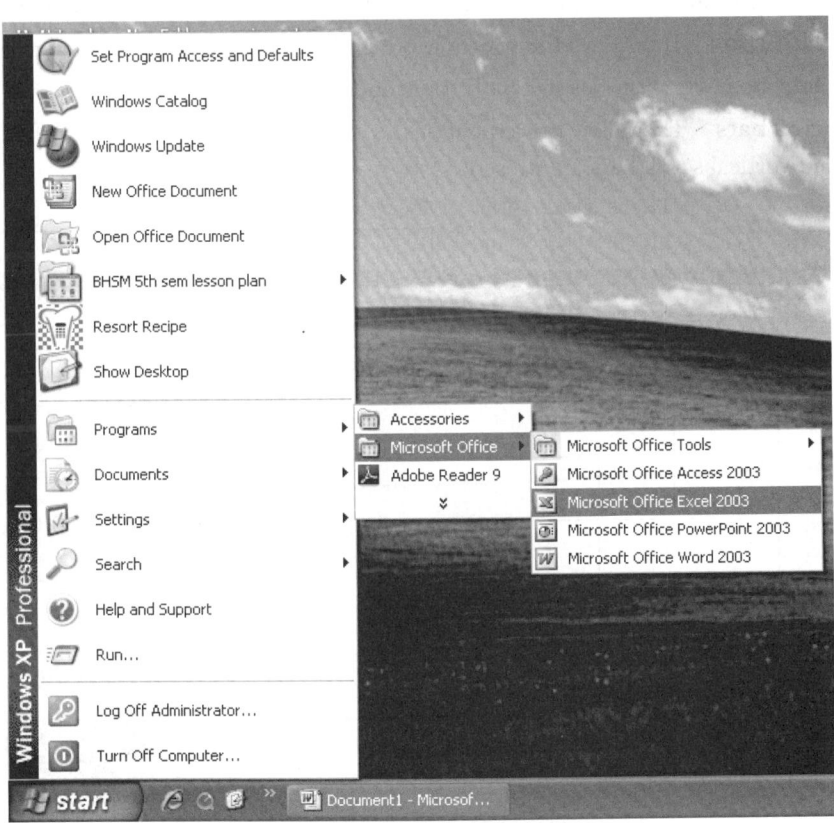

Fig. 3.24 Starting MS Excel in Windows XP

- By using the Run command as follows:

  Step1: Click the Start button on the taskbar.

  Step 2: Choose the Run option to display the Run dialog box.

  Step 3: Type excel in the text box and then click OK.

Components

A typical Excel 2003 Window is shown here. It may look different in a different installation as the components of the Excel window can be customized. The Excel window has the following parts (as shown in Fig. 3.25).

Title bar   It shows the name of the application (Excel) and the name of the workbook, and is situated at the top of the window.

Menu bar   It contains commands under different topics to perform special tasks and is located below the Title bar.

Toolbar   It is a collection of short cut buttons, represented by icons, for the options under the Menu bar.

Status bar   It displays the current status of the application.

Formula bar   It displays and allows editing of formula in the current cell.

Name box   It displays the row number and column name of the active cell.

Help box   It allows us to type a topic and search for help.

Scroll bars   They allow movement of the worksheet in and out of the window view. Horizontal and vertical scroll bars are present.

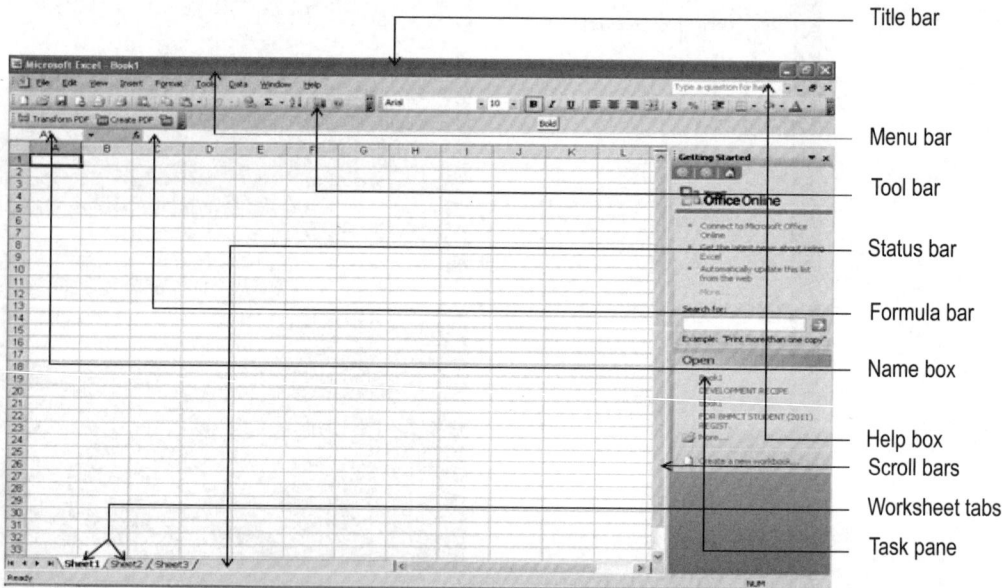

Fig. 3.25   Typical MS Excel 2003 starting window

Worksheet tabs　They make the desired worksheet active.

Task pane　It displays commands that are currently available.

Operations Performed

Excel helps us sort numerical data, prepare tables, perform financial calculations, and analyse data. The basic operations performed in MS Excel are as follows:

- Creation of a worksheet
- Saving a worksheet
- Insertion of functions in a worksheet
- Insertion of charts in a worksheet

Creating Worksheets

We can create a worksheet in MS Excel by simply inserting data in the cells of the worksheet. To create a worksheet, perform the following steps:

Step 1: Open the MS Excel – Book 1 window.

Step 2: Insert data into the cells according to requirement (as shown in Fig. 3.26).

Saving Worksheets

After entering data in the worksheet, we need to save the worksheet at a suitable location in the computer system. To save a worksheet, perform the following steps:

Step 1: Click the File menu.

Step 2: Select the Save option.

Step 3: Choose a destination to save the worksheet.

Step 4: Enter the file name.

Step 5: Click the Save button.

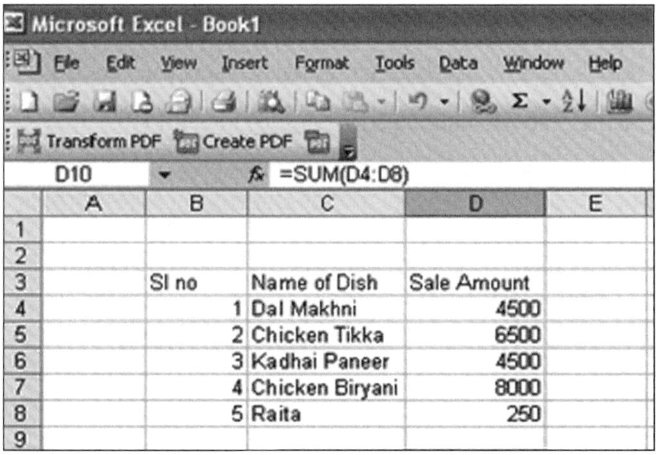

Fig. 3.26　Entering data in MS Excel

Inserting Functions in Worksheets

A function is a predefined formula that takes zero or more input values (also known as arguments) and produces (or returns) a result. The arguments can either be constant quantities or references to cells or ranges as long as data types of all the participating values are consistent. The general structure of a function is as follows:

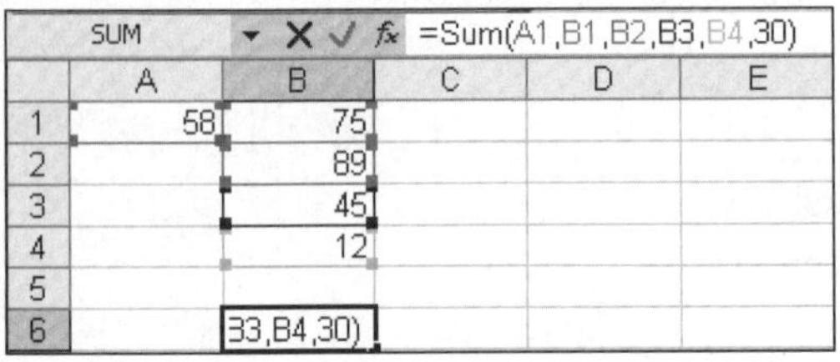

Fig. 3.27   Summation in cells in MS Excel

Function_name(comma separated argument list)

For example, SUM is an inbuilt function that may be used as follows:

SUM(A1,B1:B4,30)

The result will be the sum of values stored in cells A1, B1, B2, B3, B4, and 30 (as shown in Fig. 3.27).

MS Excel includes a wide variety of inbuilt functions that can be used in spreadsheets. They are grouped into logical categories as follows:

Mathematical   SUM, EXP, POWER, LOG, etc.

Statistical   AVERAGE, COVAR, FISHER, etc.

Logical   AND, OR, NOT, etc.

Date and time   DATE, DAY, HOUR, etc.

Text   LEFT, LEN, MID, etc.

In most cases these functions will suffice. However, sometimes we may not find an inbuilt function to meet our requirement. In this case, we can create our own function in Visual Basic and later use it in a worksheet.

Inbuilt function   It is very easy to insert a function in a cell as Excel assists us while using these. There are broadly two ways of inserting functions, manually and by using the function wizard.

We will now take an example of the sale amounts of a few dishes that were sold in a restaurant. Let us calculate the total sale amount using the SUM function. To insert a function in a cell, follow this procedure:

Step 1: Click the cell in which we want to insert the function.

Step 2: Enter the formula =SUM(D4:D8) in cell D10.

Step 3: Now press Enter.

The total appears in cell D10 (as shown in Fig. 3.28).

We can also calculate the average sale of all the dishes in D11. Here we can either use a formula or an inbuilt function.

Step 1: Click cell D11.

Step 2: Type =SUM(D4:D8)/5

or

The function will show on the formula bar as =Average(D4:D8). Let us see how Excel helps in the insertion of inbuilt functions.

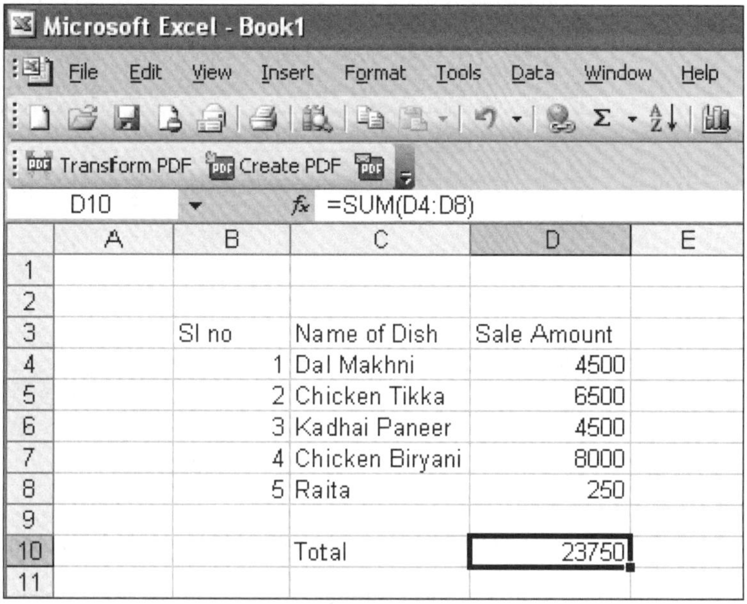

Fig. 3.28   Summation in MS Excel

Step 1: Type =AVERAGE (in the cell D11). Excel immediately shows us information about this function (as shown in Fig. 3.29).

Notice how Excel promptly offers to help. It also shows that we can use as many numerical arguments as we want but they should be separated by a comma. In this example (as shown in Fig. 3.30), we have used five arguments.

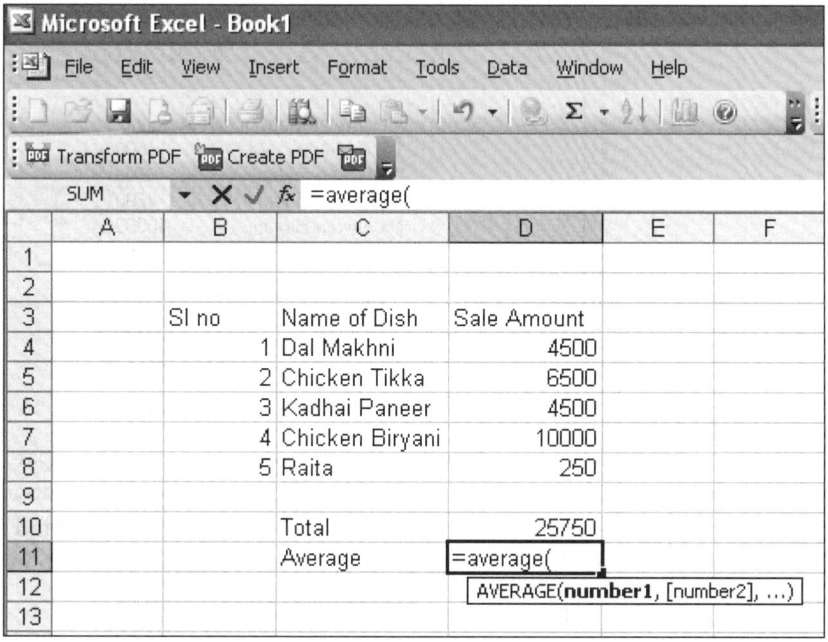

Fig. 3.29   Average calculation showing separation by comma

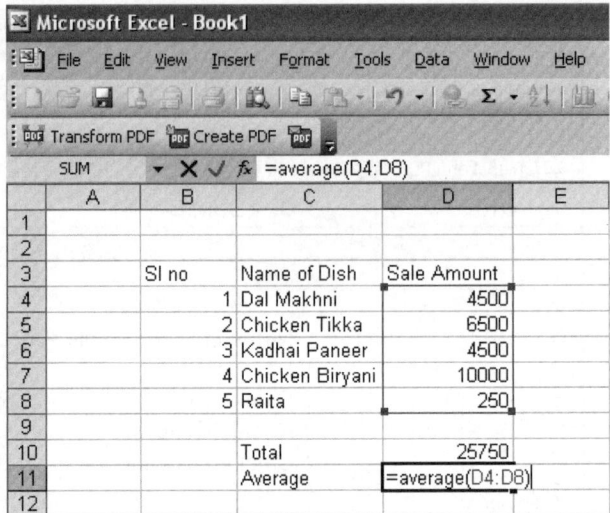

Fig. 3.30   Average calculation using the inbuilt function

If we want to include any more ranges, type a comma and repeat the process. Close the parenthesis to indicate that you have finished the range selection.

Step 2: Press Enter to see the result (as shown in Fig. 3.31).

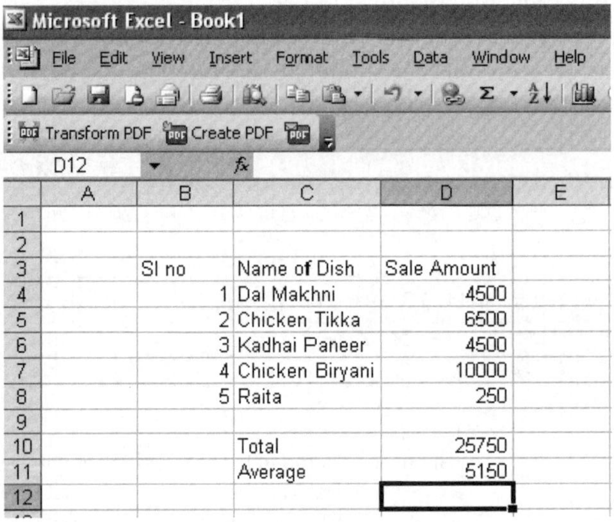

Fig. 3.31   Result of average function

We can also insert functions using the function wizard. This is particularly convenient when we do not have much idea of a function. To invoke the function wizard, select the cell in which you want to type the formula. Then click the function command in the Insert menu.

Let us insert a function for finding the standard deviation of a dish in cell D12.

Step 1: Select cell D12.

Step 2: Click the Insert menu.

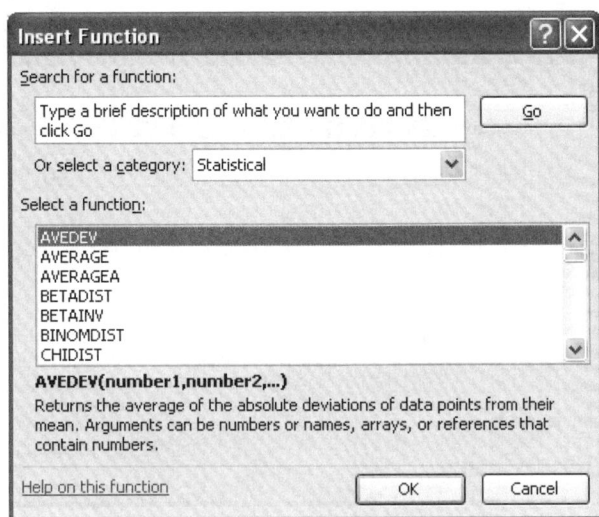

Fig. 3.32   Insert Function dialog box in MS Excel

Step 3: Select the function wizard.

The Insert Function dialog box will appear (as shown in Fig. 3.32).

Study the dialog box carefully. We can search for the details of the function we want to use by typing the function name in the *Search for a function* box and then clicking the Go button.

If we don't remember the function's name or are not sure about the function we want to work with, select a category in the *Or select a category* box. All the functions of the selected category are displayed in the *Select a function* box. Select the desired function, after which Excel will guide you.

For the example being discussed, select the Statistical category and observe the functions listed (as shown in Fig. 3.33).

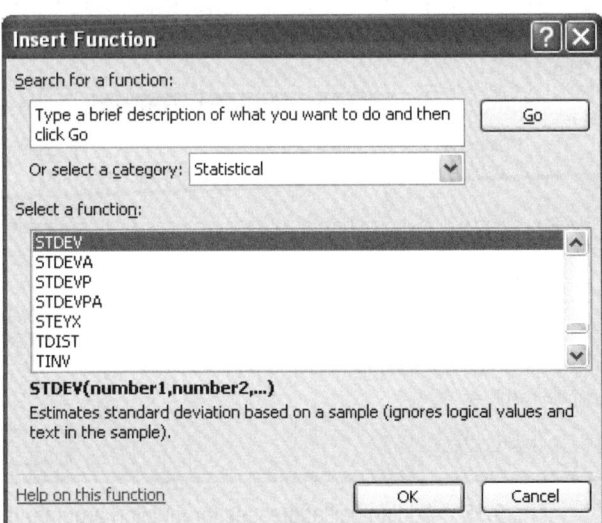

Fig. 3.33   Choosing standard deviation from the dialog box in MS Excel

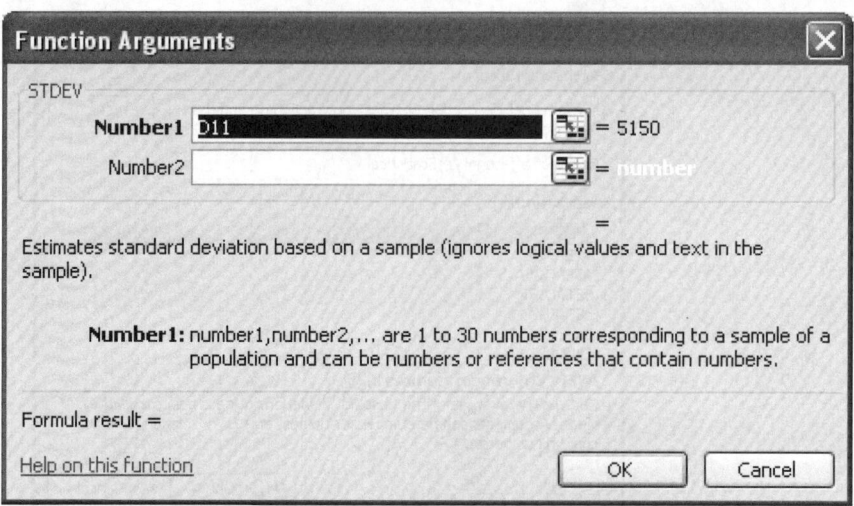

Fig. 3.34   Function Arguments dialog box in MS Excel

Step 4: Select the STDEV function.

Step 5: Click OK to proceed. The *Function arguments* dialog box opens up (as shown in Fig. 3.34).

Step 6: Select the numbers to be included as arguments in the function.

Step 7: Click the number selection icon near Number1. The dialog box rolls up exposing the cells of the worksheet.

Step 8: Select the range. The function argument dialog box pops back.

Step 9: Once the desired selection is completed, click OK (as shown in Fig. 3.35).

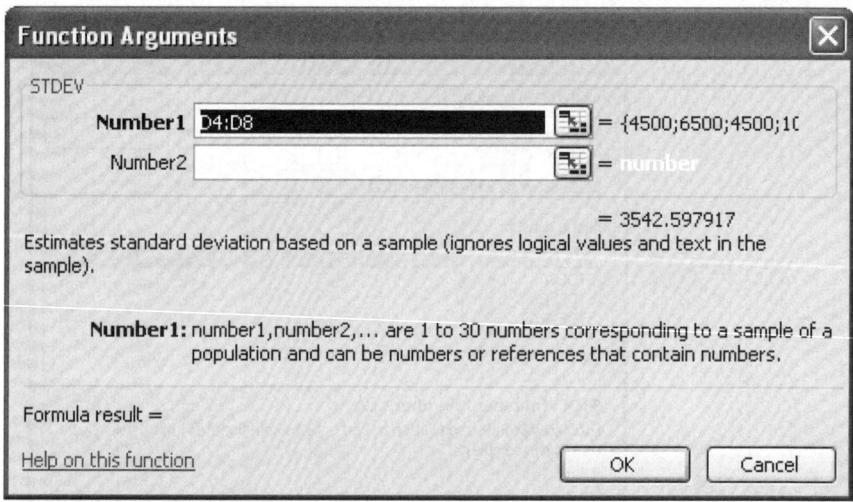

Fig. 3.35   Adding cell addresses for function arguments in MS Excel

Excel enters the function in the cell (as shown in Fig 3.36).

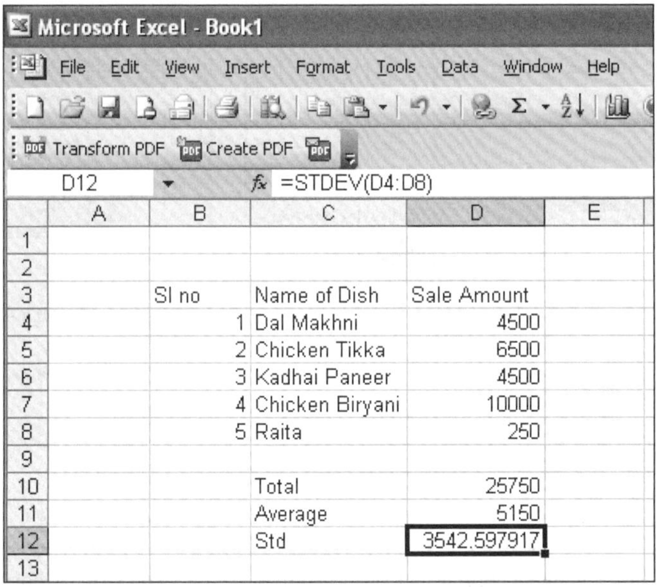

Fig. 3.36 Result of standard deviation through STDEV( ) function

Inserting Charts

Pictures are more understandable than words. Charts are visually appealing and make it easy for users to see comparisons, patterns, and trends in data. For example, instead of analysing several columns of a worksheet, we can see at a glance if sales are falling or rising over quarterly periods and how the actual sales are with respect to projected sales.

We can create a chart on a separate sheet or as an embedded object on a worksheet. We can also publish a chart on a web page. To create a chart, we must first enter data for the chart on a worksheet. Then select that data and use the Chart Wizard to go through the process of choosing a chart type and chart options.

Using Chart Wizard One of the easiest ways of generating a chart from the data entered in a worksheet is by using the Chart Wizard. It appears in the toolbar as an icon and simplifies the process of creating a chart. The procedure involved is as follows:

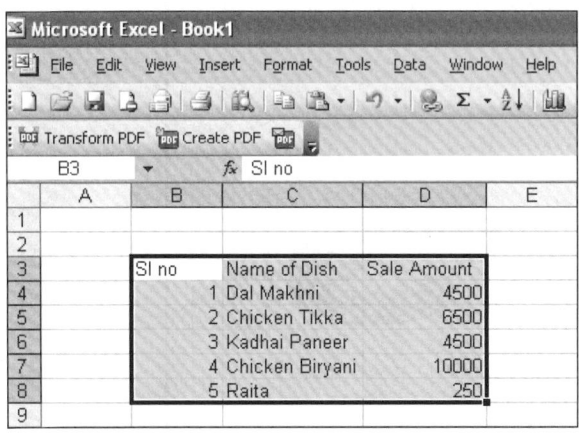

Fig. 3.37 Making a chart in MS Excel

Step 1: Select the range of data for which we want to make a chart (as shown in Fig. 3.37).

Step 2: Select the Chart option from the Insert menu.

Step 3: Select the type of Chart required. Click the *Press and Hold to View Sample* button to preview the chart. As an example, let us choose a pie chart (as shown in Fig. 3.38).

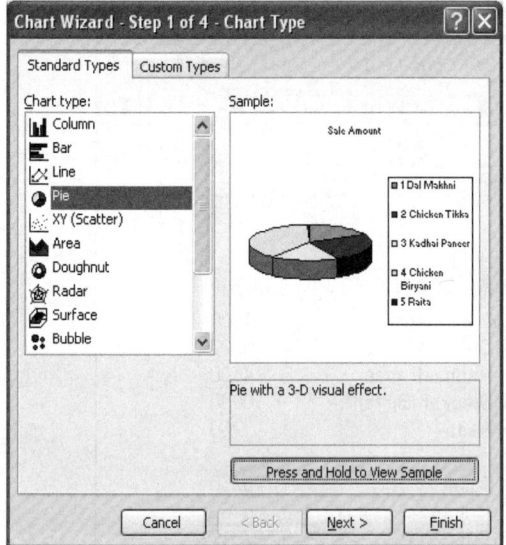

Fig. 3.38   Choosing chart from Chart Wizard in
MS Excel

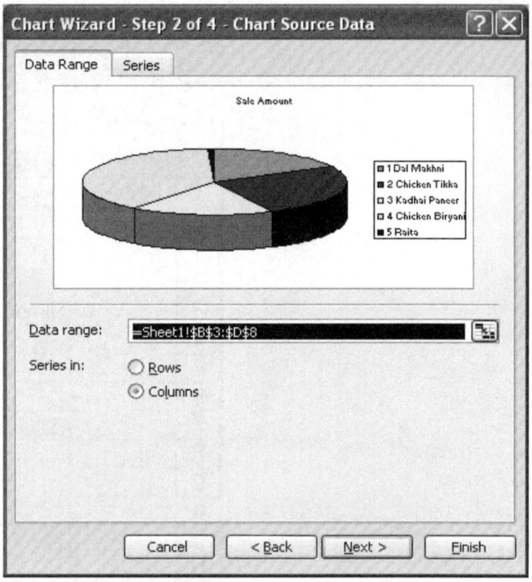

Fig. 3.39   Changing data range for chart
preparation in MS Excel

Step 4: If the chosen chart is not of our choice, click the *Custom Types* tab.

Step 5: Select the chart type and Click the Next button.

Step 6: If any data range has to be changed, then it can be done now (as shown in Fig. 3.39).

Step 7: The chart title or additional information on the X and Y axes can be added
(as shown in Fig. 3.40).

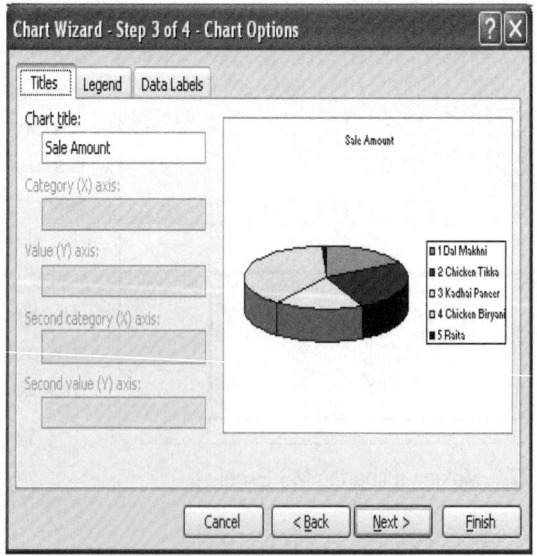

Fig. 3.40   Adding title to chart in MS Excel

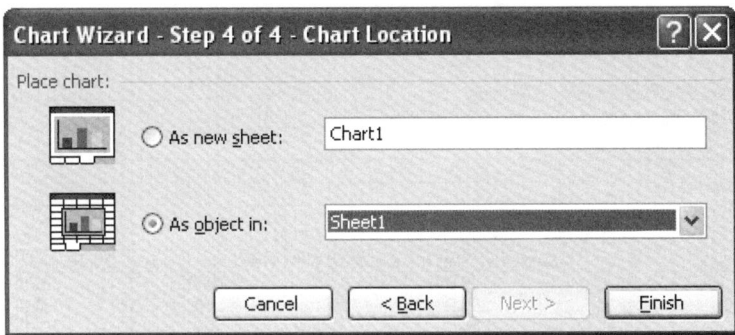

Fig. 3.41   Option regarding placement of chart in MS Excel

Step 8: On clicking the Next button, a dialog box appears asking us to specify where the chart is to be placed (as shown in Fig. 3.41).

Step 9: Click the Finish button to insert the chart into the worksheet (as shown in Fig. 3.42).

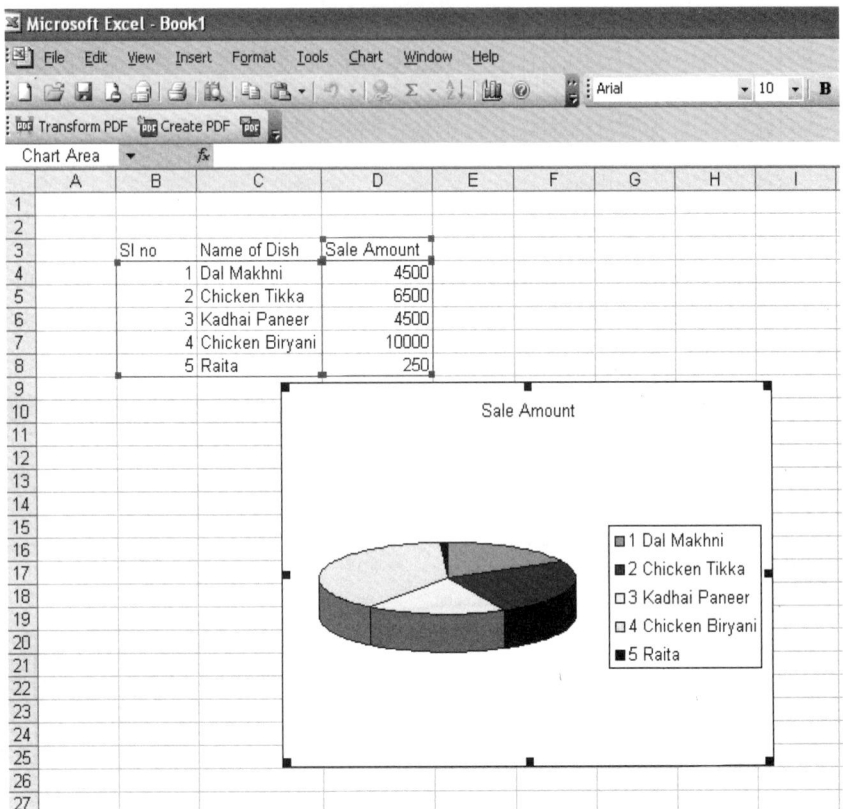

Fig. 3.42   Inserting chart in MS Excel sheet

Graphs can also be made by comparing more than a single set of data (as shown in Fig. 3.43).

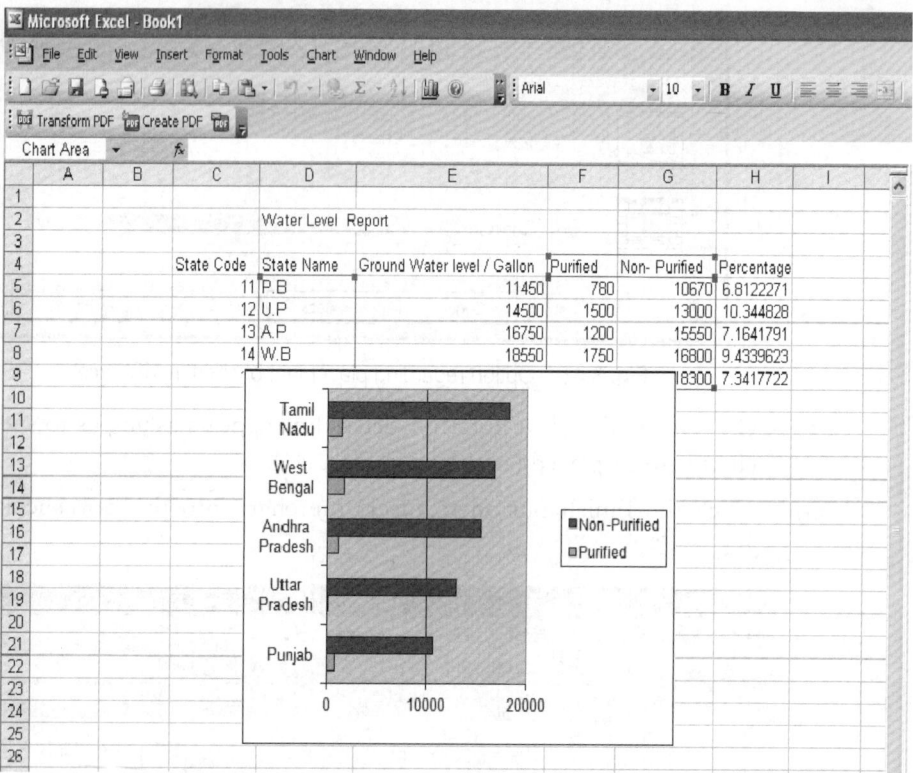

Fig. 3.43   Comparing data in Excel with the help of a chart

Using the steps mentioned in the section on inbuilt functions, we can find the sum of a set of data along a row or column. In the following example, we have found out the average marks obtained by students in three subjects (as shown in Fig. 3.44).

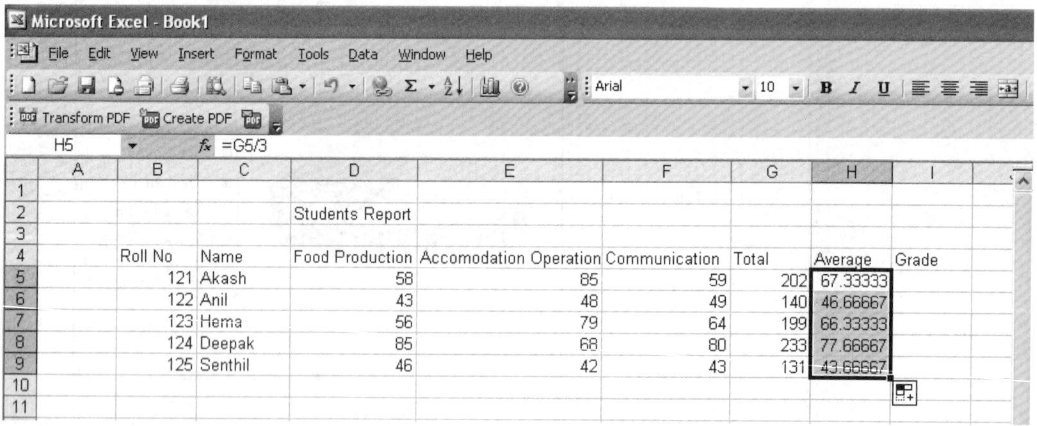

Fig. 3.44   Calculation of average marks of students

A function can be used to decide the result or grade for the example shown in Fig. 3.44. To do this, click the Insert toolbar and choose the Function option (as shown in Fig. 3.45).

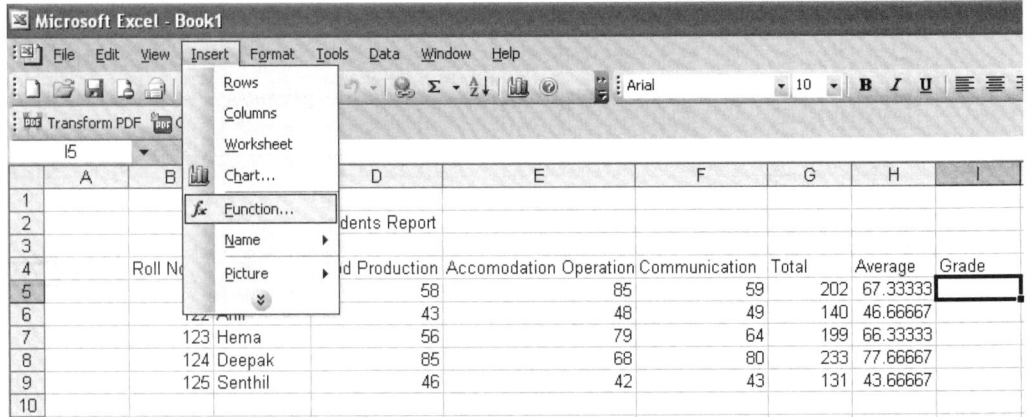

Fig. 3.45   Choosing Function from Insert option

Step 1: Click the Function option from the Insert menu. The *Function Arguments* dialog box appears.

Step 2: We have defined a logic that states that if an individual obtains marks greater than or equal to 50, it is 'pass', else it is 'fail'. This is shown in Fig. 3.46.

Step 3: After the logic is entered and we click OK, the result will be shown in the designated cell (as shown in Fig 3.47).

The result of each individual entry can be seen by dragging the cursor down (as shown in Fig. 3.48).

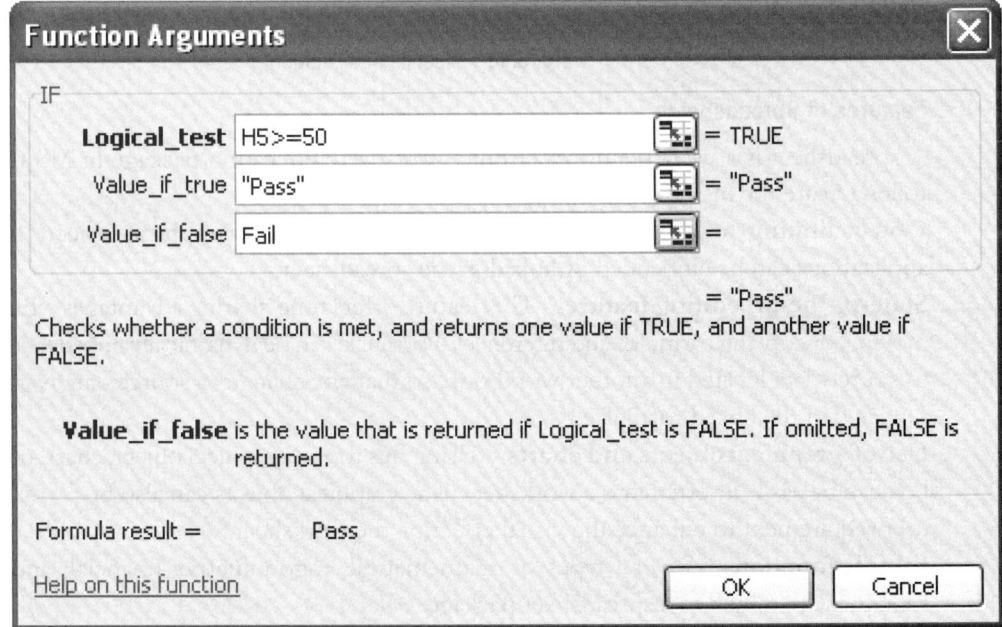

Fig. 3.46   Function Arguments using logical test in MS Excel

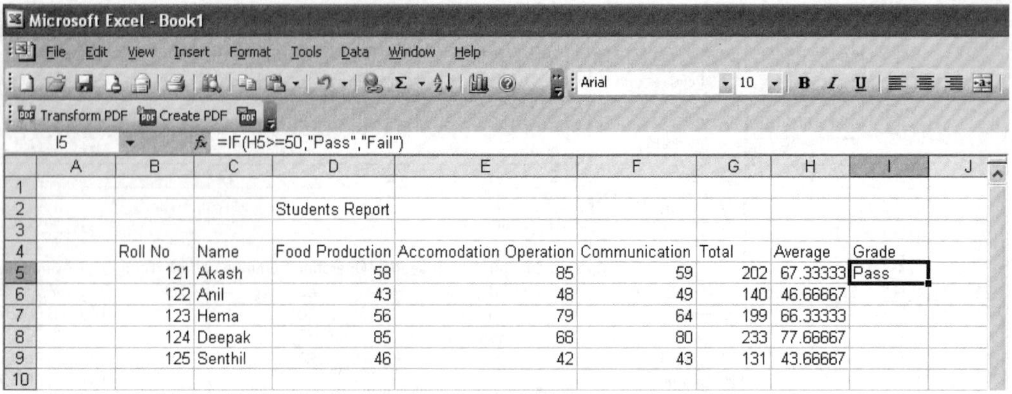

Fig. 3.47 Results of logical test

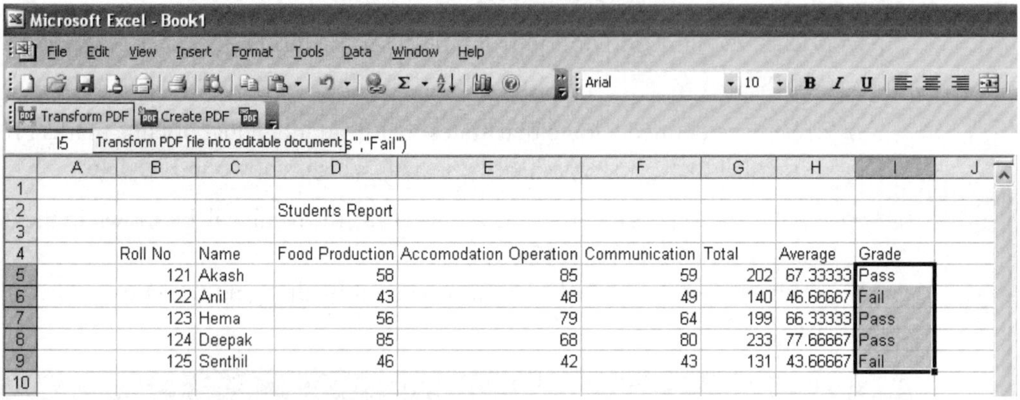

Fig. 3.48 Results of all students

Features of Spreadsheets

A spreadsheet is a user-friendly electronic worksheet application package in MS Office. The following are the main features of spreadsheets:

Object linking and embedding   Using the OLE facility, graphical objects from other applications can be embedded or linked in the spreadsheet.

State-of-the-art editing feature   This feature offers time-sharing advantages while creating, moving, or copying formulae and references. Formulae can be dynamically linked in a workbook to source data located in another workbook, so that any change to source data is immediately reflected in the linked formulae.

Use of graphical objects and charts   Using this tool, a graphical object, chart, or comment box can be easily inserted into a worksheet. The graphical objects can also be formatted as per user requirement to enhance the visual effect of the worksheet.

Inbuilt functions   Several types of mathematical, trigonometric, financial, and statistical functions are inbuilt in a spreadsheet package.

Accuracy   With the electronic worksheet program, we can easily and quickly produce reports and get answers by means of the 'what-if' facility so as to generate accurate results.

MS Excel helps us in tabulating and representing data in rows and columns. It also aids us in carrying out standard numerical calculations with the help of functions that are inbuilt in MS Excel. Data can also be represented in a graphical manner.

Thus, with the help of MS Excel, data can be interpreted and conclusions can be drawn. However, it cannot be presented in a sequential order to be viewed by superiors or subordinates. MS PowerPoint, which helps us prepare a presentation in a formal manner, will be discussed in the next section.

MS POWERPOINT 2003

A presentation is a powerful managerial communication tool through which we can compile and deliver ideas, concepts, plans, or products to the audience in a structured, effective, and impressive manner. The presentation may include slides, printed handouts, notes, outlines, graphics, and animations. A slide in MS PowerPoint is a combination of images, graphics, charts, and text that is used to convey information. MS PowerPoint presentations are currently used for businesses, schools, colleges, and training programmes.

Starting MS PowerPoint

MS PowerPoint can be opened in either of the following ways:

- Using the Start button

Step 1: Click the Start button on the taskbar.

Step 2: Select the Programs option.

Step 3: Click Microsoft Office.

Step 4: Select Microsoft Office PowerPoint 2003 (as show in Fig. 3.49).

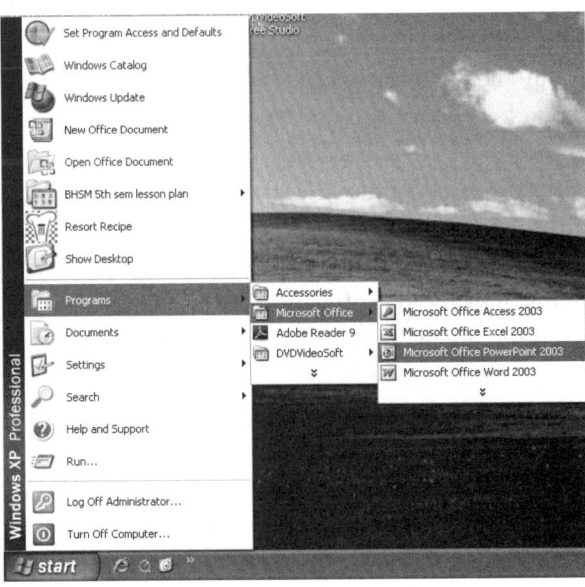

Fig. 3.49   Opening MS PowerPoint in Windows XP

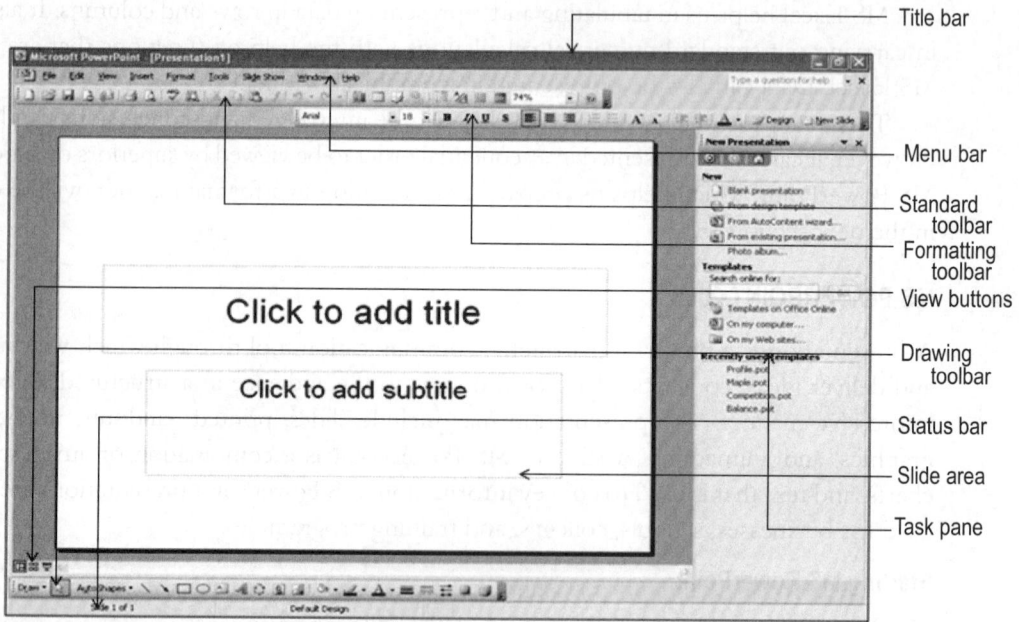

Fig. 3.50 Opening window of MS PowerPoint 2003

- Using the Run command

Step 1: Click the Start button on the taskbar.

Step 2: Choose the Run option to display the Run dialog box.

Step 3: Type powerpnt in the text box and then click OK.

As soon as PowerPoint is selected, the screen shown in Fig. 3.50 appears.

Components

The following are the components of an MS Office PowerPoint window, shown in Fig. 3.50:

Title bar Found at the top of the PowerPoint window, it shows the name of the current presentation.

Menu bar It is located under the Title bar and contains various types of PowerPoint commands, functions, and options.

Standard toolbar This is found just below the Menu bar and contains buttons for easy access to standard commands of the software.

Formatting toolbar This toolbar contains buttons and list boxes, and is located under the Menu bar.

View buttons The three kinds of view buttons are slide view, slide sorter view, and slide show view.

Drawing toolbar It is located just above the Status bar. Using this toolbar, we can create and manipulate text, graphics, shapes, and pictures.

Status bar The Status bar is located at the bottom of the window and contains a message area indicating the current slide number and the total number of slides in the presentation.

Slide area   The slide area is the working area for creating and editing a slide.

Task pane   It displays the commands that are currently available.

Operations Performed

MS PowerPoint is generally used for creating both professional and informal presentations in an efficient manner. The following operations can be carried out in PowerPoint:

- Creating presentations
- Formatting presentations
- Working with design templates
- Inserting graphics in presentations
- Adding slide transition effects
- Using slide show
- Preparing organization charts

Creating Presentations

PowerPoint offers three ways of creating a new presentation. This can be done using the wizard, template, or by creating a blank presentation (colour scheme fonts and design features set to default value). PowerPoint is provided with inbuilt professional designs known as Auto layouts and templates.

Step 1: Click the <u>N</u>ew option in the <u>F</u>ile menu; click the New button represented by the icon; or press Ctrl+N.

Step 2: The New Presentation window will appear on the right-hand side of the window (as shown in Fig. 3.51. In this window, we have the option of creating a new presentation in one of the following ways:

- Blank presentation
- From design template
- From AutoContent wizard
- From existing presentation
- Photo album

As an example, we will start a presentation using the Blank presentation option.

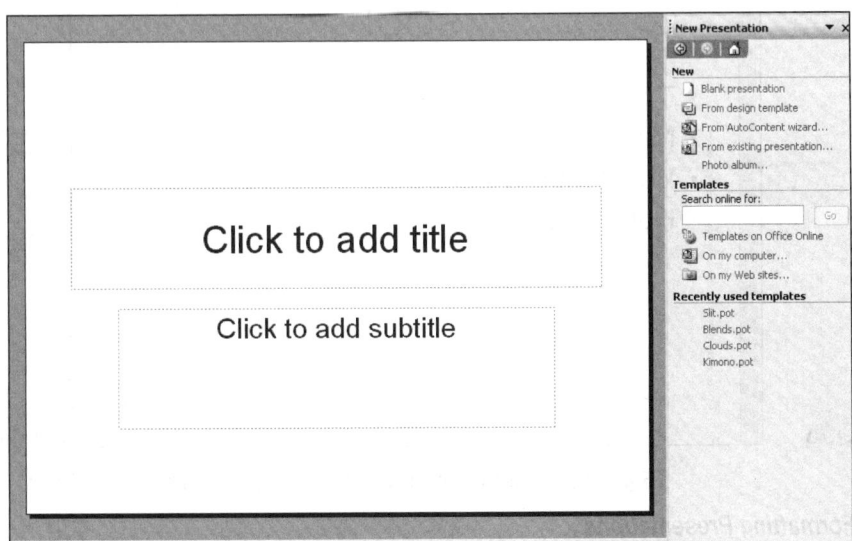

Fig. 3.51   New presentation window in MS PowerPoint

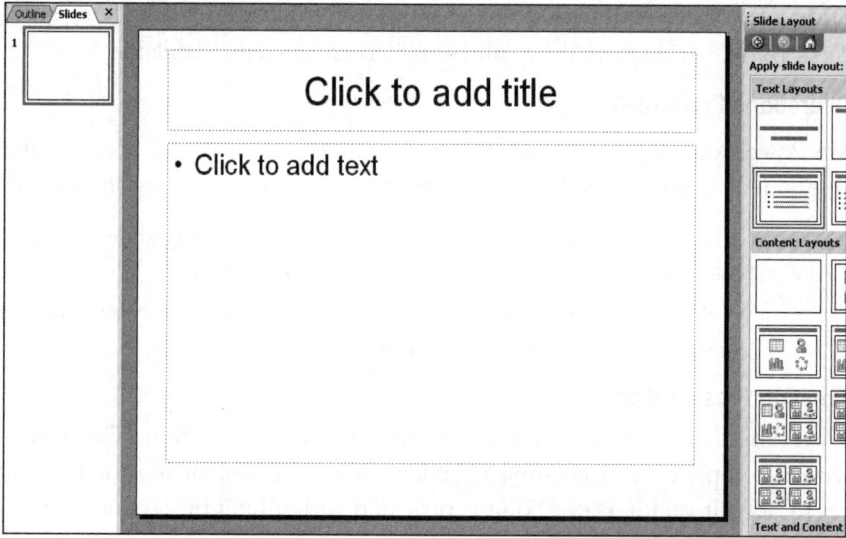

Fig. 3.52   Applying slide layout in MS PowerPoint

Step 3: We can select any of the available slide layouts by clicking it, upon which, that layout is applied immediately to the open slide. Next, choose text and content layouts as shown in Fig. 3.52.

Step 4: We can add a title or heading to the slide.

Step 5: Add necessary bullet points. In the following example (as shown in Fig. 3.53), the different types of soups have been listed out.

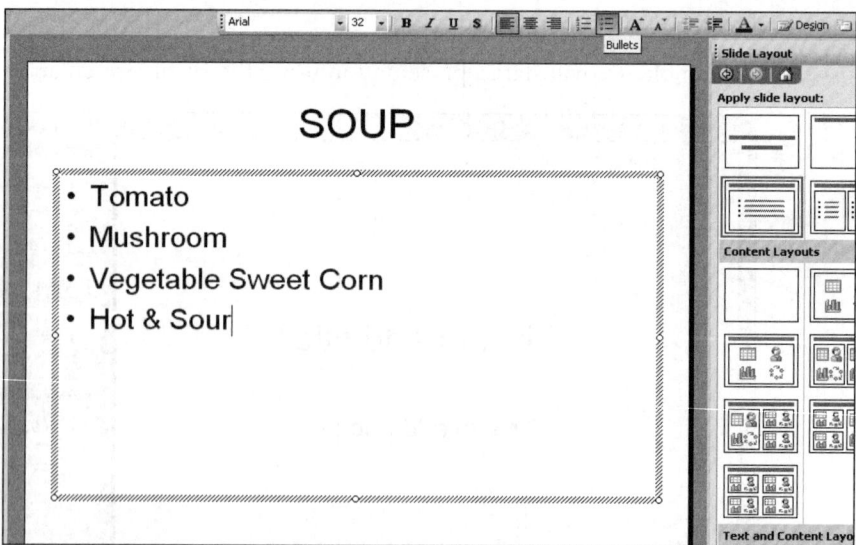

Fig. 3.53   Addition of title and bullet to a slide in MS PowerPoint

Formatting Presentations

The presentation can be designed and formatted as per requirement.

- Changing font style
  - To change the font style into Bold, select the text and click the Bold tool on the toolbar, as shown in Fig. 3.54.

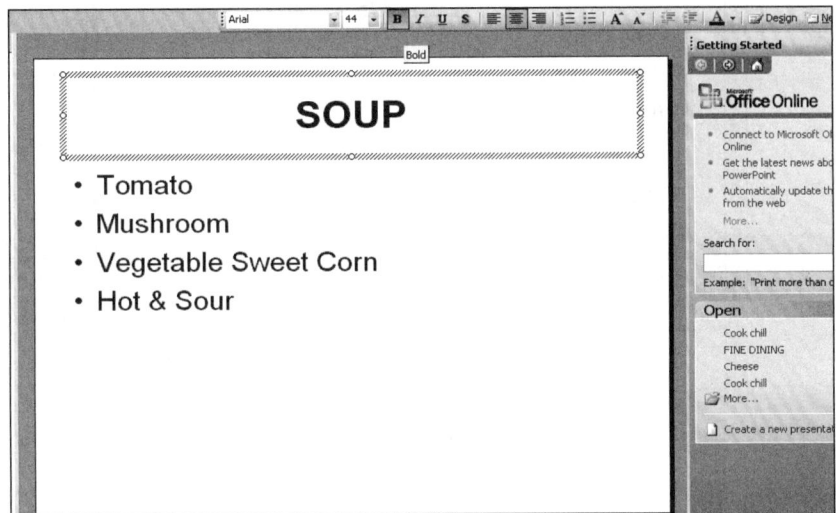

Fig. 3.54   Changing font to bold in MS PowerPoint

  - To change the font style into Italics, select the text and click the Italics tool on the toolbar, as shown in Fig. 3.55.

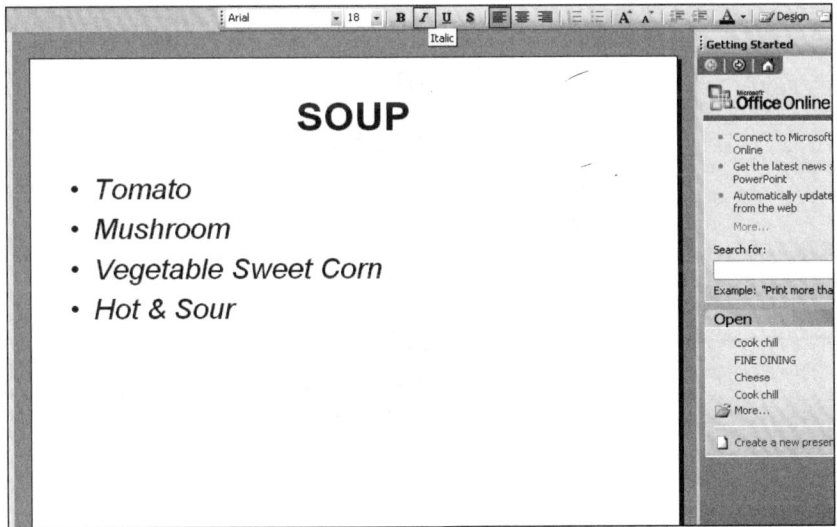

Fig. 3.55   Changing font to italics in MS PowerPoint

- Changing font colour
  - To change the font colour of the text, select the text and click the font colour tool of the toolbar, as shown in Fig. 3.56. We can choose the colour from the More Colors option.

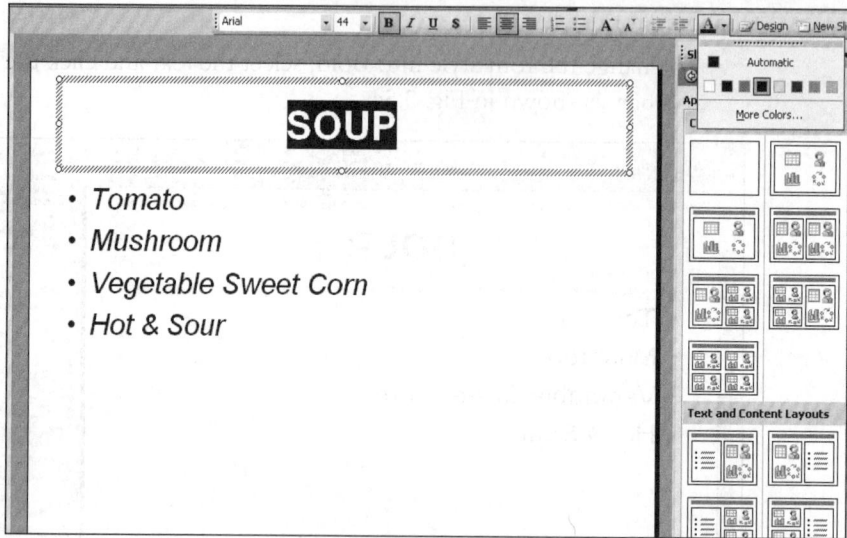

Fig. 3.56   Changing colour of heading in MS PowerPoint

—   We can also change the font colour of the content by choosing an appropriate colour (as shown in Fig. 3.57).

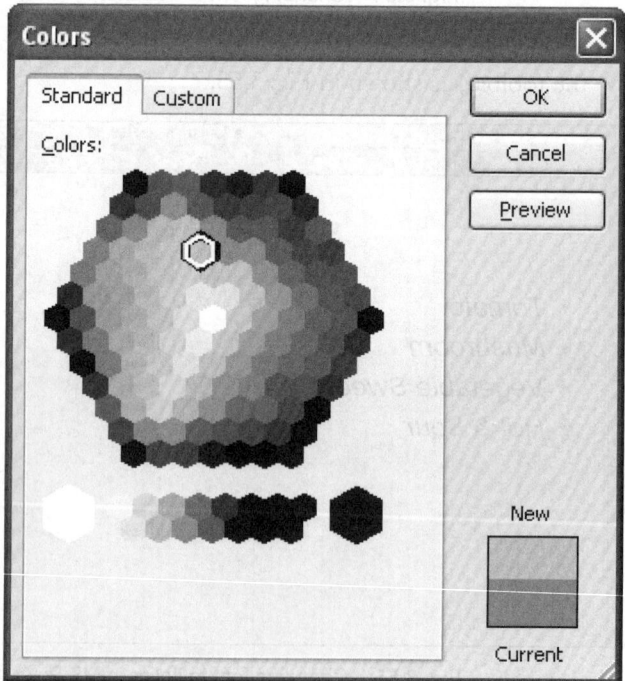

Fig. 3.57   Choosing font colour in MS PowerPoint

—   After choosing the desired colour for the content, click the OK button. The colour changes to the selected choice (Fig. 3.58).

•   Changing the background colour

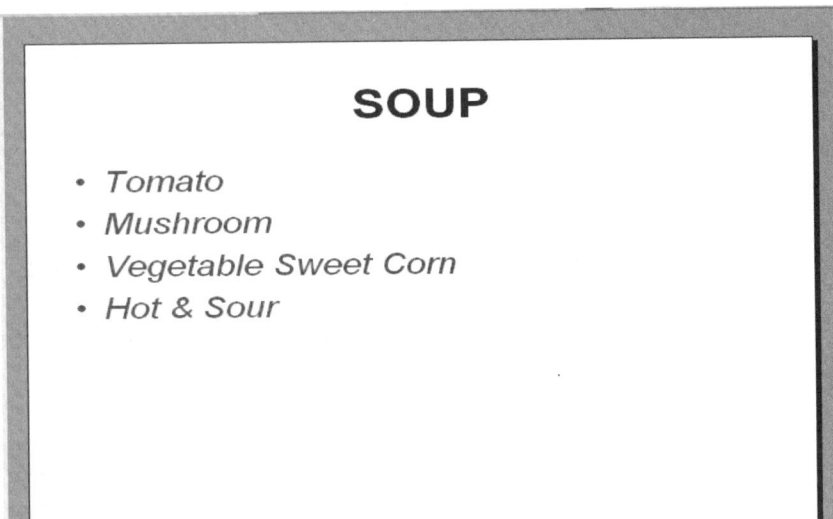

Fig. 3.58   Colour added to title and content in MS PowerPoint

Slide colour can be added by clicking the Design tool of the toolbar and choosing the Colour Schemes option (as shown in Fig. 3.59). On the task pane, various colour schemes will be displayed. Select one of the choices.

Fig. 3.59   Choosing and applying a colour scheme in MS PowerPoint

Working with Design Templates

Design templates can be employed in the presentation using the following steps:

Step 1: Click the Design tool of the toolbar

Step 2: Different kinds of design templates will be displayed on the task pane. Choose one among these. Fig. 3.60 shows the design templates and Fig. 3.61 shows a slide with the applied design template.

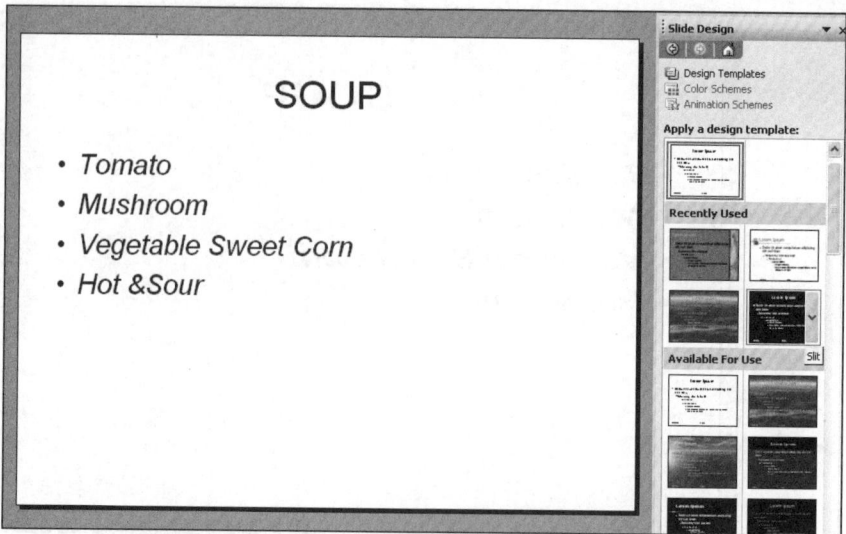

Fig. 3.60   Choosing design templates in MS PowerPoint

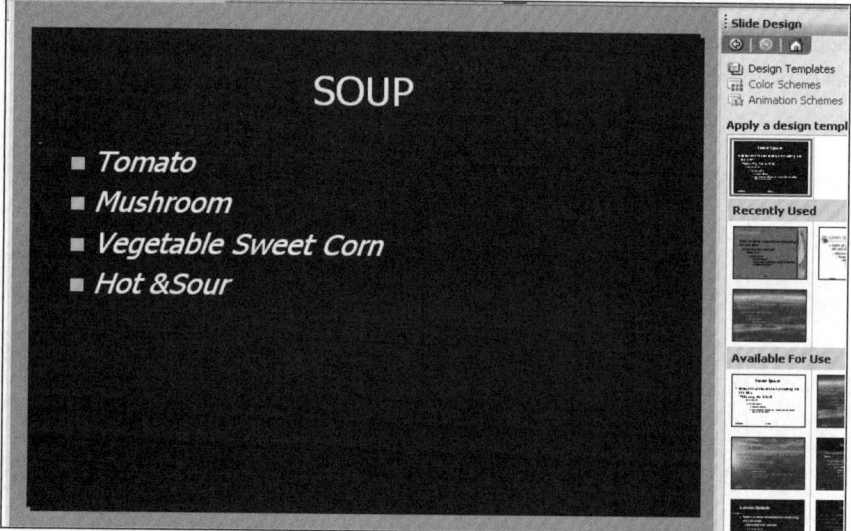

Fig. 3.61   Applying a design template in MS PowerPoint

Inserting Graphics in Presentations

MS PowerPoint provides us with a wide range of predefined Clip Art graphic libraries. To insert Clip Art, the following procedure is followed:

> *Step 1*: Choose the Clip Art command from the Picture option on the Insert menu (as shown in Fig. 3.62).
>
> *Step 2*: Type the desired Clip Art in the media clip search box (as shown in Fig 3.63). In this example, on entering 'soup' in the search option, images of soup are displayed.

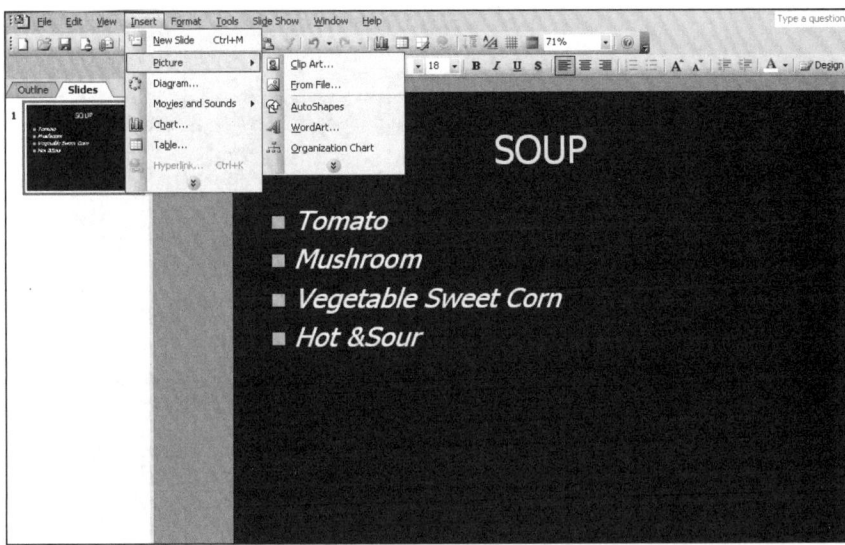

Fig 3.62   Choosing Clip Art from Insert menu in MS PowerPoint

Fig. 3.63   Inserting picture in the slide in MS PowerPoint

Step 3: Select a suitable clip (as shown in Fig. 3.64). The Clip Art picture is inserted in the slide.

Adding Slide Transition Effects

The audience might find it boring to view the monotonous way in which slides change in a presentation. Transition helps in changing slides gradually. To add transition effects to slides follow the given procedure:

Step 1: Click the Slide Transition command in the Slide Show menu to open the list in the task pane.

Fig. 3.64   Clip Art picture inserted in the slide in MS PowerPoint

Step 2: In the Slide transition dialog box, select a transition from the Effect drop-down list (as shown in Fig. 3.65).

Step 3: We can choose the speeds slow, medium, or fast.

Step 4: Set the advance time and sound if desired.

Step 5: Click the Apply button.

Step 6: To apply the transition to all the slides, click the Apply to All button.

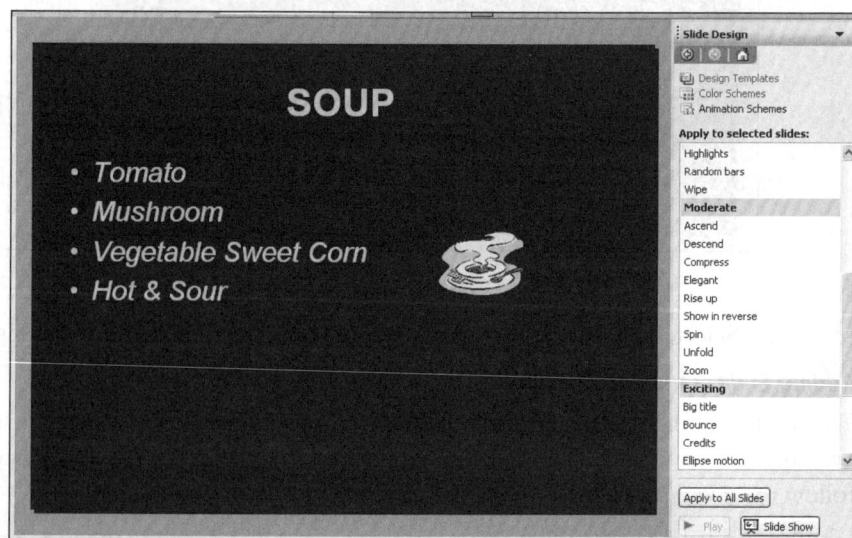

Fig. 3.65   Slide transition in a slide in MS PowerPoint

Using Slide Show

After the slide has been prepared, to view the slide as a slide show, go to the Slide Show option and choose the View Show option or press F5 (as shown in Fig. 3.66).

Fig. 3.66   View slide show in MS PowerPoint

Preparing Organization Charts

Many other features can be added to the PowerPoint presentation by choosing a suitable slide layout. We can prepare an organization chart and fill the required details (as shown in Fig. 3.67).

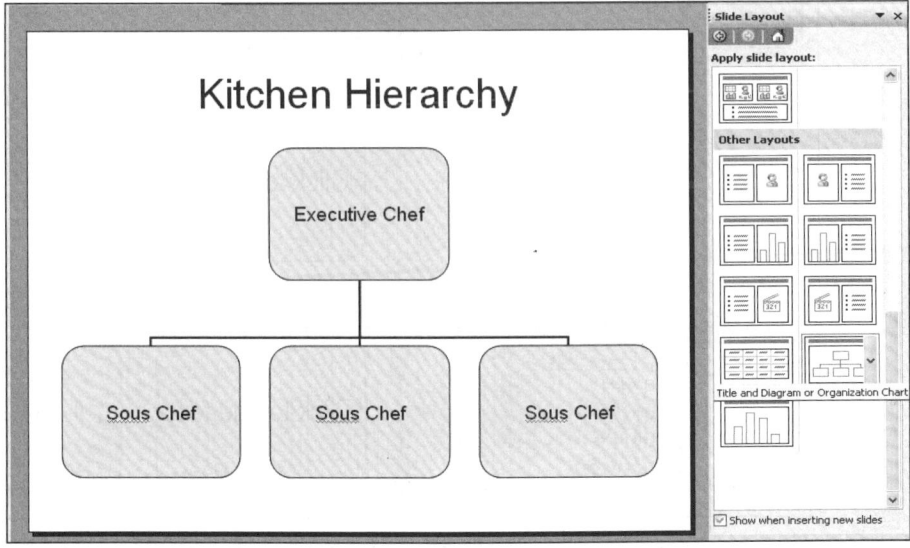

Fig. 3.67   Adding hierarchical structure in MS PowerPoint

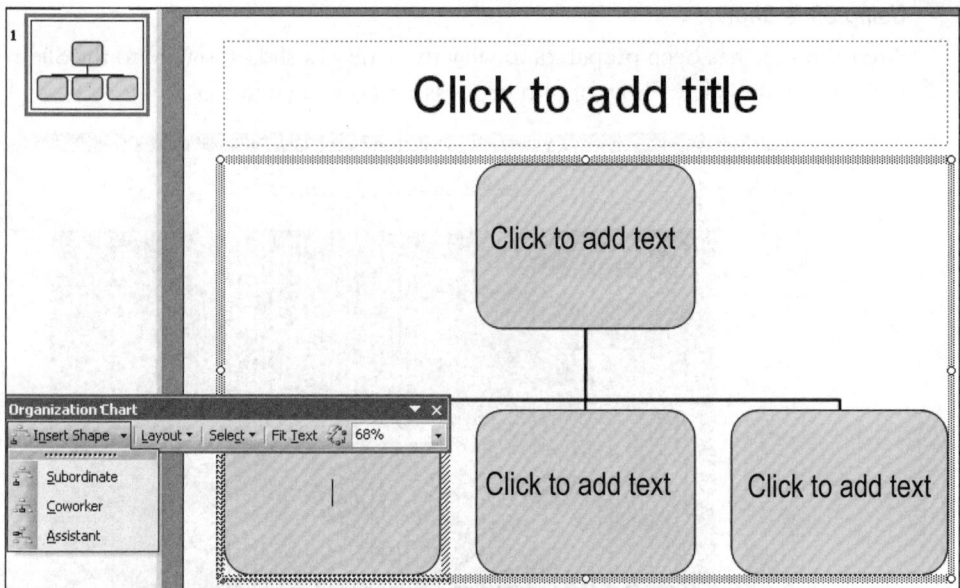

Fig. 3.68 Inserting shapes in an organization chart in MS PowerPoint

In the organization chart shown in Fig. 3.68, we can also insert shapes such as subordinate, coworker, and assistant.

The layout of an organization chart (as shown in Fig. 3.69) can be changed as per the needs of a user. Standard, both hanging, left hanging, right hanging, and auto layout are the different kinds of layouts available.

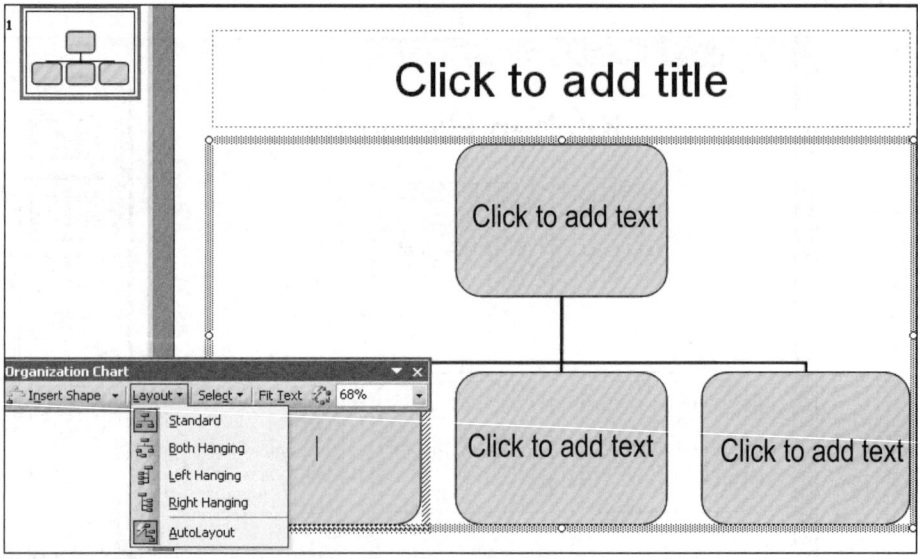

Fig. 3.69 Layout options in an organization chart in MS PowerPoint

A sample organization chart showing subordinates and coworkers in the F&B service of a hotel is shown in Fig. 3.70.

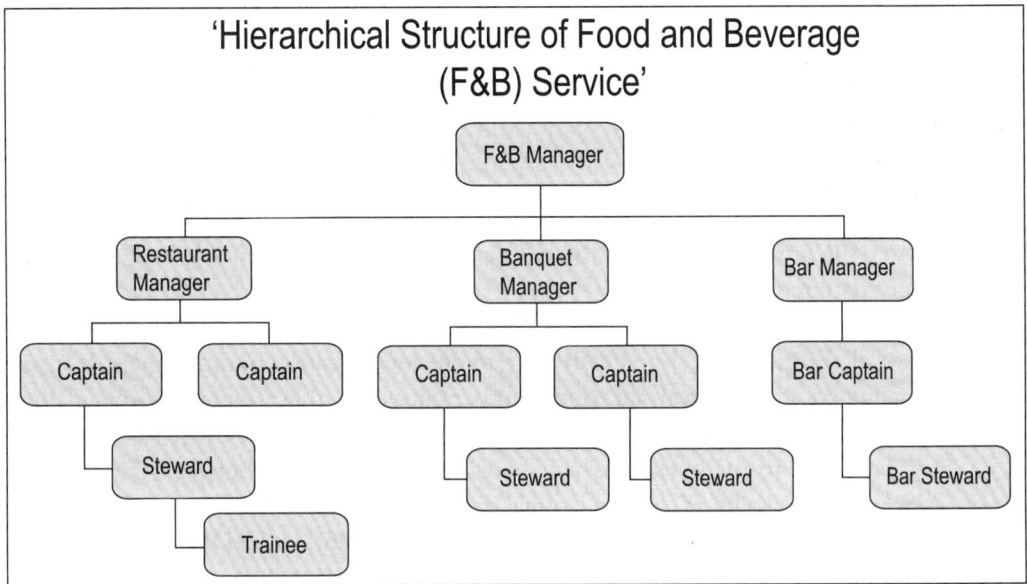

Fig. 3.70   F&B organization structure in MS PowerPoint

Features

MS PowerPoint is a presentation software package included in the Microsoft Office suite and is rich in features. It is used in preparing professional-quality presentations as it helps us structure ideas and information that we want to convey to our audience. It also helps in creating the content for our presentation by typing and inserting text, pictures, sounds, and animations. With the help of this package, we can also add visual images, supporting documents, and audio recordings to enhance the effectiveness of the presentation.

PowerPoint makes the creation of presentations simple by providing inbuilt professional designs called Auto layouts and templates. The following are the main features of PowerPoint:

- It consists of templates that are helpful in designing presentations.
- It provides four views—slide view, outline view, normal view, and slide sorter view— that are very useful during the development of a presentation.
- Graphics, charts, audio, video, and text can be added in a single presentation.
- Presentations prepared using PowerPoint can be directly projected as a slide show.
- Different transition and building effects may be included in the slides to make the presentations attractive.
- An entire presentation can be printed on paper using the printing features.
- Different colours can be added to columns and text from a large palette of colour schemes.

PowerPoint helps in making impressive presentations with the help of text, graphics, images, clip art, audio, and video. However, for representing large amounts of data like guest details of a hotel, making a presentation becomes difficult. We would thus require a database program. We will next discuss database management in MS Access.

MS ACCESS 2003

It would be easy for the staff to remember guests' names and room numbers if the hotel was small and had around 20 rooms. However, if the hotel had around 250 rooms it would be very difficult to relate a guest with a room number. For this, we would require a database (systematic organization of data).

To create a database structure, many fields are required. A field can contain data in the form of text, number, or date/time. Once the fields have been selected, we can create tables as per requirement.

In MS Access, multiple tables can be created. For example, we can have a table containing a list of all the guests in a hotel with their unique identification number (guest ID), another table containing information about room numbers, another table about the company they represent (for company guests), and yet another one about the guests' country of origin. All the tables can be linked to the guest ID, hence creating a relationship between them. For this reason, MS Access is also known as a relational database management system (RDBMS). We will now discuss database, database management system (DBMS), and RDBMS.

Some of the terms associated with MS Access are as follows:

Database   A database is an organized collection of information related to a particular subject or purpose.

Database management system   A DBMS is a collection of programs or inter-related data that helps in the creation, organization, and management of data.

Relational database management system   An RDBMS is a method of getting information from different tables that are connected to one another through some information.

Starting MS Access

To start a session of Access, carry out the following steps:

- Using the Start button

Step 1: Click the Start button.

Step 2: Select Microsoft Office from Programs.

Step 3: Click MS Office Access (as shown in Fig. 3.71) or click the icon for MS Access. As a result, an Access session begins.

Note that the Access icon may be available at a location other than the one shown here depending on the current configuration of the system.

- Using the Run command

*Step*1: Click the Start button on the taskbar.

Step 2: Choose the Run option to display the Run dialog box.

Step 3: Type msaccess in the text box and then click OK.

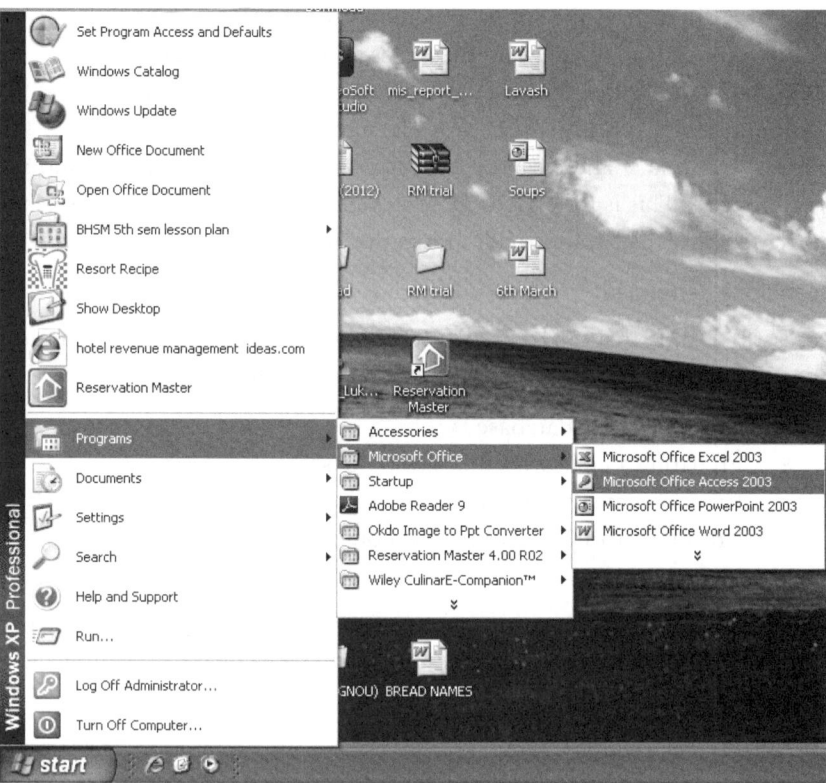

Fig. 3.71   Starting MS Access in Windows XP

Components

MS Access is another Windows-based application and hence its window has components common to other MS Office applications(as shown in Fig. 3.72).

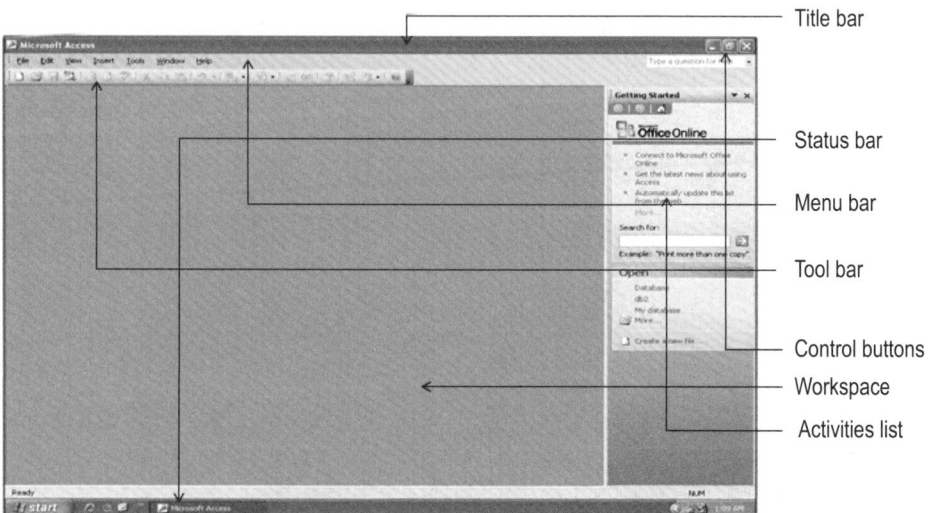

Fig. 3.72   MS Access window

Title bar   It displays the application name and file name if one is currently opened.

Status bar   It displays relevant messages and the special key status.

Menu bar   It displays the seven menus: File, Edit, View, Insert, Tools, Window, and Help.

Toolbar   It displays the short cut tools in the form of icons.

Control buttons   These can be used for minimizing, maximizing, and resizing the window.

Workspace   This is the space where databases are opened for users to work on.

Activities list   This list provides tasks that we can ask Access to perform.

Operations Performed

Access is a DBMS that can be used for creating a database. It thus helps in storing data efficiently. The operations that MS Access can perform are as follows:

- Creating a new database
- Preparing a table with the help of design view
- Entering data into a table
- Creating a form using the auto content wizard

Creating Databases

A well-designed database ensures convenient and quick access to information. An optimized database helps us get the required results from the database. The procedure for creating a new database is as follows:

Step 1: Click the File menu.

Step 2: Click the New option; click the New tool button on the tool bar; or press Ctrl+N.

A list of options appears in the task list of the Access window (as shown in Fig. 3.73).

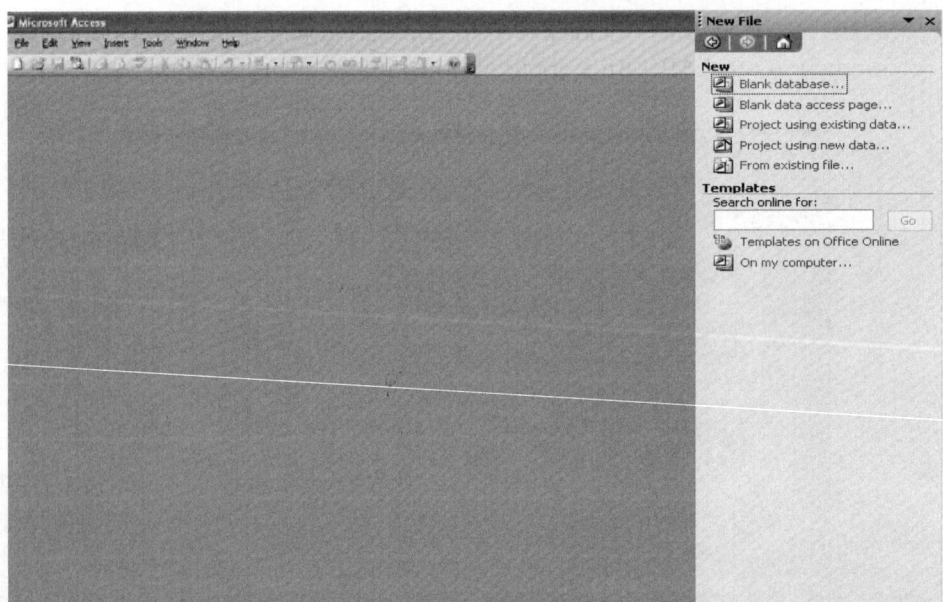

Fig. 3.73   Creating a database

Access helps us create documents such as a blank database, a blank data access page, and projects. To create a blank database, the following procedure is carried out:

Step 1: Select the Blank database option in order to create a new blank database. Access asks us to choose a location and name for the database.

Step 2: We have entered the database name as 'Database' and chosen the folder 'My Documents' as its location (as shown in Fig. 3.74).

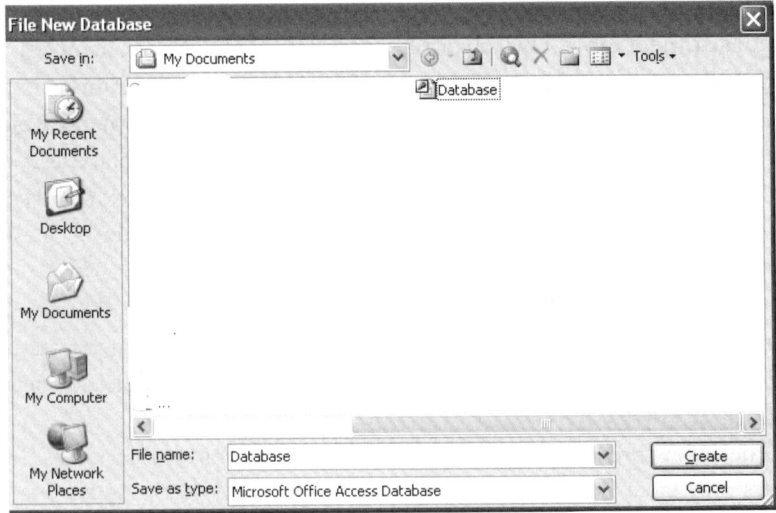

Fig. 3.74   Saving a database in MS Access

Access creates a blank database having a specified name (e.g., Database) and places it in the specified folder (e.g., My Documents). Click the Create button to save the database. The database opens in its window (as shown in Fig 3.75).

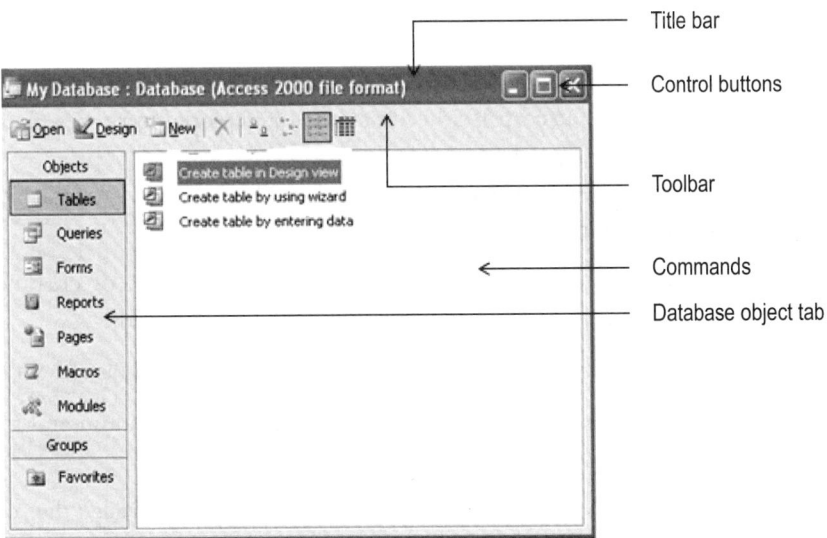

Fig. 3.75   Parts of the database window in MS Access

We can open more than one database at a time. Each one will have its own window. The database is empty as it does not have any table or form in it. The different parts of a database window are shown in Fig. 3.75.

Title bar   It displays the database name.

Control buttons   It allows the maximization, minimization, and resizing of a window.

Toolbar   It displays the short cut buttons for tools.

Commands   These commands allow users to issue data definition commands, for example, create table in design view.

Database object tab   This tab displays seven database objects, namely Tables, Queries, Forms, Reports, Pages, and Modules. Each tab offers relevant data definition commands. We will work with tables and forms in this section. The following are some of the objects of MS Access:

Table   A table is a fundamental building block of an Access database. A database must have at least one table since data is stored here. A table is made up of rows and columns. Each column in a table is called a field. A field could contain information about guest IDs, first names, or last names.

Query   A query is a filter through which data is evaluated.

Form   A form helps us simplify data entry work and displays information in a specific manner.

Report   A report helps us summarize data in a format suitable for viewing on the screen or printing.

The other three objects available are pages, macros, and modules. In this chapter we will discuss only table and form.

Creation of Database Table

We can easily create a table in MS Access by following any of these three options (as shown in Fig. 3.75).

- Create table in design view: This option allows us to create the design and fields of the table. It also allows us to enter data type and data into the table. Here, we will only discuss design view.
- Create table by using wizard: This options provides some sample tables to the user. They could be used for either business or personal purposes. A user can input data after selecting a particular table.
- Create table by entering data: This option allows the user to simultaneously design the table and enter data into it.

Creating a table in design view is the most common method of creating a table. To create a table in MS Access (in design view), follow the given procedure:

Step 1: Double-click the Create table in Design view option (as shown in Fig. 3.75).

Step 2: Enter the name of the fields in the Field Name column and the data type in the Data Type column (as shown in Fig. 3.76).

We may wish to create a database of the guests who have availed the services of the hotel with field names such as guest ID, first and last names, and contact number. As an example, the completed table of guest details is shown in Fig. 3.77.

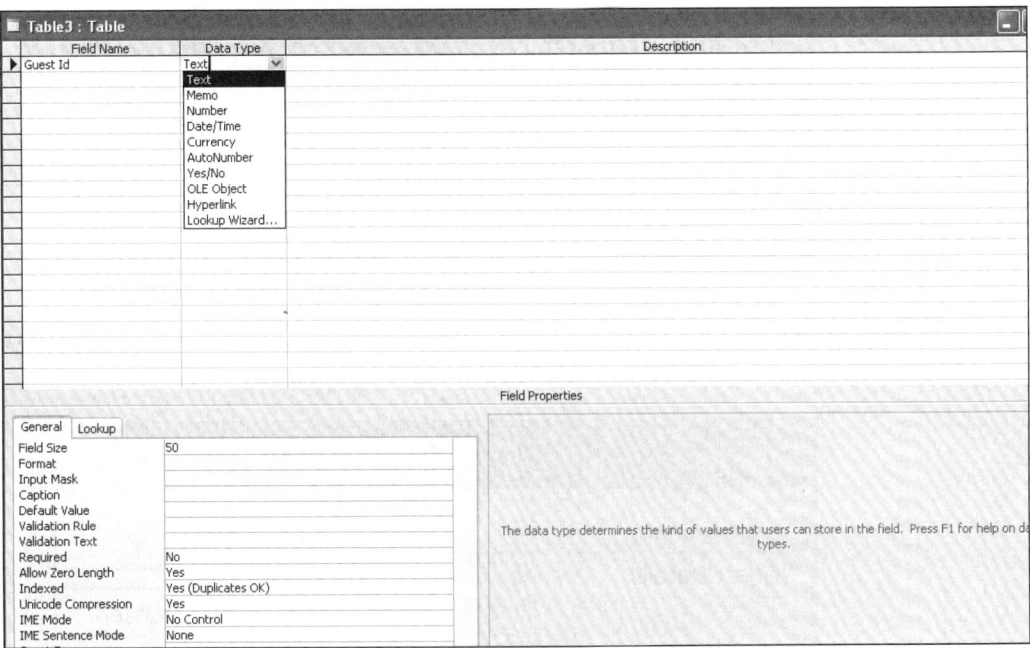

Fig. 3.76   Addition of field name and data type in a table in MS Access

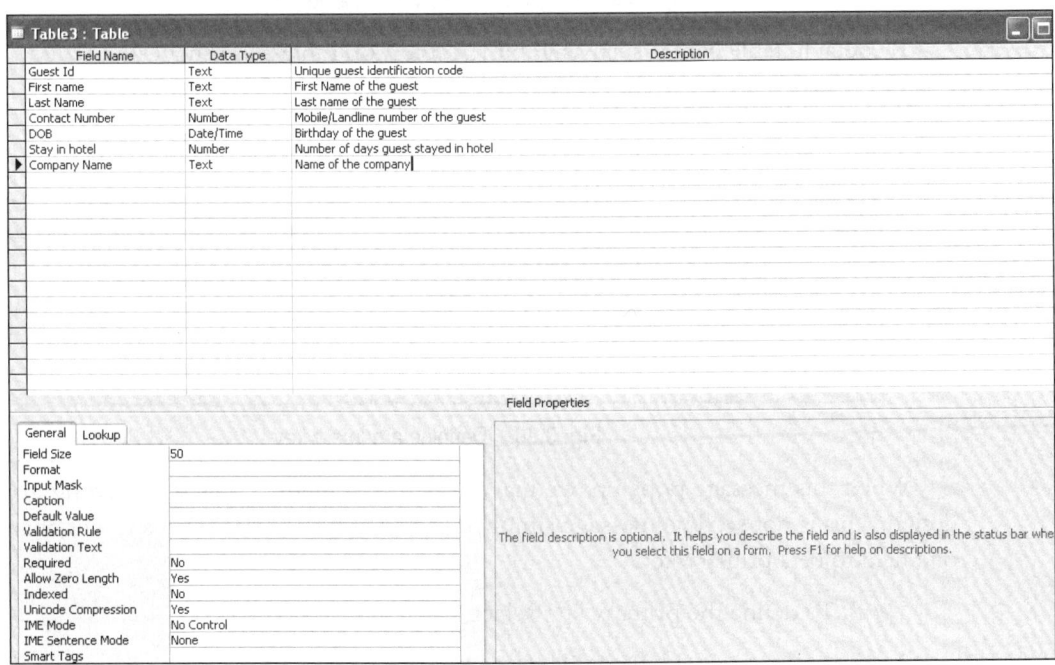

Fig. 3.77   Creation of a database with field name, data type, and
description in MS Access

Step 3: After we have completed defining the structure of the database, click the Close
button of the table window. A dialog box, as shown in Fig. 3.78, appears.

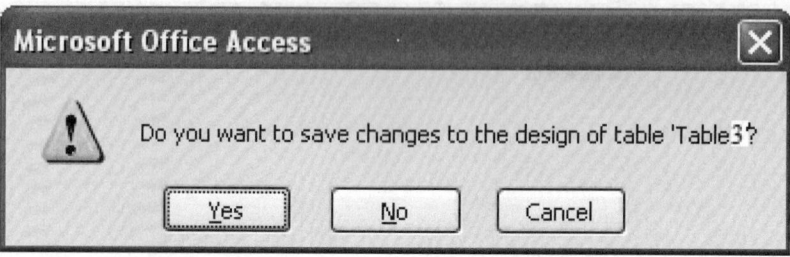

Fig. 3.78   Saving a table in MS Access

Step 4: The message is self-explanatory. If we want to save the table, click the Yes button. Another dialog box, as shown in Fig 3.79, appears.

Fig. 3.79   Saving the table with a name in MS Access

By default, table names such as Table1, Table2, or Table 3 are suggested.

Step 5: Write the name of the table (e.g., Guest Details).

Step 6: Click OK. Then another dialog box, as shown in Fig. 3.80, appears.

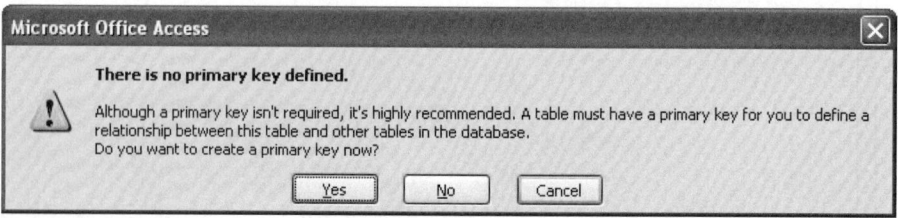

Fig. 3.80   Defining a primary key

When this message pops up, we may think that something has gone wrong in the table design. This is not so. The message simply means that Access expects us to designate at least one field to be a primary key.

Step 7: To define the primary key, select Yes; else select No.

A primary key is a field (it could be a combination of fields as well) that uniquely identifies each record in the table. It also implies that such a field cannot have empty values and no two records can have identical entries in such fields (guest ID field could be a primary key). After a primary key is chosen, Access does not allow us to enter identical values in it or leave it empty.

Entering Data in Tables

In order to manipulate the contents of a table, we can open the table in the following way:

Step 1: Double-click the table name. We can double-click the 'Guest Details' table to open it (as shown in Fig. 3.81).

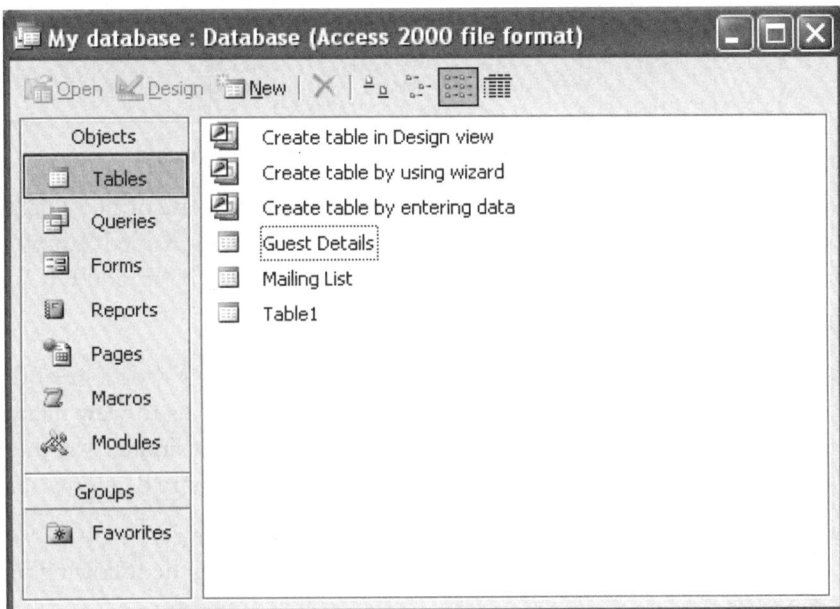

Fig. 3.81 Opening the guest details table in MS Access

Step 2: A blank table with the field names will appear (as shown in Fig. 3.82).

Guest Details : Table

	ID	Gust id	F Name	L Name	Contact Numbe	Date Of Birth	Stay in Hotel	Company Name
	1				0		0	
*	(AutoNumber)				0		0	

Fig. 3.82 Blank table in MS Access

Step 3: Enter data in the table as shown in Fig. 3.83.

Guest Details : Table

	ID	Gust id	F Name	L Name	Contact Numbe	Date Of Birth	Stay in Hotel	Company Name
	1	Guest-00001	Pankaj	Aggarwal	58958952	5/10/1960	5	Raj enterprise
	2	Guest-00002	Venkat	Raghavan	78945456	5/7/1999	2	S.N associates
▶	3	Guest-00003	Anshuman	Chakraborty	94854899	12/8/2000	10	C.K machines
	4	Guest -00004	Laxman	Pande	85496566	5/12/1995	6	L.R Group
*	(AutoNumber)				0		0	

Fig. 3.83 Data entered in the table in MS Access

Manipulating a table Manipulation of a table involves addition of new rows and columns to the table where a row represents a record in the table and a column represents a field. We can enter data in the empty record marked with an asterisk (*), as shown in Fig. 3.83. The asterisk shifts to the next record as soon as data is entered in the new record.

Editing a table   Editing a table involves changing the existing data in the table. For this, select the cell in which we have to change the phone number. As an example, we have changed the last digit in 78945456 to 78945457 in the contact number of Guest-00002 (as shown in Fig. 3.84).

	ID	Gust id	F Name	L Name	Contact Numbe	Date Of Birth	Stay in Hotel	Company Name
	1	Guest-00001	Pankaj	Aggarwal	58958952	5/10/1960	5	Raj enterprise
	2	Guest-00002	Venkat	Raghavan	78945457	5/7/1999	2	S.N associates
	3	Guest-00003	Anshuman	Chakraborty	94854899	12/8/2000	10	C.K machines
	4	Guest -00004	Laxman	Pande	85496566	5/12/1995	6	L.R Group
*	(AutoNumber)				0		0	

Fig. 3.84   Editing a table in MS Access

Similar to the procedure followed for preparing a database of guest details, we can prepare a separate database of the employees. This table can been named 'Employee database' with fields such as employee ID, first name, last name, address, basic, and HRA.

Working with Forms

A set of data can be viewed with the help of forms that provide a great degree of flexibility for viewing and entering data. Though a data sheet allows us to view many records at a time, the number of fields that can be seen are limited. On the other hand, forms can help us rearrange fields and view more data on a single screen.

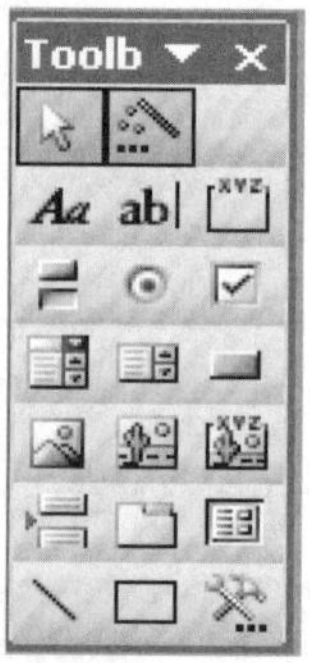

Planning Forms

When planning a form, we should keep the following points in mind:

*   Determine the purpose of the form
*   Determine the record source
*   Gather the source document used to design the form
*   Determine the type of control for each element of the form

Fig. 3.85   Control bar in MS Access

Clearly labelled fields and appropriate formatting are important for designing forms. The control bar is provided with form design tools. The controls are grouped in a toolbox bar (as shown in Fig. 3.85).

Types of Forms

The data in forms can be viewed in two ways: columnar and tabular.

Columnar forms   The fields are arranged as columns and resemble a manual data entry form.

Tabular forms   The fields are arranged as columns and the records are entered in rows, which resemble a data sheet.

Using Design View to Create Forms

We will now create a form using Auto form Wizard by employing the following steps:.

Step 1: Select the Forms tab in the database window.

Step 2: Click Create form by using wizard command.

Step 3: Select the table 'Employee Database'.

Step 4: Select the fields of the table that we would like to include in the form.

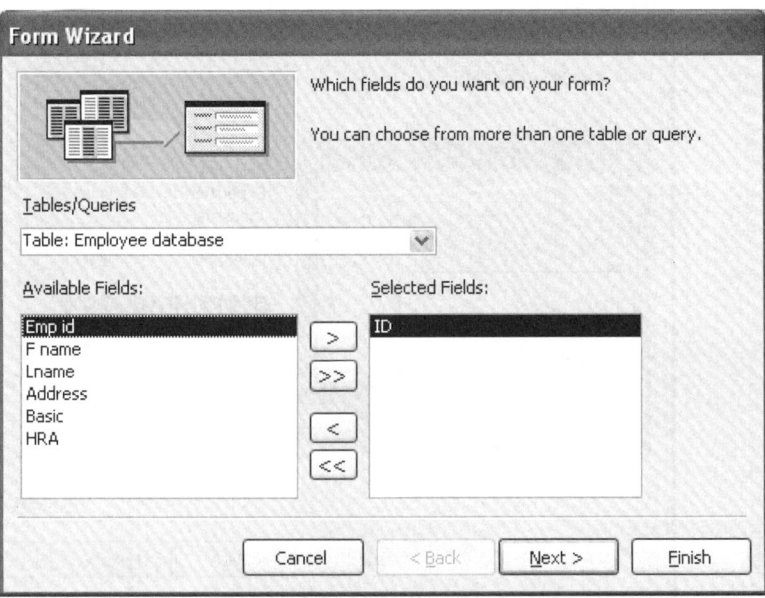

Fig. 3.86   Selecting fields in Employee database in MS Access

Step 5: Use < or > buttons to move a single field and << or >> buttons to move all the fields. We have selected a single field (as shown in Fig. 3.86).

Step 6: Click Next. When the next dialog box appears, select the desired form type. As an example, we select the Columnar type (as shown in Fig. 3.87).

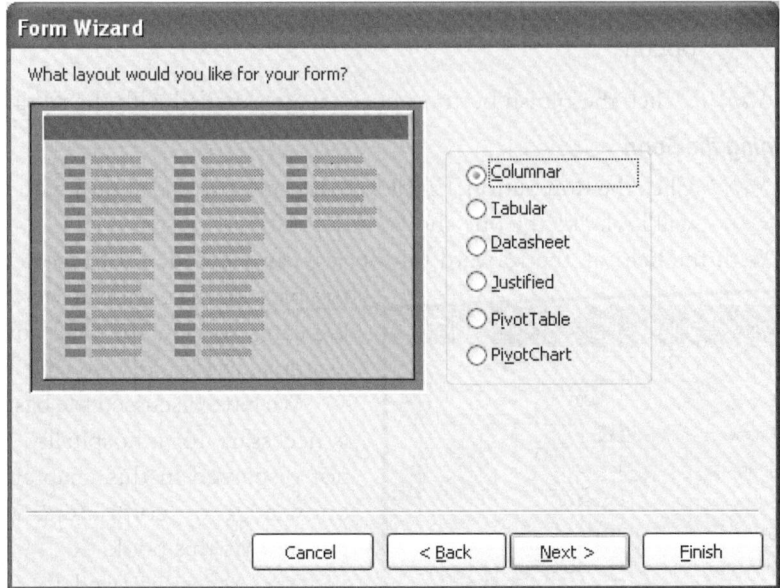

Fig. 3.87   Choosing a layout in MS Access

Step 7: Click Next. In the next dialog box, select the style of the form from the given list

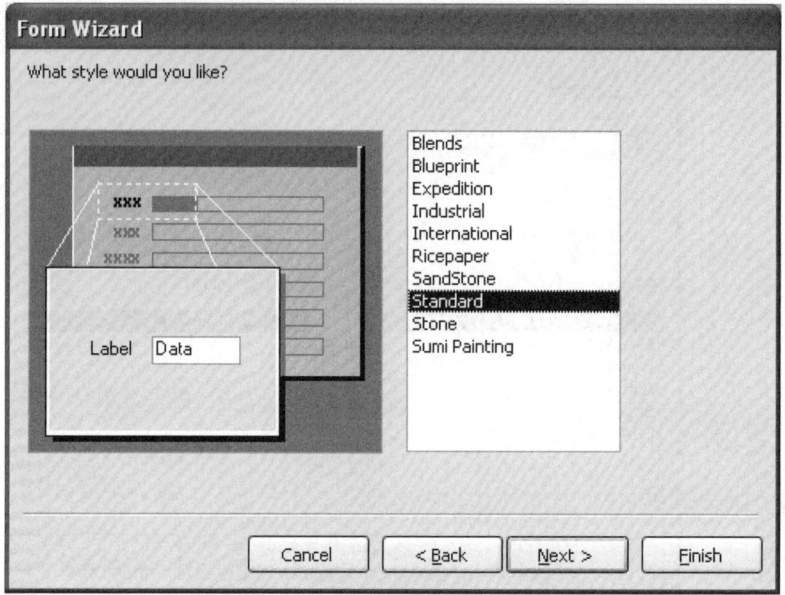

Fig. 3.88   Choosing a style in MS Access

of inbuilt styles. Here, we have selected the Standard style (as shown in Fig. 3.88).

Step 8: Click the Next button. In the next window, give a title to the form (e.g., employee database1).

Step 9: In the next step, we can opt to either open the form in the design view to modify it or run the form. At any point, we can go back to the previous screen and change the option.

Step 10: Click the Finish button once you are satisfied with the settings.

Running the Form

Step 1: Click the Run button. Employee database1 would appear with the record. This form will look like the one shown in Fig. 3.89.

With the help of the forward button, we can view the next record and with the help of the back button, we can view the previous record. We can also view the total number of records. For example, in Fig. 3.89 we have four records totally.

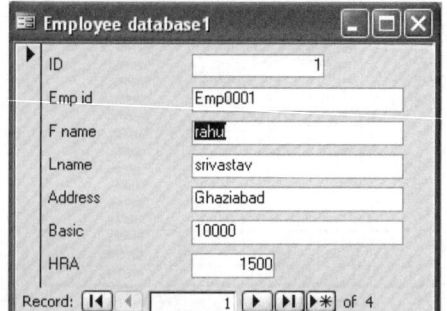

Fig 3.89   Run the Employee database form in MS Access

We have discussed the basics of MS Access that is necessary for a hospitality student. The sections not discussed in this chapter require an in-depth knowledge of computers, which is beyond the purview of this book.

MS Access is used to create and prepare a database in a table. Database in tables help us store large quantities of data, which could either be of hotel guests or employees. The use of query in MS

Table 3.3   Overview of MS Office 2003 software packages

	Word	**Excel**	**PowerPoint**	**Access**
Purpose	Prepares personalized documents	Prepares numerical datasheets	Prepares formal and informal presentations	Prepares relational databases
Features	Automatically checks spelling and grammatical errors	Numerical and logical functions can be performed	Other than text, audio-visual images can be added to enhance the presentation	Prepares forms and relates it to two tables
	Mail merge	Graphical presentation of data	Organization charts can be prepared	Forms and reports can be prepared
	A document consists of one or more pages	A workbook consists of one or more worksheets	A presentation consists of one or more slides	A database consists of one or more tables

Access allows us to find data in the table and sort it accordingly. This function of MS Access is not available in MS Word, Excel, or PowerPoint. In MS Excel, we can create a table and store data in rows and columns, but we cannot prepare a database or sort data. Excel is a spreadsheet application, whereas Access is an RDBMS. Excel is usually used for calculations, whereas Access is used to store information that can be accessed by people. Table 3.3 provides an overview of the four MS Office 2003 software packages discussed in this chapter.

MS Office is a user-friendly suite that helps in the efficient management of an organization. The MS Office suite helps us maintain records of individuals, edit documents and send letters, calculate and tabulate data as per necessity, and also represent data graphically. Its features enable an individual to present data attractively in the form of a presentation. Thus, we can conclude that an individual familiar with MS Office can perform jobs with competence.

SUMMARY

Microsoft Office 2003 has been introduced in this chapter to initiate us into the process of learning and using the various features of the package.

In this chapter, we have discussed the basic concepts of MS Word, MS Excel, MS PowerPoint, and MS Access. In Word, we learnt how the Edit menu helps in editing, how a document is saved, and how Mail Merge helps in sending personalized documents to many individuals by personally addressing them.

In Excel, we talked about functions such as Sum, use of formula, and percentages. We also learnt how data is used for preparing charts and why charts are used for representing data differently. In PowerPoint, we familiarized ourselves with the basics of preparing presentations. We learnt how to change font and colour, add pictures, change layout, and customize our presentation. In Access, we touched upon the preparation of databases using tables and forms.

KEY TERMS

Blank web page   It is a web document that is used when we want to display the document's contents on an Internet web browser. It opens in a web layout view and the files are saved with .html extensions.

Cell   It is the smallest addressable data container in Excel.

Chart   It is a graphical representation of a worksheet data. Charts can be used to compare different worksheet values. When a user changes the data source on a worksheet, the chart is automatically updated.

Clipboard   After a Cut operation is performed, the selected object is removed from the present document to a memory area called clipboard.

Formula   In Excel, the user can analyse data by using a formula. A formula begins with an equal to (=) sign. A cell containing a formula normally displays the resulting value of the formula in the formula bar of a worksheet.

Function   It is a predefined formula that takes zero or more input values (also known as arguments) and produces (returns) a result.

Mail Merge   It allows us to take a document (e.g., a form letter)

and a set of information (e.g., the names and addresses of all our clients) and merge them together. Using this feature, we can merge the form letter and the list of names to generate mass mails.

Template   It is a general pattern that will remain the same throughout the presentation.

Workbook   A workbook is a single file in Excel having multiple worksheets in them.

Worksheet   A worksheet is a rectangular collection of many cells. By default, a user can use worksheets in a workbook.

REFERENCES

Balagurusamy, E., *Fundamentals of Computers,* Tata McGraw Hill, New Delhi, 2009.

Gupta, S. and S. Gupta, *Computer Aided Management,* Excel Books, New Delhi, 2004.

Mukherjee, P. and S. Bandhopadhay, *Introduction to Computer*

Science, Vol I, Deep Prakashan, Kolkata, 2001.

Saxena S., *A First Course in Computers,* Vikas Publishing House, New Delhi, 2003.

Sinha, P.K. and P. Sinha, *Computer Fundamentals,* Fourth edition, BPB Publications, New Delhi, 2007

EXERCISES

Concept Review Questions

1. What is MS Word used for?
2. What is Mail Merge? What are its advantages?
3. What are the ways of starting MS Excel from our computer system?
4. Describe the following functions:
   (a) SUM()   (c) AVERAGE()
   (b) IF()
5. What is the method of inserting a picture in a slide?
6. What is MS Access? How can we create a database in MS Access?
7. Explain the method of creating tables in MS Access using the design view.
8. What are the different types of forms? Describe them.
9. Describe the following terms:

   (a) DBMS   (c) Report
   (b) RDBMS   (d) Query

10. How can Mail Merge be used for sending letters that are personally addressed to the recipients?
11. In a student result database, how can we categorize the aggregate results into first class, second class, and pass taking into account the fact that 60 per cent and above is first class, 50 per cent and above is second class, 35 per cent and above is pass, and below 35 per cent is fail.
12. What points need to be considered while preparing a formal presentation for a business purpose?
13. While preparing a database of guest details, the guest ID is usually used as a primary key. Why is it that fields such as guest name and room number are not preferred?

Multiple Choice Questions

1. The Microsoft Word menu bar consists of _____ menus.
   (a) seven   (c) eight
   (b) nine   (d) six
2. What is the keyboard short cut to open the Format menu?
   (a) Alt+F   (c) Alt+O
   (b) Ctrl+O   (d) Ctrl+F
3. Which of the following toolbars provide many drawing tools?
   (a) Menu bar   (c) Formatting toolbar
   (b) Drawing toolbar   (d) None of these
4. Which of the following is a powerful presentation package?
   (a) Microsoft Word   (c) Microsoft PowerPoint
   (b) Microsoft Excel   (d) Microsoft Access

5. A worksheet is a rectangular collection of many
   (a) rows   (c) columns
   (b) cells   (d) workbook
6. A formula entered in a cell begins with
   (a) a bracket sign   (c) a hash sign
   (b) a question mark sign   (d) an equal to sign
7. From which function can we get the standard deviation value?
   (a) STDEV   (c) STDEVA
   (b) STANDARDIZE   (d) AVEDEV
8. Which of the following is a graphical representation of data?
   (a) Picture   (c) Formula
   (b) Chart   (d) Function

Project Work

1. The club manager of ABC club wants to invite his platinum members (121 members) to an exclusive Christmas Eve celebration on 24th December at the club premises. How will he prepare and send letters to the invitees?

2. Create two databases in Access for managing information of guests in a hotel restaurant.

3. With reference to the databases mentioned in Question 2, design suitable forms to represent the data.

4. Create a worksheet on the sales report of Hotel Surya for December 2012. The fields of the worksheet will be room number, room rent, luxury tax (5 per cent of room rent), VAT (13.5 per cent of room rent), total tax from room, and net sale from rooms.
   (a) Save the file with the name Suryadec'12.xls.
   (b) Create a column chart with the aforementioned data. Compare the net sales with the rooms sold, giving appropriate titles and labels for the chart.

5. Prepare a presentation on your course curriculum. Consider the subjects in your syllabus semester-wise and year-wise.

6. Prepare an organization chart of the front office department of a hotel.

CHAPTER **4**

Internet

LEARNING OBJECTIVES

After going through this chapter, you will be able to understand the following:

- Internet, and its features and uses
- Types of network and topologies
- Data communication
- E-commerce
- Latest wireless technology
- Internet security

The Internet is a network of computers linking many systems across the world. It is a network of various networks, sharing a common mechanism for addressing (identifying) computers and a familiar set of protocols for communication between two computers on a network.

HISTORY OF INTERNET

The Internet began in the Advanced Research Project Agency network (ARPANET) of the US defence department. ARPANET was the first wide area network (WAN) with only four locations in 1969. It slowly evolved from the basic idea of ARPANET being used for connecting

---

EXHIBIT 4.1 Internet for Everyone

Mrs Dixit, a housewife, wants to prepare chicken *makhni* for her family but does not have the recipe. Mr Dixit, on the other hand, wants to know the value of his shares. Their son Karan wants to know more about tropical rain forests and his little brother Pratap wants to play car racing games. Though the needs of the family are different, the solutions can be found within the comfort of their home by simply logging on to the Internet. The Internet is one of the most wonderful inventions of the industrialized world. You would simply need a search engine such as Google or Bing to get

the answers to your queries. A person can also refine his/her search by typing suitable key words. Websites related to the words typed in the search engine are shown on the screen. A user can look into these websites, copy content if permitted, download material, and even take printouts.

From airline to railway reservations, and hotel to cruise liner bookings, everything is possible through the Internet. Nowadays we cannot think of life without the Internet, be it entertainment, fun, or even education.

computers, and was used by various research organizations and universities in the beginning to share and exchange information. Since 1990, the Internet has grown rapidly to become the world's largest network. It now interconnects thousands of networks with more than 10 million computers and 50 million computer users across 150 countries around the world. The Internet is growing at a swift pace by adding new users every day.

EQUIPMENT NEEDED TO CONNECT TO THE INTERNET

The following equipment is required to use the Internet:

Computer   Any IBM compatible, Macintosh, or UNIX computer that has a memory of 2 GB or more, and at least 16 MB RAM or a 200 MHz processor is needed.

Modem   A modem is a device that is used to transmit data over the network. A computer sends data in binary code to the modem, which converts the binary-coded data to an analog signal. The data then usually travels along the telephone network and reaches the destination computer. The modem connected to the computer converts the analog signal back into binary-coded data, which can be read by the computer.

Telephone line   It was earlier essential to have a telephone connection to get connected to the Internet. Data and information used to pass through telephone lines in the form of analog signals. Nowadays, with the advancements in technology, the Internet is available on mobile phones, and signals have also become digital.

Internet service provider   We get Internet connection from Internet service providers (ISPs). Earlier in India, Internet connection was available through Videsh Sanchar Nigam Limited (VSNL). Now we can choose from 120 companies that have been granted ISP licences. Some of the most popular ISPs are Mahanagar Telephone Nigam Limited (MTNL) in Delhi and Mumbai, Bharat Sanchar Nigam Limited (BSNL), and Mantra Online. Nowadays, many private players (e.g., Reliance, Airtel, Tata, etc.) provide Internet services to subscribers.

Web server   A web server is a computer that can store various web pages and contains web server software. The web pages on the server are mostly hypertext markup language (HTML) documents. The web client (i.e., the browser) makes a request to the web server. The web server software running on the server accepts the request, makes a search, and then returns the result to the web client.

Web browser   The software used to navigate through the web is known as a browser. The most popular browsers are Microsoft Internet Explorer, Mozilla Firefox, Google Chrome, and Safari (for Mac computers). MS Internet Explorer is installed along with MS Office 2000/XP.

BASIC INTERNET SERVICES

The Internet has revolutionized the world of communication by allowing us to send and retrieve information across countries. It helps us to not only send text, but also images, audio, and video files.

Electronic Mail

Electronic mail service (known as e-mail, in short) helps an Internet user send a mail (message) to another Internet user in any part of the globe in an almost real-time manner. An e-mail

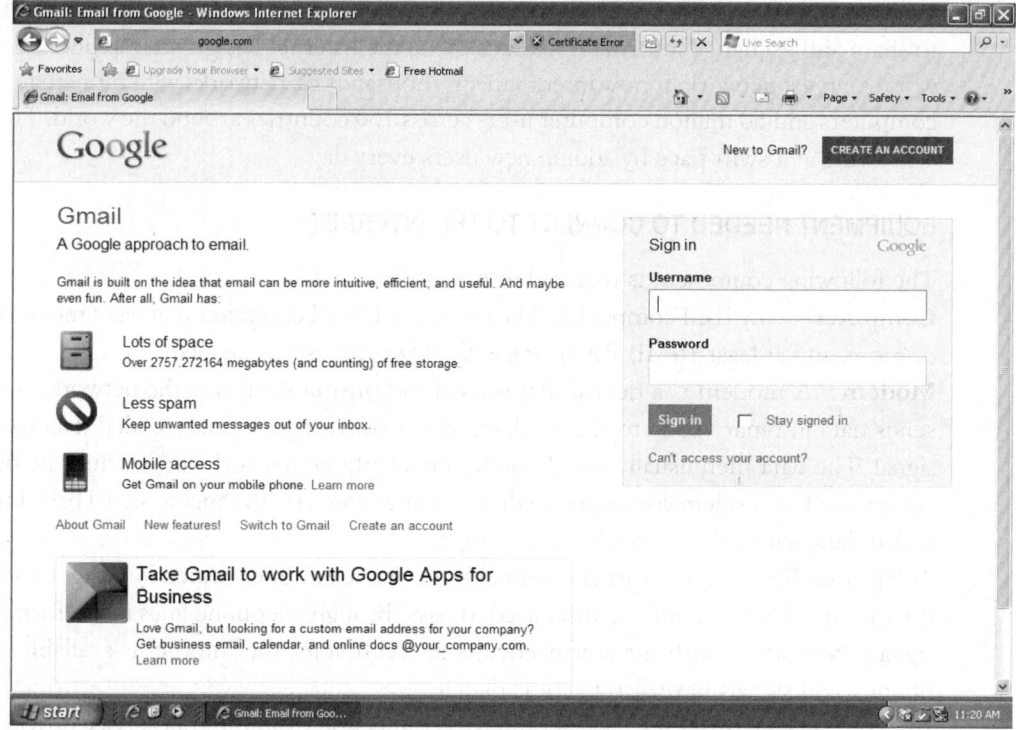

Fig. 4.1 Gmail, an e-mail service website

Source: https://mail.google.com/mail/?tab%3Dwm&scc=1&ltmpl=default&ltmplcach

message might take a few seconds to a couple of minutes to reach the destination(s) as it may have to travel from one network to another. One of the most popular e-mail service providers is Gmail (shown in Fig. 4.1).

Messages in an e-mail can not only contain text documents, but also images, audio, video, or a combination of these. The only constraint is that the image has to be digitized and later converted into a computer-readable format.

File Transfer Protocol

The file transfer protocol (FTP) service enables an Internet user to shift a file from one computer to another on the Internet. A file can contain any type of digital information—text, image, video, audio, or even software. Shifting/Copying a file from a remote computer to one's personal computer is known as downloading, whereas shifting/copying a file from a personal computer to a remote computer is known as uploading a file.

Telnet

The telnet service allows an Internet user to log into another (host) computer on the Internet from the local computer. This process is also referred to as remote login. Telnet also operates on the client–server principle. The local computer uses a telnet client program to establish the connection and displays data on the local computer's monitor. The remote or host computer uses a telnet server program to accept the connection and sends requests for information back to the local computer.

Limitations

Telnet is no longer popular due to the following reasons:

- It was originally designed for Unix-based systems but now, most of these sites are not maintained and many do not work at all.
- It does not work with embedded pictures and graphics.
- It is used for only point-to-point communication. Therefore, only two sites can communicate at a time. In addition, remembering the port number and machine type is a cumbersome process.

Usenet

Usenet is a global network of discussion groups, better known as newsgroups. We can join any discussion group of our choice. This is an electronic forum where people discuss topics of mutual interest. By reading and posting in newsgroups, we can ask questions from experts, find suitable job postings, get necessary advice, and even download photos and music.

Google groups are Usenet clients. http://groups.google.com is an example.

World Wide Web

The world wide web (WWW) is the most popular method of accessing the Internet. The main reasons for the popularity of the web are the point-and-click nature of the access, its graphical interface service, and the fact that it combines graphics, music, and animation into an improved communication medium. Later in this chapter, we will discuss the basic components of the WWW. It consists of a group of web servers containing several web pages belonging to different websites.

Search Engine

A web search engine is an interactive tool that helps users locate information on the world wide web. It is a collection of programs that gathers information from the web, indexes it, and puts it in a database so that it can be searched for. Google is one of the most popular search engines.

A search engine takes keywords or phrases entered in the search form and then searches the database attached to it, with the search phrase. If it does not find a match it returns a message that says 'no match found', otherwise it returns hyperlinks to sources that contain the search phrase. Search engines contain computer programs called *spider, crawler, bot,* or *robot* that are responsible for building up the database attached to each search engine. Other popular search engines include Bing, Yahoo!, AltaVista, and Lycos.

Internet Relay Chat

Another popular application of the Internet is live chat whereby people can chat with others online. Chat, in Internet terminology, means to hold online discussions on various topics with people around the world. Internet relay chat (IRC) is one of the oldest and fastest chat clients, developed by Jarkko Oikarinen of Finland in the late 1990s. IRC has been an effective source of communication during natural disasters and calamities, for example, the 1994 Los Angeles earthquake.

IRC works on a client–server technology that requires client software to perform the chatting process. In order to chat, one has to subscribe to a chat client. There are many IRC clients, the most popular one being Microsoft Internet Relay Chat (mIRC).

There are many websites that are dedicated to chat, and others where chat is just one of the features offered. The difference between an IRC chat and a web chat is that an IRC chat is faster and more number of people can be connected as it has many servers that are interconnected. However, web chats are more attractive as most of them are free of cost. Web chats allow voice conversations and communication with others through live video over the Internet. One of the most popular web chat services is Google talk.

USES OF INTERNET

The Internet currently caters to millions of subscribers worldwide, who use it for either commercial or non-commercial purposes. Some uses of the Internet are discussed here:

Software sharing   The Internet provides access to a large number of shareware software, development tools, and utilities. A few examples of such shareware tools are compilers, code libraries, mail servers, and operating systems.

Online communication   Computer users across the world use e-mail services on the Internet to communicate with each other.

Customer support service and promotions   Organizations use the Internet to provide customer support. The use of e-mail and other services on the Internet enhances our knowledge about various products. Many organizations also use the Internet for promoting their products.

Discuss topics of common interest   The Internet has a number of groups where members exchange their views on topics of common interest. The Internet presently has thousands of electronic subscriptions and books that are either available free of cost, at a low price, or at market price. E-books, journals, and magazines can also be downloaded from websites. For example, Cornell Hospitality, a hotel management journal, is available electronically.

Online shopping   The Internet has facilitated the concept of virtual shops. These shops remain open 24 hours, 365 days a year, and are accessible to purchasers around the world. eBay and Amazon are examples of popular online shopping websites.

BASIC COMPONENTS OF WORLD WIDE WEB

We have discussed the history, basic services, and uses of the Internet. Let us now take a look at some of the components of the WWW.

Hypertext

A hypertext is a special type of database system in which objects (text, pictures, music, and programs) can be linked to each other. When we select an object, we can see all the other objects that are linked to it. We can also move from one object to another even though they might have different forms. Hypertext systems are particularly useful for organizing and browsing through large databases.

Hyperlink

A hyperlink is a connection between two documents that allows users to navigate from one document to another that may even be located on a different Internet server. They are usually distinguished by font colours that are different from the regular text and are sometimes even underlined to make the user aware of their presence.

Web Page

A web page is a single unit of information (often called a document) that is available via the WWW. A web page can consist of text, moving pictures, graphics, image forms, sound, animation, hyperlinks, and hypertext links.

Website

A website is a set of related web pages that contains content such as text, images, video, and audio. A website is hosted by at least one web server, accessible via a network such as the Internet or by a private local area network (LAN) through an Internet address, also known as a uniform resource locator (URL).

Accessing a website   In order to access the web and view websites, we require a WWW browser. For a web client, a computer needs to be loaded with a special software tool known as a WWW browser.

Designing a website   Web pages can be designed using HTML. The WWW is the brainchild of Tim Berners Lee, who had the idea of creating an electronic web to search for information. In 1980, he developed a programming language called HTML and a web based on HTML.

HTML is a standardized and portable language. HTML is a text that is designed on one computer and will look exactly the same when viewed on different computers having different operating systems and configurations.

HTML is a subset of standard generalized markup language (SGML), which is an ISO standard for defining the structure and managing the contents of the HTML. Presently, we use dynamic hypertext markup language (DHTML) to make websites more interactive and user-friendly.

Browser

In the client–server model, the browser functions as the client program. It acts on behalf of the user. The browser contacts a web server and sends a request for information, receives the information, and then displays it on the user's computer screen. Browsers generally provide the following navigation facilities to help users save time while surfing the Internet:

- Browsers enable users to visit a server computer's site by specifying its URL address. A URL is an addressing scheme used by WWW browsers to locate sites on the Internet.
- Browsers also enable a user to create and maintain a list of favourite URL addresses of server computers that the user is likely to visit in the future.
- Browsers have a 'history' feature. The browser retains a history of the server computers visited during a particular surfing session.

Browsers are of two types: text-based browsers and graphical browsers. Linux is an example of a text-based browser while Microsoft Internet Explorer, Google Chrome, Mozilla Firefox, Mosaic, Opera, Amaya, and Hot Java are examples of graphical browsers. In a text-based browser, graphics are not visible. On the other hand, in a graphical browser all sorts of graphics/images can be viewed. In a graphical browser, we are provided with a point-and-click facility. Thus, by using a mouse, we can select drop-down menus, hypertext link, and toolbar buttons to navigate and access resources on the Internet.

The most popular web browsers are Internet Explorer, Mozilla Firefox, and Google Chrome.

Internet Explorer

Internet Explorer is a product of Microsoft and is usually supplied with every version of Windows. Internet Explorer 6.0 is supplied with Windows XP and can be easily downloaded for free. To open the browser, double-click the Internet Explorer icon.

Mozilla Firefox

Mozilla is a global, non-profit organization that is dedicated to making the web better. Mozilla works with a worldwide community to create open source products like Mozilla Firefox and innovations for the benefit of individuals and the betterment of the web. Firefox improves web performance with its improved start-up and page load times, and speedy web application performances. Firefox offers a 'Do Not Track' feature that lets us choose that we would not like to be tracked by websites. Firefox has been manufactured for the rich, interactive websites of the future. To open Mozilla Firefox, click its icon.

Google Chrome

Google Chrome has the unique feature of sandboxing, which prevents harmful web pages from leaving behind some programs on the computer or even stealing private information from it. Google Chrome gets updated regularly with the latest security features without any action from our end.

Google Chrome is launched within a few seconds of double-clicking the icon. It alerts us with a warning message if the site we wish to visit may cause harm to the computer via a virus (which is discussed in the section on Internet security).

Portal

Large search engine sites such as Yahoo! and Lycos that have branched off to offer a wide variety of services are known as portals. The idea is that a web user could access the web by using only one website, the portal. A portal can be used for searching, getting stock queries, buying products, etc. There are two types of portals depending on the type of information required:

Hortal Google India is an example of a horizontal portal, known as a hortal. Using this search engine, a user can get information on a wide range of subjects depending on the query.

Vortal With vertical portals known as vortals, we can go deep into the Internet to research a particular topic. The portals are usually subject-oriented so that any information that is required can be obtained easily. For example, the site www.criconline.com is dedicated to cricket.

Keeping in mind the Indian angle, portals can be classified as national level or regional level, as shown in Table 4.1 and Table 4.2 respectively.

Table 4.1 Examples of national-level portals

Portal names	URLs
Yahoo India	http://www.yahoo.co.in
123India	http://www.123India.com
Indiatimes	http://www.indiatimes.com

Table 4.2 Examples of regional-level portals

Portal names	URLs
Cafe Mumbai	http://www.cafemumbai.com
Hello Kerala	http://www.kerala.com
A guide to Bangalore (Bengaluru)	http://www.bangaloreguide.com

Uniform Resource Locator

The URL is a form of address that all web browsers can understand. The basic structure of a URL is [protocol://server-name.domain.top_level domain:port/directory/filename].

For example, http://www.kolkatahotels.org.uk/ is a comprehensive guide to hotels in Kolkata, which provides information on booking and reservation.

In the aforementioned example, the first part shows what type of protocol is being used (e.g., FTP, IP, and Telnet). In this case, it is http. The second part indicates the resources (in this case, WWW). The third part consists of the web server (in this case, kolkatahotels) to which we wish to get connected. The fourth part is a domain name (.org). Other domain names that are in use include .in, .edu, .gov, .ac, .com, and .net. The most common way of referring to websites on the Internet is by using URLs.

Protocol

It refers to a set of rules that we must observe while using the Internet. We discussed FTP in the section on basic services of the Internet. Other commonly used protocols are mentioned here:

Transmission control protocol/Internet protocol   TCP/IP is a collection of public protocols or rules that helps us copy files from one computer to another over the network. An IP address is a 32-bit identifier assigned to a host that uses the Internet protocol. The IP address is represented by four octal numbers.

Hypertext transfer protocol   HTTP is based on the client–server principle. Hypertext is the text that is specially coded using a standard system called HTML.

Simple mail transfer protocol   SMTP is responsible for transmission of mail messages from one computer to another. SMTP is a reliable and efficient method of mail transmission. It receives spooled e-mail and then checks the messages in the queue to transfer them.

Multi-purpose Internet mail extensions   MIME is a set of protocols developed in 1992 that allows different kinds of documents, especially multimedia, to be exchanged among different computer systems.

Gopher   Gopher is a protocol designed to search, retrieve, and display documents from various remote sites on the Internet. It is a menu-driven application developed at the University of Minnesota. It possesses an Internet navigation tool that allows users to find and retrieve information using a hierarchy of menus and files.

Domain Name

A domain name is a way of identifying and locating computers connected to the Internet. Domain names have been designed on the basis of organization type or geographical locations. An example of a domain name is yahoo.com, where .com indicates that Yahoo is a commercial organization.

In general, while typing an Internet address, it is better to use lower case letters. There are two types of top-level domains. A top-level domain is the domain that is at the highest level in the hierarchical domain name system of the Internet.

- Non-geographical domains are those that indicate the type of organization. The major categories of top-level non-geographical domain names are given in Table 4.3.

Table 4.3 Top-level non-geographical domain names

Domain name	Description	Example
com	For commercial organizations	yahoo.com
edu	For higher educational institutions	manipal.edu
gov	For government organizations	whitehouse.gov
mil	For US military organizations	army.mil
net	For network resources/ organizations	pacific.net
int	For international organizations	tpc.int

Table 4.4 Geographical domain names

Domain	Country
.lk	Sri Lanka
.pk	Pakistan
.np	Nepal
.uk	United Kingdom of Great Britain and Northern Ireland

Geographical domains are the two-character codes that represent individual country codes, for example www.vsnl.net.in. Here, the .in indicates that the network connection is in a country, India. A few geographical domains are listed in Table 4.4.

Client–Server Architecture

With the increasing popularity of computer networks, it has become possible to interconnect several computers so that they can communicate with each other over the network.

The client–server architecture has now become an important designing system that rests on the network. In this system, one or more clients (processes) and servers rest on various host sites of a network. Client–server communication is also supported by facilities for communication within it and with hosts.

NETWORK AND DATA COMMUNICATION

Computers were initially developed as standalone, single-user systems. A standalone system received the user's data, processed it, and provided results. Initially, the data processed was of the user's only, but later when computers spread across the office and data had to be exchanged, various software and hardware were developed that enabled computers to share and communicate information among them.

Network

A computer network is a network of computers that is spread out but connected in a manner that enables meaningful transmission and exchange of data between them. The sharing of information, resources (both hardware and software), and processing load is the prime objective of a computer network.

In this chapter, we will learn about the various types of data communication mediums. We will also learn how these technologies can be used for building different types of network topologies.

Data Communication

Communication is the process of transferring a message from one point to another. The three basic elements of any communication system are as follows:

- A sender (source), which creates and sends a message

- A medium, which carries the message
- A receiver (sink), which receives the message

In a data communication system, the sender and receiver are normally machines such as computers, terminals, and peripheral devices (printers, plotters, and disks). The transmission medium could be telephone lines, microwave links, or satellite links. Hence, a data communication system is an electronic system that is used to transfer data from one point to another.

Data Communication Medium

A transmission medium is a design factor of any communication system. It is a physical channel (path) between the transmitter and receiver in the data transmission system. It links the source with the destination. The transmission medium is broadly divided into two classes (as shown in Fig. 4.2). Data rates and distances are two important factors of data transmission.

The data rate is dependent on bandwidth and usually increases with bandwidth. On the other hand, signal power decreases exponentially with distance. For this reason, repeaters are used after a few kilometres to increase signal strength.

The classification of transmission media is as follows:

Bounded media (guided media)   In guided media, the waves are guided along a solid medium like a copper wire. Examples of bounded media are twisted pair wires, coaxial cables, optical fibres, etc. In bounded media, electrical cables carry information as electrical signals and optical cables carry information as optical signals. Bounded media is best suited for short-distance communication.

Twisted pair wire   A twisted pair wire consists of two bundles of thin copper wires, each bundle separately enclosed in a plastic insulation. The bundle is then twisted around each other to reduce interference by adjacent wires. It is an unshielded twisted pair (UTP) cable. Other than the plastic insulation around the two individual bundles of copper wires, nothing else shields it from outside interference.

UTP cables are commonly used in local telephone communications and short-distance (up to about 1km) digital data transmission. They are used for connecting terminals to the main computer if they are placed a short distance apart. It is inexpensive and easy to install but has limited use since it picks up noise signals easily (resulting in high data error rates) when the line length extends 100 meters.

Coaxial cable   A coaxial cable consists of a group of specially wrapped and insulated wire line cables that are used for transmitting data at high rates without distortion or loss of signal.

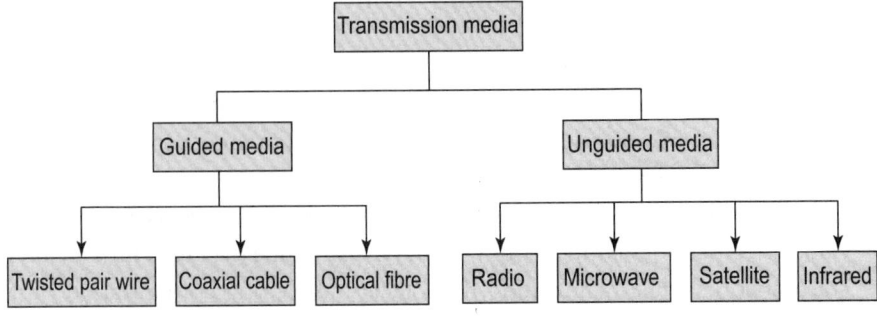

Fig. 4.2   Classification of transmission media

It consists of a central copper wire surrounded by a polyvinyl chloride (PVC) insulation over which there is a case of copper mesh and a foiled shield. The copper mesh case is shielded by an outer shield of thick PVC material. The signal is transmitted by the inner copper wire and is electrically shielded by the outer copper mesh. They are extensively used in long-distance telephone lines, as cables for cable television, and by telephone companies to transmit data.

Optical fibre   Optical fibres are thin, hair-like threads of glass, plastic, or silica. They can transmit data at speeds much higher than copper wires or coaxial cables with no significant loss of intensity over long distances.

A fibre optic cable consists of three concentric layers: inner core, a cladding around it, and an outer protective coating.

Unbounded media (unguided media)   In unguided media, the waves pass through the atmosphere and outer space. It is known as wireless transmission. It is suitable for long-distance communication. Examples of unbounded media include radio, satellite, and microwave.

Microwave system   A microwave system uses very high frequency radio signals to transmit data through space (wireless communication). Microwave systems use electromagnetic waves that cannot pass through or bend around obstacles such as tall buildings or hills. Thus, the transmitter and receiver of a microwave system have to be mounted on very high towers since they should be in the line of sight. Hence, this system may not be possible for very long distance transmissions. Taking into consideration the problem of line of sight and power amplification of weak signals, microwave systems use repeaters at intervals of about 25–30 km between the transmitting and receiving stations.

Communication satellite   A communication satellite consists of microwave relay stations that are placed in outer space. The satellites are launched by either rockets or space shuttles and are positioned precisely 36,000 km above the equator with an orbit speed that matches the Earth's rotational speed. The satellite is positioned in a geosynchronous orbit so that it remains stationary relative to the Earth and always stays over the same point on the ground. It thus becomes easy for the ground station to aim its antenna to a fixed point in the sky. Presently, there are hundreds of satellites in orbit to handle domestic and international data, and voice and video communications. The Indian satellite (INSAT) series is positioned in outer space in such a way that the signals can be received and sent from any place in India.

TERMINOLOGY USED IN NETWORKS

The terminologies and concepts that will help us understand computer networks have been discussed here:

Node   The personal computer and the server are known as network devices or network nodes. In a network, a node is a connection point, either a redistribution point or an end point for data transmissions. In general, a node has programmed or engineered capability, which recognizes, processes, or forwards transmissions to other nodes.

Hub   A hub is a physical layer device that connects multiple user stations, each via a dedicated cable. Electrical interconnections are established inside the hub. Hubs are used to create physical star networks while maintaining the logical bus or ring configuration of the LAN. In some cases, a hub functions as a network hub in Ethernet LANs. These allow transmission lines to be connected together easily. Figure 4.3 shows the connections of a hub network. Hubs come in a variety of options and names. Token ring networks use devices known as multiple access units (MAUs) instead of hubs, but they function in a similar way.

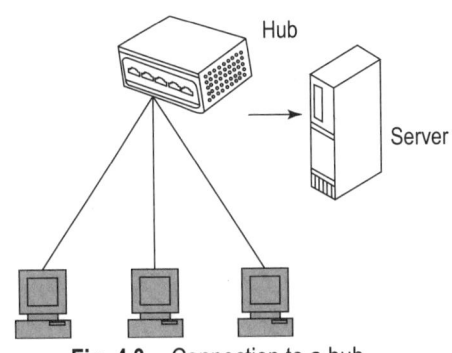

Fig. 4.3   Connection to a hub

Source: http://www.windowsnetworking.com/articles_
tutorials/autoslct.html

Server   The term server refers to any device that offers a service to network users. A server is a computer program that provides services to other computer programs (and their users) within the same computer or to different computers. A server can be hardware, software, or both. The most common servers are file servers, print servers, and gateways (also known as communication servers).

Host   The term host refers to the microcomputer attached to a network device. It may also refer to the native operating system or workstation.

Workstation   A workstation is a computer that is used as a node on a network, which is primarily used to run application programs. For example, in the IBM token ring network, any IBM personal computer can be used as a workstation.

NETWORK TOPOLOGY

The topology of a network relates to the way in which the network's nodes (computer or other devices that need to communicate) are attached together. It also determines the data paths available between any pair of nodes in the network. Though the number of possible network topologies could be limitless, the five major ones are bus network (a bus is a distinct set of conductors within a computer system to which pieces of equipment might be connected in parallel), star network, ring network, completely connected network, and multi-access bus network.

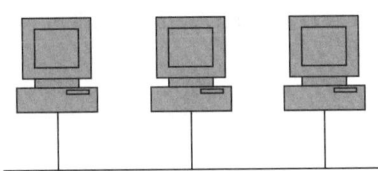

Fig. 4.4   Bus network

Bus Network

In a bus network, devices are connected to a common cable (as shown in Fig. 4.4). If one device wants to access another device on the network, it puts a message addressed to the device on the bus. If one of the devices attached to the bus fails, the rest of the network is not affected. The network fails only when the bus fails. Bus failure is rare because the bus is simply a wire with no active components; it has to be cut into two for it to fail. A new implementation of this network is the use of a fibre optic cable as the bus. This network is commonly used along with the LAN.

Advantages

- Bus topology requires a short cable length and a simple wiring layout. Thus, the installation and maintenance costs are reduced.
- Additional nodes can be connected to an existing bus network at any point along its length. Thus, this network is easy to extend.

Disadvantages

- In most LANs based on a bus, the control of the network is not centralized by any particular node. Hence, it is required to check text at different points to detect a fault.
- It is difficult to isolate faults in this network. If a node on a bus is faulty, it must be rectified at the point where the node is connected to the network. Once the fault has been located, the node can be removed. If the fault is in the network medium itself, an entire segment of the bus has to be disconnected.

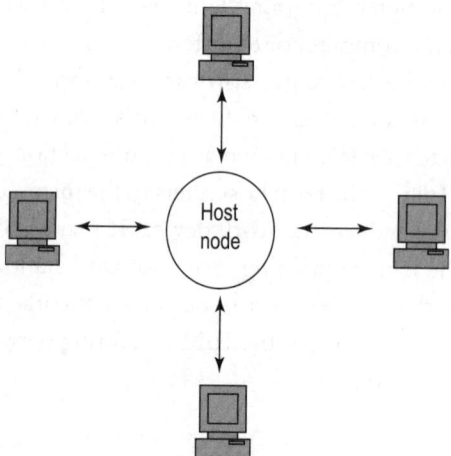

Fig. 4.5   Star configuration of a computer network

Star Network

It has multiple nodes connected to a single host node. The nodes in the network are linked to each other through the host node and can communicate only through the host node (as shown in Fig. 4.5). The routing function is performed by the host node (a centrally controlled communication point between any two other nodes) by establishing a logical path between them.

Advantages

- Star topology has very low line cost since only $n - 1$ lines are required for connecting n nodes.
- If a node other than the host node fails, the remaining nodes remain unaffected.

Disadvantage   The system essentially depends on the host node. If it fails, the whole network fails.

Ring Network

In a circular or ring network, every node has two communicating subordinates (neighbouring nodes with which it can communicate directly) but there is no master/central node for controlling other nodes (as shown in Fig. 4.6).

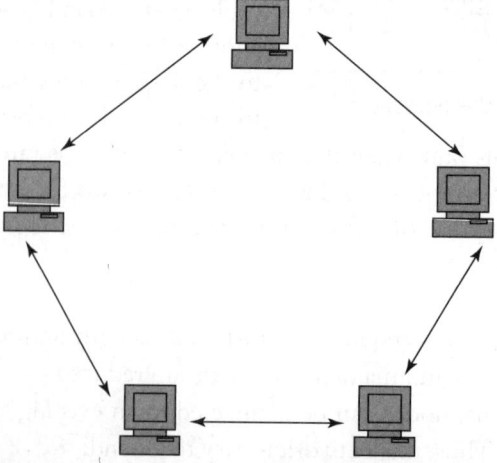

Fig. 4.6   Ring configuration of a computer network

Advantage The ring network works well if there is no central node for making the routing decision.

Disadvantages

- It requires complex control software as compared to the star network.
- In a ring network, the communication delay is directly proportional to the number of nodes in the network. Therefore, addition of new nodes in a network increases the communication delay.

Completely Connected Network

In this network, a separate physical link is present for connecting a node to any other node. In this way, each node has a direct link, called point-to-point link, with all other nodes in the network. Thus, the control is distributed among the nodes, which in turn decides its communication properties (as shown in Fig. 4.7).

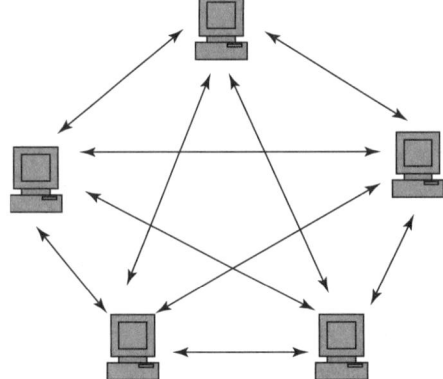

Fig. 4.7 *Completely connected computer network*

Advantages

- Each node in the network need not require individual routing capability.
- Communication between any two nodes is considerably fast.
- It is reliable as any link failure might affect only the direct communication between the nodes connected by that link.

Disadvantage It is an expensive network due to the link cost. If there are n nodes in a network, $n(n - 1)/2$ links are necessary. Hence, the cost of linking the system increases with the square of the number of nodes.

Multi-access Bus Network

In a multi-access bus network, all the nodes share a single transmission medium. All the nodes are attached to the same communication line (channel), as shown in Fig. 4.8. If a node wants to send a message to another node, it supplements the destination address to the message and then checks if the communication line is available. As the message travels along the line, each node checks if the message has been addressed to it. The message is then picked up by the addressee node, which sends an acknowledgement to the source, after which the node frees the line. This type of network is also known as multipoint, multidrop, or broadcast network.

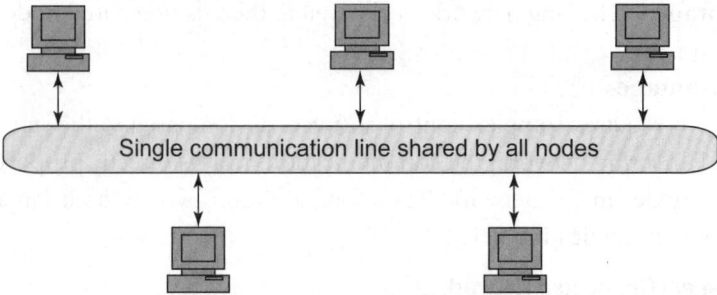

Fig. 4.8   Multi-access bus network

It may also be appropriate for use in a local area network where a high-speed communication channel is used and the computers are confined to a small area. It is also suitable when satellite communication is used as many computers at different geographical locations share one satellite channel.

Advantages

- It is easy to add new nodes to the network.
- If a particular node fails, communication among the other nodes in the network is not affected.

Disadvantages

- For some reason, if the shared communication link fails, the entire network will fail.
- All the nodes in the network should have good communication and decision-making competence.

Hybrid Network

Different network topologies have their own advantages and disadvantages. A pure star, ring, or completely connected network is seldom used. Instead, organizations use a hybrid network, which is a combination of two or more different network topologies. The exact configuration of the network varies with the needs and structure of the organization. A hybrid network may have components of star, ring, and completely connected networks (as shown in Fig. 4.9).

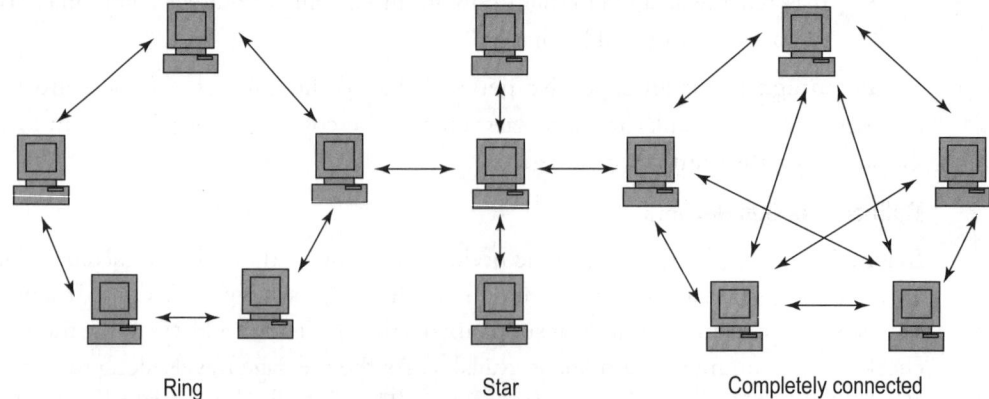

Ring   Star   Completely connected

Fig. 4.9   Hybrid network having a combination of ring, star, and completely connected network topologies

CLASSIFICATION OF NETWORKS

Networks can be broadly classified into two types: local area network (LAN) and wide area network (WAN). WANs are also said to be long-haul networks.

Local Area Network

A LAN refers to a group of computers in a localized area. It consists of two or more computers directly linked within a small, well-defined area such as a hotel, room, office building, or campus. We will now discuss the characteristics of LAN.

Geographical distribution   A LAN is restricted to a limited geographical coverage of a few kilometres.

Data transmission rate   Data transmission rates are higher in LAN when compared to other network transmissions. The data transmission rates in LANs usually range from 10 megabits per second (Mbps) to 1 gigabit per second (Gbps).

Error rate   LANs usually have only a few errors during data transmission over a communication or network connection.

Communication link   The most common communication links used in LANs are twisted pairs, coaxial cables, and fibre optics because the nodes are physically distributed over a small geographical area.

Ownership   A LAN is usually owned by a single organization because of its limited geographical coverage.

Communication cost   The cost of transmitting data in a LAN is negligible since the transmission medium is owned by the user's organization.

Famous LANs available for microcomputer users include Ethernet, Novell, PC Network, ARCNET, and Omninet. We will discuss Ethernet, ARCNET, and Novell.

Ethernet

Nowadays, Ethernet is widely used as it is comparatively faster than other LAN technologies and economical. It was introduced by Digital Equipment Corporation (DEC), Intel, and Xerox in the year 1980. Later, however, a modified version of the same was adopted by the Institute of Electrical and Electronics Engineers (IEEE) as a standard LAN technology. Ethernet uses a network topology that is a multi-access bus topology. The data transfer rate is 10 Mbps. The message is transmitted from one machine/node to another by first breaking it into packets (called frames in Ethernet) and then broadcasting those packets to the bus. There is an address designator linked with each packet. When the packet travels on the bus, each node or machine checks if the packet is addressed to it, and then the address machine picks up the message.

ARCNET

Attached resource computer network (ARCNET) is designated as a high-speed LAN well suited for real-time control applications in both industrial and commercial markets. The ARCNET resource centre strives to provide as much assistance as possible to allow users around the globe to work smartly and efficiently with this technology. Some examples of ARCNET applications are industrial monitoring and control systems for a variety of industries including nuclear power plant systems and networking of switching systems used in pay telephone products.

Devices for local area network connections   Bridges, routers, and gateways are used to interconnect different LANs. They could be used to build the backbone network that ties together many LANs to a business location.

Bridge   Bridges are simple devices that connect LANs and use exactly the same network architecture. They pass filtered messages and help reduce unnecessary network traffic.

Router   Routers are used to connect LANs that may either be of the same or different type of network architecture. It may choose the best route to send a message when several possible routes are available in a complex network topology. It maintains a routing table, which identifies the optimal path for crossing the interconnected network.

Gateway   A gateway is a network point that allows us to enter other networks. It is often a major component for interconnecting various backbone networks to form a larger regional wide area network (WAN).

Metropolitan Area Network

A network that shares some of the characteristics of both LANs and WANs is sometimes referred to as metropolitan area network (MAN). MANs generally cover a wider geographical area (upto 50 km in diameter) than LANs and frequently operate at speeds very close to those of LANs. The main objective of MANs is to interconnect LANs located in an entire city or metropolitan area. The communication links used for WANs are coaxial cables and microwave links.

Wide Area Network

A wide area network (WAN) consists of two or more geographically dispersed computers, linked by communication facilities such as telecommunication or microwave links. A WAN is a computer network covering a relatively large geographical area such as large organizations and government agencies. We will now discuss the characteristics of WANs.

Geographical distribution   A WAN may extend over several thousand kilometres and may operate nationwide or even worldwide.

Data transmission rate   Data transmission rates usually range from 1200 bits per second (bps) to 2 Mbps. With advancement in technology, data transmission rates are being increased continuously.

Error rate   The error rate in WANs is high when compared to LANs.

Communication link   Typical communication links used in WANs are telephone lines, microwave links, and satellite channels since the nodes are physically distributed over a large geographical area.

Ownership   A WAN is usually formed by interconnecting multiple LANs that may belong to a different area.

Communication cost   With a WAN, the cost involved is very high as the transmission mediums used are leased lines or public communication systems such as telephone links, microwave links, and satellite channels.

Famous WANs available for microcomputer users include ARPANET, ERNET, and NICNET.

The ARPANET of the Advanced Research Projects Agency of the United States Department of Defence was the first WAN. It was connected to about forty universities and research institutions all over the United States and Europe and had about fifty computers ranging from minicomputers to supercomputers, in its network. Education and research network (ERNET) is a WAN linking several educational and research institutions in India. National Informatics Centre's network (NICNET) is a WAN linking several offices of the National Informatics Centre (NIC) with the head office. The Indian railways has its own WAN, linking booking offices across the nation to help passengers book tickets on all major trains from anywhere in India.

Network Software

The term network software is a general phrase that refers to software that is designed to help set up, manage, and monitor computer networks. Networking software applications manage and monitor networks of all sizes, from the smallest home networks to the largest enterprise networks.

Windows NT

Windows NT is a network operating system software that has been designed by Microsoft Corporation. This has all the features and applications necessary for a complete network operating environment. The Windows NT server provides a complex security system for controlling access within and between the domains. It is a multitasking, multi-threaded network operating system.

Novell

Novell was the first to realize that a personal computer could be connected to boards and a special operating system was designed to enable single or more personal computers to act as a server, which could be connected to client personal computers. In 1982, Ray Noorda established Novell in Provo, Utah, to design and market personal computer networking software called Novell Netware. With the netware network operating system (NOS), personal computers could be used to share files, printers, and other system peripherals among a group of users.

Network Applications

Computer networks offer a number of advantages:

Sharing files and information   Networks allow us to share information and files with other computers on the network.

Sharing resources (hardware)   We can set up certain computer resources such as a disk drive or printer, so that all the computers on the network can access them. For example, a printer attached to a server computer is a shared resource. It means that anyone on the network can use it. We can also share other resources such as compact disk read-only memory (CD-ROM) drivers or modems.

Sharing software   A network allows users to share resources with others on the network. This can be software related, for example, a database application like a word processor. System administrators in a server-based network can configure or install new software on remote network computers.

Sharing programs   Sometimes it is best to place programs that everyone uses on a shared disk, rather than keeping separate copies of the programs on each person's computer.

For example, if there are ten computer users who use WordPerfect, we could store either ten copies of WordPerfect (one on each computer) or just one copy of WordPerfect on a shared disk.

Backup   Since all data is stored on the server, backing up critical data becomes a simple process.

Communication　The benefits of using computer networks are improved communication via electronic mails and groupware applications. Through e-mail, members of a network can send messages and ensure safe delivery of data to other members even when the receiver is not present at the destination. Groupware applications allow users to collectively work on the same document.

WIRELESS TECHNOLOGY

Over the years, several wireless technologies have been introduced. Communication via wireless technology works by modulating radio waves or pulsing infrared light. Wireless communication is linked to the wired network infrastructure by a stationary transceiver. The area covered by an individual transceiver signal is known as cell. The cell size varies widely. For example, an infrared transceiver can cover a small meeting room, a cellular (mobile) phone transceiver has a range of a few miles, whereas a satellite beam can cover an area more than 400 miles in diameter. We will now discuss the uses of wireless technology.

4G Technology

4G is the abbreviation of fourth generation wireless technology, the stage of broadband mobile communications that would supersede third generation (3G) wireless technology. A 4G network requires a mobile device to be able to exchange data at 100 Mbps. A 3G network, on the other hand, can offer data speeds as low as 3.84 Mbit/sec. 4G is not as popular as 3G technology. However, when fully implemented, 4G is expected to enable pervasive computing in which simultaneous connections to multiple high-speed networks will provide seamless handoffs throughout a geographical area. Coverage enhancement technologies such as femtocell and picocell are being developed to address the needs of mobile users in homes, public buildings, and offices. This will free up network resources for mobile users who are in roaming or in remote service areas.

2G and 3G Technology

Early mobile phones used first generation technology, which was analog, circuit-based, narrow band, and used only voice communication. After that came the next generation or second generation wireless technology (generally known as 2G technology), which was digital, circuit-based, narrow band but suitable for voice. However, it had limited data communication capability. Presently, we are using third generation (generally known as 3G technology) wireless networking technology, which is suitable for voice and advanced data applications as well as online multimedia and even e-commerce.

Wireless Local Area Networks

Wireless local area networks (WLANs) are like conventional LANs and have a wireless interface that enables wireless communication between equipment that are part of LAN. The main component of WLAN is the wireless interface card, which possesses an antenna. This interface card is then connected to the mobile unit as well as to the fixed network. It has a limited range and is designed for local environments such as colleges, parks, buildings, hotels, or office complexes. The advantages of wireless LANs are its flexibility and mobility.

Wireless fidelity (Wi-fi) is the embedded technology of WLAN. It was actually developed to be used for mobile computing devices (e.g., laptop) in LANs. It is presently used for services such as the Internet, voice over Internet protocol (VoIP), phone access, gaming, and for basic connectivity in consumer electronic products (television, digital cameras, and DVD players).

Wi-Max

Wi-Max stands for worldwide interoperability for microwave access. It has proved to be a useful technology for providing wireless broadband access to areas that do not have a good network or cable TV network.

Multi-hop Wireless Network

In wireless network environments, radio channels are sometimes used to connect mobile equipment to a base station in a single hub, though the base station itself is connected to a wired infrastructure.

Applications of Wireless Systems

- Mobile e-commerce applications, popularly known as m-applications
- Web surfing using wireless devices
- Mobile video-on-demand applications
- Access to corporate data by sales people and frequent travellers while they are travelling.
- Location-sensitive services such as programs that help finding nearby theatres, restaurants, and hotels in unfamiliar locations.

E-COMMERCE

The Internet has eliminated the need for standing in long queues for paying electricity bills or purchasing railway tickets. All this can now be done from an individual's residence or office. We just have to log in to the appropriate website and the task is completed in a few seconds. Many banks also offer the facility of net banking. In net banking, we can carry out online transactions with the bank, thus saving a lot of time and energy.

Commerce on the Internet

In today's web-oriented society, the most popularly discussed term is e-commerce. In simple words, it can be considered to be an electronic form of commerce. The term commerce is defined as transactions or exchange of goods and services for their worth. It applies to e-commerce as well. E-commerce involves the selling and purchasing of commodities and services using a computer network, preferably via the Internet. An individual can launch an online business venture by conceptualizing the business and having a computer with Internet connection.

Popularity of E-commerce

E-commerce is gaining popularity for the reasons mentioned here:

Low set-up cost   An individual can easily set up his/her own website. There are organizations and institutions that help customers develop and launch websites. It is not necessary to have large retail showrooms to market a product. A website showing the characteristics of the product including cost details is more than sufficient.

Global free market   No one can actually dominate the global market as a presence on the Internet is easy not only for global giants, but also for small organizations that can actively participate at low costs and compete with tough competition from the global giants.

Global access   Since hundreds of countries are connected to the Internet, a person who can afford a television and telephone can also access the Internet and benefit from the information available.

Availability of technology   Since the technology used in web servers, browsers, search engines, and Internet connections is similar throughout the world, businesses can be conducted easily.

Multiple opportunities   By using e-commerce, multiple activities (e.g., selling, renting, purchasing, etc.) can be performed. Thus, a whole variety of transactions can be provided under the same roof. Some popular e-commerce websites are mentioned here:

- *www.amazon.com*: On this website, buyers can place orders for books, computers, video games, kitchen and housewares, toys, games, and DVDs. The credit card is the most convenient and preferred means of payment.
- *www.etoys.com*: This website sells toys for infants who are 0–12 months old and children who are 5–7 years old. Other than toys, the website also sells art and craft items, musical instruments, games, puzzles, drawings, etc.
- *www.flipkart.com*: This online shopping website is used to shop for cameras, computers, books, mobile phones, home and kitchen equipment, and MP3 players.

E-commerce in India

Many e-commerce sites have come up over the last few years, thus changing the ways in which we shop. These sites have removed the myth that Indians need the touch-and-feel experience and prefer to go to a store and buy goods across the counter. Indians are now ready and contented to view items on their computer screens and make purchases using their credit cards. Nowadays, Indians shop for everything (including jewellery) and make real estate deals online. In addition, even bus tickets are purchased online.

Sites such as Mydala or SnapDeal help us pick discounted deals for the day at a restaurant, spa, or any nearby store. Policybazaar advises users on insurance policies. Letsbuy.com is a young Internet company where people can buy items ranging from mobiles to computers and even gadgets such as microwaves and toasters. It is also possible to purchase jewellery online from the cyber jeweller, CaratLane, and deal with real estate on Groffr. Redbus makes bus tickets to some places in the country available online.

Customers in small cities and towns who have the money may wish to buy goods that are usually available in big cities. E-commerce can easily reach these customers too.

INTERNET SECURITY

There are concerns regarding privacy, theft, liability, and loss of productivity of individuals if information is not secured on the Internet. Thus, organizations should assign high priority to online security.

The common threats to Internet security are as follows:

Virus

A virus is a piece of programming code that usually disguises itself and causes an unexpected and usually undesirable event. A virus is often designed so that it automatically spreads to other computers. Viruses can be transmitted from e-mail attachments, file downloads, or be present on a diskette or CD. We may not be aware of the viruses present in the files downloaded. They may be quite harmful, erasing data or causing the computer's hard disk to be reformatted. Some examples of viruses are given in Table 4.5.

Table 4.5   Examples of viruses

Virus name	Commonly known as
W32.Kaza A. Benjimin	Benjimin
W32.Sircam	Sircam
W32.Nimda	Nimda32

If we follow simple safety measures, we can easily protect our computer from getting corrupted by virus attacks. The following points should be kept in mind:

- Be well-informed about viruses, their symptoms, behaviour, general working, and common sources of origin.
- Update antivirus software regularly from the Internet to empower it against new viruses.
- Always scan a CD or pen drive before use as it might contain viruses.
- Do not open any suspicious or unfamiliar e-mail or website. It is preferable to delete such e-mails.
- Take regular backup of files so that a possible damage is minimized if a virus attack does occur.
- Regularly remove files having extension .tmp and .chk.

Worm

Similar to a virus, a worm is a program that is designed to harm the computer and destroy data stored in a computer. Worms are more harmful than viruses because they are designed to copy themselves from one computer to another computer over a network (e.g., by using e-mails).

Firewall

It is a wall that isolates the computer from the Internet using a 'wall of code', which inspects each individual 'packet' of data as it arrives on either side of the firewall, inbound to or outbound from the computer. The firewall decides if the 'packet' should pass or be blocked. It can 'filter' the arriving packets based on the originating machine's IP address and port number and the destination machine's IP address and port number.

ANTIVIRUS SOFTWARE

Antivirus software protects our computer from unexpected virus attacks. Apart from protecting our system, they perform the following tasks:

- It detects the name of the virus and its type.
- It comes with alert features that warn users about the virus.

- With regular updates, antivirus software can instruct the computer on new viruses.

Norton Antivirus, McAfee, and Kaspersky are some examples of antivirus software.

McAfee

McAfee is the world's largest dedicated security technology company. It delivers proactive and proven solutions and services that help secure systems and networks around the world. McAfee protects consumers and businesses of all sizes from the latest malware and emerging online threats. It is designed to work together, integrating antimalware, antispyware, and antivirus software with security management features that deliver unsurpassed real-time visibility and analytics, reduce risks, ensure compliance, improve Internet security, and help businesses achieve operational efficiencies. McAfee security technologies uses a unique, predictive capability that is powered by McAfee Global Threat Intelligence, enabling home users and businesses to stay one step ahead of online threats.

Kaspersky

Kaspersky Antivirus 2012 is the backbone of a personal computer's security system, delivering real-time protection from the latest malware and viruses. It works behind-the-scenes with intelligent scanning and small, frequent updates, while proactively protecting us from known and emerging Internet threats. The features of Kaspersky Antivirus 2012 are as follows:

- Scans all websites and e-mails for malicious software
- Detects and helps repair program vulnerabilities
- Checks files, applications, and websites with one click
- Rolls back most harmful malware activity
- Maximizes PC performance with speed and efficiency

Norton Antivirus System

Norton Antivirus system provides four unique layers of powerful protection to proactively stop online threats before they can infect a computer. It checks where the files came from and how long they have been around to stop new online threats before they can cause trouble. It protects us while we surf the web by warning and blocking unsafe and fake websites in our search results. This antivirus is cloud based so we can download, install, transfer, update, or renew Norton products from anywhere over the Internet. It detects threats as they travel over a network and eliminates them before they can reach our computer. This antivirus secures, remembers, and automatically enters our website user names and passwords to prevent cybercriminals from stealing our information as we type. It also includes an online version so we can share logins among additional PCs protected by Norton Internet Security 2012. It keeps our mailbox free from unwanted, dangerous, and fraudulent e-mails. This antivirus protects the system by checking for and blocking online threats as our browser loads, so as to stop them before they can damage the computer. It scans our news feed for dangerous downloads and links to unsafe websites and warns us and our Facebook friends about them.

SUMMARY

In this chapter, we learnt the basics of Internet and network. We also discussed the various types of networks and network topologies, illustrated with suitable features and examples. In addition, we got a brief idea about the latest and most popular wireless technologies and the role of e-commerce in the business field. The Internet has now become absolutely essential for all individuals. We can use the Internet to chat and send e-mail and thus stay in touch with people across the world. We also discussed data communication mediums with figures. Many a time, we face problems when working with computers because of virus attacks. In this chapter, we discussed the importance of Internet security and the types of antivirus that are available.

KEY TERMS

Antivirus software   They protect our computer from unexpected virus attacks. With regular updates, they also safeguard our computer from new viruses.

Bridges   They are simple devices that connect LANs that use the same network architecture and hence reduce unnecessary network traffic.

Bus network   In such a network, each computer is connected to a single communication cable through an interface and every computer can directly communicate with every other computer or devices on the network.

Client   They interact with users, providing an interface that allows the user to request for services from the server and displays the results returned by the server.

Client–server   A network of computers consists of a server computer and multiple client computers that share programs and data from the server.

Coaxial cable   It consists of specially wrapped and insulated wire line cables that are used for transmitting data at high rates without distortion or loss of signal.

Communication satellite   They are microwave relay stations placed in the outer space. Presently, there are hundreds of satellites in orbit to handle international and domestic data, and voice and video communication.

Completely connected network   It has a separate physical link for connecting each node to every other node. The control is distributed, with each node deciding its communication properties.

Domain name   It is a way of identifying and locating computers connected to the Internet.

Downloading   It is the process of moving a file from a remote computer to our own computer.

E-commerce   It encompasses the entire online process of developing, marketing, selling, delivering, servicing, and paying for various products and services.

Electronic mail   An e-mail service enables an Internet user to send a mail (message) to another Internet user in any part of the world.

File transfer protocol (FTP)   It is used to make files and folders publicly available for transfer over the Internet. We can upload and download files using FTP.

Firewall   This software inspects every packet of data that arrives at our computer. It has total power over what the computer receives over the Internet.

Gateways   They are complicated interconnecting devices. They are a type of communication server.

Gopher   It efficiently uses Internet resources and helps the user find and use information scattered across the world.

Host/Remote computer   It is the computer from where data, information, or programs have to be accessed. It could be in the near vicinity or far away.

Hypertext transfer protocol (HTTP)   It is a set of rules or protocol related to the transfer protocol.

Hypertext markup language (HTML)   Web pages can be designed using HTML.

Hub   It is a physical layered device that connects multiple user stations, each via a dedicated cable.

Hybrid network   It is a network that is a combination of two or more network topologies.

IP   The Internet has more than a million computers attached to it. It requires a proper addressing system for communication. This system of addressing is known as Internet Protocol (IP) addressing system.

Internet relay chat (IRC)   This feature of the Internet lets us instantly communicate with people sitting at any location around the world by sending messages to and receiving messages from them.

Local area network (LAN)   It is a computer network of interconnecting computers and other peripheral devices within a limited geographical area of a few kilometres.

Local computer   The computer on which a user can work easily is called a local computer.

Metropolitan area network (MAN)   These networks share some of the characteristics of both LANs and WANs. They usually interconnect computers spread over a geographical area of about 50 kilometres.

Multi-purpose Internet mail extension (MIME)   It is a set of rules that allows different kinds of documents to be exchanged among many computer systems.

Microwave system   This system uses very high frequency radio signals to transmit data through space (wireless communication).

Modem   It is a device that is used to transmit data over the network. It converts digital signals to analog signals and vice versa.

Multi-access bus network   In this network, all nodes share a single transmission medium, that is, all nodes are attached to the same communication line (channel).

Network   It is a simple connection between two or more computers, thus interconnecting them.

Network topology   It refers to the geometrical arrangement of computer resources, remote devices, and communication facilities.

Node   Each computer or device in a network is called a node.

Optical fibre   Hair-like thin threads of glass, plastic, or silica that are used for long distance data transmission are called optical fibes. They use light signals for data transmission.

Protocol   It is a set of formal operating rules, procedures, or conventions and is a language that enables computers to speak to one another.

Ring network   It can be as simple as a circle or point-to-point connections of computers at dispersed locations. It has no central host computer. All the nodes are connected in a closed loop.

Routers   They are used to connect LANs that may either be of the same or different LAN architecture. They choose the best and optimal route to send a message in a complex network topology.

Search engine   It is a collection of programs that gathers information from the web, indexes it, and puts it in a database so that it can be searched for.

Server   It is a high-capacity, high-speed computer with a large hard disk capacity. The server also contains network versions of programs (software) and large data files.

Simple mail transfer protocol (SMTP)   It ensures a reliable and efficient method of mail transmission.

Star network   In this network, several devices or computers are connected to a centralized computer.

Transmission control protocol/Internet protocol (TCP/IP)   It helps us copy files from one computer to another computer across the world.

Telnet   It is a terminal emulation protocol that enables us to remotely log in to other computers on the Internet, using a command line interface.

Twisted-pair cable   It is a simple, inexpensive, and wired slow data transmission medium used for short distances.

Uploading   It is the process of moving a file from our computer to another host computer.

Uniform resource locator (URL)   Internet addresses for all servers and resources are given in the form of URLs.

Usenet news   It is a bulletin board service featuring a large number of discussion groups involving millions of people around the world.

Virus   It is a programming code usually disguised as something else that causes an unexpected and undesirable event.

Web browser   It is a piece of software that acts as an interface between the user and the Internet, especially the world wide web.

Wide area network (WAN)   It is a computer network that interconnects computers spread over a large geographical area. This type of network may be developed to operate nationwide or worldwide.

World wide web (WWW)   It is a network of computers across the world that are interconnected on the Internet and uses the concept of hypertext to link Internet sites and information on the Internet.

Worm   Like a virus, it is a program that is designed to harm a computer and destroy data stored on it. It spreads itself to many computers over a network.

REFERENCES

Balakrishnan, Paran, 'Click and buy', Graphiti, *The Telegraph Magazine*, Kolkata, 11 September 2011.

Balagurusamy, E., *Fundamentals of Computers,* Tata McGraw Hill, New Delhi, 2009.

Gupta, S. and S. Gupta, *Computer Aided Management*, Excel Books, New Delhi, 2004.

Gupta, Vikas, *'O' Level Course Internet and Web Design*, Dreamtech Press, New Delhi, 2003.

Internet for Business Managers, Study Material for Distance Learning, IMT Ghaziabad.

Mukkhopadhay, A.K. and A. Das, *Elements of Computer Science*, Vol. II, Kalimata Pustakalaya, Kolkata, 2002.

Sinha, P.K. and P. Sinha, *Computer Fundamentals*, Fourth edition, BPB Publications, New Delhi, 2007.

Web References

http://andreprtik.blogspot.in/2010/10/macam-macam-topologi.html, last accessed on 2 February 2012.

http://en.wikipedia.org/wiki/Top-level_domain, last accessed on 17 January 2013.

http://en.wikipedia.org/wiki/Website, last accessed on 17 January 2013.

http://encyclopedia2.thefreedictionary.com/bus%2Bnetwork&docid=BBrwkAlleSmO, last accessed on 2 February 2012.

http://http://images.yourdictionary.com/fiber-optics-glossary, last accessed on 21 January 2013.

http://in.norton.com/internet-security/, last accessed on 20 March 2012.

http://kristin-itgs.wikispaces.com/LAN%2Band%2BWAN71, last accessed on 2 February 2012.

http://searchnetworking.techtarget.com/definition/node, last accessed on 19 August 2012.

http://trustmeher.net/freeware/free_anti_virus_tools.htm&docid=, last accessed on 20 March 2012.

http://wccftech.com/internet-explorer-9-ui/the-new-internet-explorer-9-ie9-logo-2, last accessed on 15 March 2012.

http://webpage.pace.edu/ms16182p/networking/cables.html, last accessed on 21 January 2013.

http://whatis.techtarget.com/definition/server, last accessed on 17 August 2012.

http://www.abtnetwork.net/2010/12/network-concepts-and-classification.html, last accessed on 2 February 2012.

http://www.computerhope.com/jargon/e/errorate.htm, last accessed on 20 August 2012.

http://www.geo-orbit.org/sizepgs/Noise.html&docid=nhYbCtf-n-XMBM&imgurl=http://www.geo-orbit.org/sizeimgs/towersr.gif, last accessed on 23 January 2013.

http://www.google.co.in/, last accessed on 20 March 2012.

http://www.google.co.in/imgres, last accessed on 23 January 2013.

http://www.google.com/talk/, last accessed on 20 March 2012.

http://www.kaspersky.com/internet-security-2012, last accessed on 20 March 2012.

http://www.linux-vs.org/Joseph.Mack/linuxexpo99/linuxexpo2.html, last accessed on 22 January 2013.

http://www.mcafee.com/us/, last accessed on 20 March 2012.

http://www.mozilla.org/en-US/press/faq/fx/, last accessed on 18 August 2012.

http://www.national-tech.com/support/technical_articles/coaxial_cable.html, last accessed on 6 February 2012.

http://www.radio-electronics.com/info/satellite/communications_satellite/communications-satellite-technology.php&docid1, last accessed on 16 March 2012.

http://www.tutorvista.com/content/physics/physics-iv/communication-systems/communication-system-elements.php, last accessed on 22 January 2013.

http://www.webopedia.com/TERM/N/network_software.html, last accessed on 17 August 2012.

http://www.windowsnetworking.com/articles_tutorials/autoslct.html, last accessed on 16 March 2012.

https://mail.google.com/mail/?tab%3Dwm&scc=1&ltmpl=default&ltmplcach, last accessed on 20 March 2012.

www.webopedia.com/TERM/H/hypertext.html, last accessed on 17 August 2012.

EXERCISES

Concept Review Questions

1. How is the Internet revolution affecting society?
2. What is the meaning of URL?
3. Define the term 'search engine'. List any two search engines.
4. Define the term 'browser' and list the various categories of a browser.
5. Expand the following abbreviations:
   (a) FTP (e) XML
   (b) TCP/IP (f) SMTP
   (c) WWW (g) MIME
   (d) HTML
6. Write about the applications of networks in our life.
7. What is a coaxial cable? What are its uses?
8. Why is communication in a communication satellite better than a microwave system?
9. What is network topology? Describe the three network topologies with appropriate diagrams.
10. Differentiate between LAN and WAN with examples supporting each.
11. Briefly explain bridge, router, and gateway.
12. What is a wireless computing system? How is it useful?
13. Write short notes on the following:
    (a) 2G and 3G technologies
    (b) Wi-Fi
    (c) Bluetooth
14. Giving examples, explain why e-commerce has become popular for businesses.
15. Briefly write about the concept of Internet security.
16. What are the major threats to Internet security? Briefly list the measures to implement Internet security.
17. Write about Windows NT in brief.
18. Differentiate between the following:
    (a) Worm and virus
    (b) Twisted pair wire and coaxial cable
    (c) Bridge and router
19. What are the factors that have led to the growth of e-commerce?

20. Draw the client–server architecture. What are the functions of the modem?

21. What are the advantages of LAN?

22. Compare the world wide web with gopher.

Multiple Choice Questions

1. This protocol is used by both HTTP and FTP:
   (a) V. 90
   (b) TCP/IP
   (c) Linux
   (d) Windows XP

2. Usually the term 'web browser' refers to
   (a) a program
   (b) a person
   (c) a file
   (d) an operating system

3. The site that is the starting point on the world wide web for a particular group/organization is
   (a) a website
   (b) a link page
   (c) a download
   (d) a home page

4. The generic term for a company that can directly connect us to the Internet is
   (a) an Internet service provider
   (b) a telephone line
   (c) a modem
   (d) none of these

5. Which of the following is a bulletin board system that allows us to pose and respond to messages on the Internet?
   (a) Telnet
   (c) Web browser
   (b) USENET
   (d) E-mail

6. Which of the following is a software that is installed on our computer and safeguards our computer by inspecting all incoming and outgoing data packets?
   (a) LAN
   (b) SMTP
   (c) Firewall
   (d) Browser

7. A network that covers a large geographical area such as a country or the entire world is called
   (a) LAN
   (b) MAN
   (c) VAN
   (d) WAN

8. Business conducted over the Internet is known as
   (a) e-conduct
   (b) e-commerce
   (c) e-intelligence
   (d) e-tech

9. Which of the following is an example of a network operating system?
   (a) Windows NT
   (b) DOS
   (c) Windows 98
   (d) DOS LINUX

Project Work

1. In a satellite communication system, a microwave signal of 14 GHz is transmitted from a transmitter on the Earth to a satellite positioned in space. How many GHz will return to the Earth?

CASE STUDY 1

Surrinder Bhalla, an entrepreneur, started Click-a-Pizza for net-savvy people to order pizzas through the computer. A customer could choose various pizza bases, three different sizes, toppings, combinations of cheese or single cheese, and also place side orders. The customer could also see how the pizza would look with the choices when prepared. The orders were home-delivered within 25 minutes to places that were located within a radius of 3 km from the base kitchen.

Though Click-o-Pizza was satisfactory initially, as the customer base grew the services took a back seat. Ordering a pizza became a slow process and customers were frustrated. The pizza received by the customers was not what they had ordered.

Accompaniments for the pizza such as oregano, chilli flakes, and mustard sauce were missing.

Bhalla, who had invested a lot in market survey, created his own website, purchased vehicles for home delivery, and set up a base kitchen, couldn't believe that the outlet was on the verge of closure in just three months.

(a) What went wrong in this case?

(b) What are the reasons for Click-o-Pizza not finding regular customers?

(c) If you were Surrinder Bhalla, what would you have done? Justify your answer.

CASE STUDY 2

Dilip Shekhawat, the front office manager of Hotel Samrat Palace, was worried about the hotel reservation system. As per the changing trends, Shekhawat had brought in new changes to the reservation system. The reservation department, other than using phone, fax, and postal mails for reservations, started accepting e-mails. For requesting room reservations by e-mail, Shekhawat had asked the owners to prepare a website of the hotel so that guests could view the facilities, tariff, proximity to various destinations, and book hotel rooms.

Though sales were high, the computer systems became slow and many files in the reservation system were found to be corrupt. Since employees also had access to the Internet, it had led to viewing of different sites, making the computers prone to virus attack.

Shekhawat called his friend, a software engineer, to find a solution to the crisis. On listening to his problem, he suggested that Shekhawat install a good antivirus in all his computer systems.

(a) What went wrong in this case?
(b) What is a virus?
(c) What is an antivirus?
(d) What are the benefits of installing an antivirus?

Answers to Multiple Choice Questions

1. (b)   2. (a)   3. (d)   4. (a)   5. (b)   6. (c)   7. (d)   8. (b)   9. (a)

CHAPTER **5**

Introduction to DBMS

LEARNING OBJECTIVES

After reading this chapter, you will be able to understand the following:

- Concept of data and information
- Types of database models
- Using database management system in Visual FoxPro 6.0
- Navigating in FoxPro 6.0 interface
- Types of data and defining the fields of a table
- Addition, deletion, and indexing of records
- Search and display of a specific record
- Creation and running of a program

Data is a vital asset of an organization. If hardware and/or software are damaged, they can be replaced; however if data is lost or damaged, it is not easy to replace. If required, an organization could procure the same hardware and software of a competing hotel but not the data pertaining to their guests.

EXHIBIT 5.1 Guest Database

The century-old Kingston Club was famous for its continental food, golf, and squash. The elite of the city were its members and there was a waiting list of new members. The club had over a thousand members and the daily footfall was around 500–600. Many times, Priyanka Saxena, the receptionist, had problems identifying members when they arrived at the club's reception as many of them didn't carry membership cards regularly.

Girish Negi, the club manager was not very satisfied with the system being followed. The names of all the members were arranged in alphabetical order in an index register, which was used as a verification tool. Guests (the members or their siblings) were asked to provide their IDs when they visited the club premises. Due to the procedure being followed, non-members who accompanied the members also knew the member identity number. Thus, non-members posed as members and availed the services of the club, even during evenings when the club's food and beverage outlets were full. It was difficult for the receptionist to verify if all guests were members. This not only caused a loss to the club but also posed security problems.

For smooth functioning, it was suggested that the club maintain a database of all its members with their photographs, so that whenever the details of an individual was typed (name, surname, or member ID), his/her profile could be viewed along with a photograph. This would save them the tedious process of cross-checking with the index and member files.

For example, a hotel might maintain a large volume of guest data and transactions on a computer's secondary storage devices. For some unforeseen reason if the data is lost or damaged, the hotel will be in trouble unless it has a backup of the data. The hotel might end up losing details regarding the history of guests and the transactions made by them.

Data is a collection of unorganized facts, but when it is organized it becomes useful information. Data can also be manipulated to produce the desired output (e.g., departmental reports, guest bills, and purchase vouchers). The output, known as information, refers to organized facts that help people make decisions. Processing is a method by which different kinds of input are transformed into useful output. Thus, data processing is a series of actions or operations that convert data into information.

Data processing involves performing diverse operations on data. For convenience, let us define these operations on different levels of arranged data. These levels form the data storage hierarchy of data processing. Figure 5.1 shows the six levels of this hierarchy.

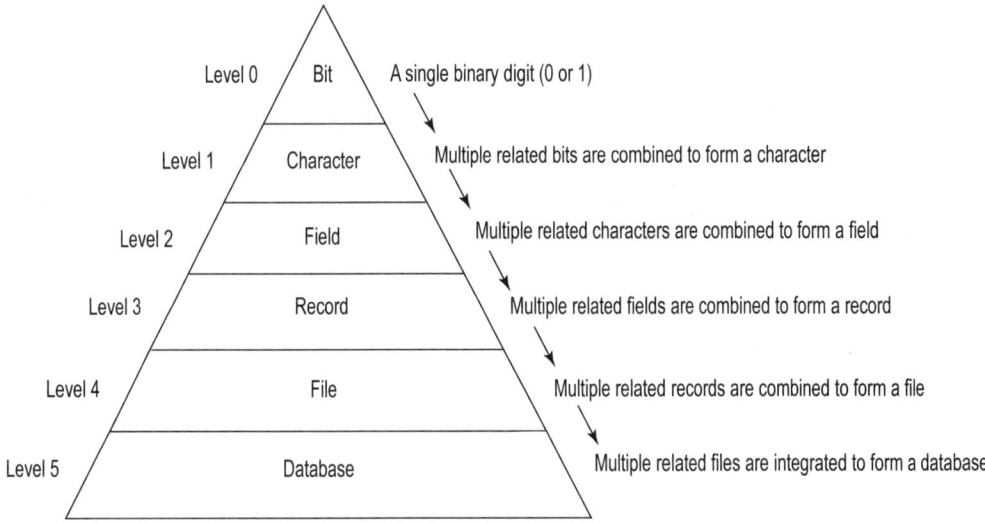

Fig. 5.1 Data storage hierarchy used in data processing

Bit The smallest item of data is a single binary digit (a bit) whose value can either be 0 or 1.

Character A character (or byte) is a combination of multiple related bits. For example, if the guest name is Sheetal Kapoor, then 'S', which is the first letter in the name is a character.

Field A field is a meaningful collection of related characters. It is the smallest logical data to be treated as a single unit in data processing. For example, in a database of guests, we may have fields such as guest name, guest ID, contact number, and company name.

Record A record is a combination of multiple related fields that are treated as a single unit. For example, Sheetal Kapoor, a guest's name, is a record.

File A file is a combination of multiple related records that are treated as a unit. Each record in the file is identified by a key field, which is unique for every record in the file. The guest name, guest ID, contact number, and company name of a guest are combined to form a file, for example, 'Guest details'.

DATABASE

A database integrates data from multiple files in such a way that data redundancy is minimized. It is an organized collection of electronic data that is amenable to efficient manipulation. Databases provide us with up-to-date and relevant information, besides following a strict central control regime.

Information can be rapidly retrieved from databases. Hence, decisions can be taken quickly and useful reports generated promptly. Some of the applications of databases are mentioned here:

- Computerized reservation system: This system has a large database related to hotel room availability, rates, and locations across the country or world.
- Guest information in a hotel: This system stores and retrieves information pertaining to guests in a hotel.

In a database-oriented approach to organizing data, a set of programs is provided to help users organize, create, update, delete, and manipulate data in a database. Together, all these programs form a database management system (DBMS).

Database Model

A database model defines the manner in which the numerous files in a database are linked together. The four commonly used database models are hierarchical, network, relational, and object-oriented. Database structures or database structuring techniques are described later in the chapter.

DATABASE MANAGEMENT

Database management is a set of activities by which a database is created, maintained, manipulated, and used. The activities include adding, updating, and even deleting records in a database.

The user employs computer programs to carry out database management activities. Thus, there is a program for adding a record in the database, another for deleting a record, and yet another for drawing reports (as shown in Fig. 5.2).

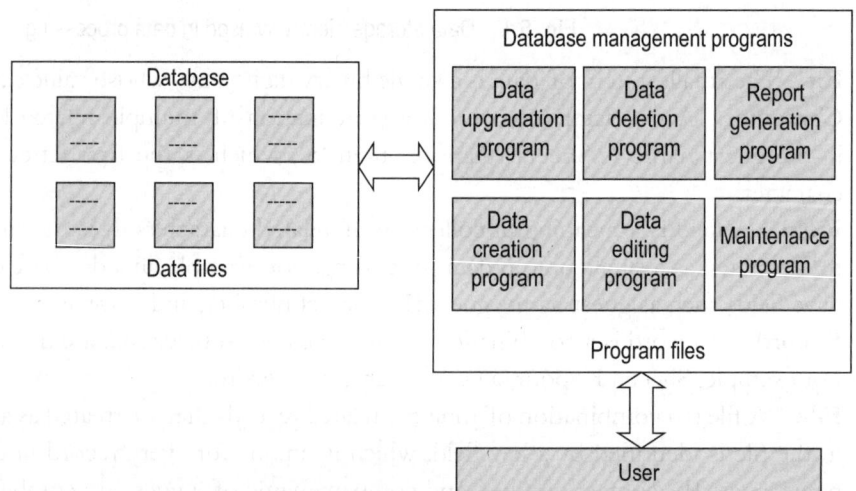

Fig. 5.2   Set of activities performed with data

DATABASE MANAGEMENT SYSTEM

A database management system or DBMS consists of a database and the necessary programs to perform the database management activities. Users do not directly interact with the data files and instead use a language interface. A DBMS is used for data storage and manipulation. Data manipulation users do not have to write programs as they are already stored in the database and can be invoked by issuing one or more commands in a database language.

Most of the currently popular DBMS have a more complex architecture than the one shown in Fig. 5.3. With this architecture, users do not have to specify how a data manipulation action has to be carried out. They simply have to state what information they want. The DBMS takes care of the rest.

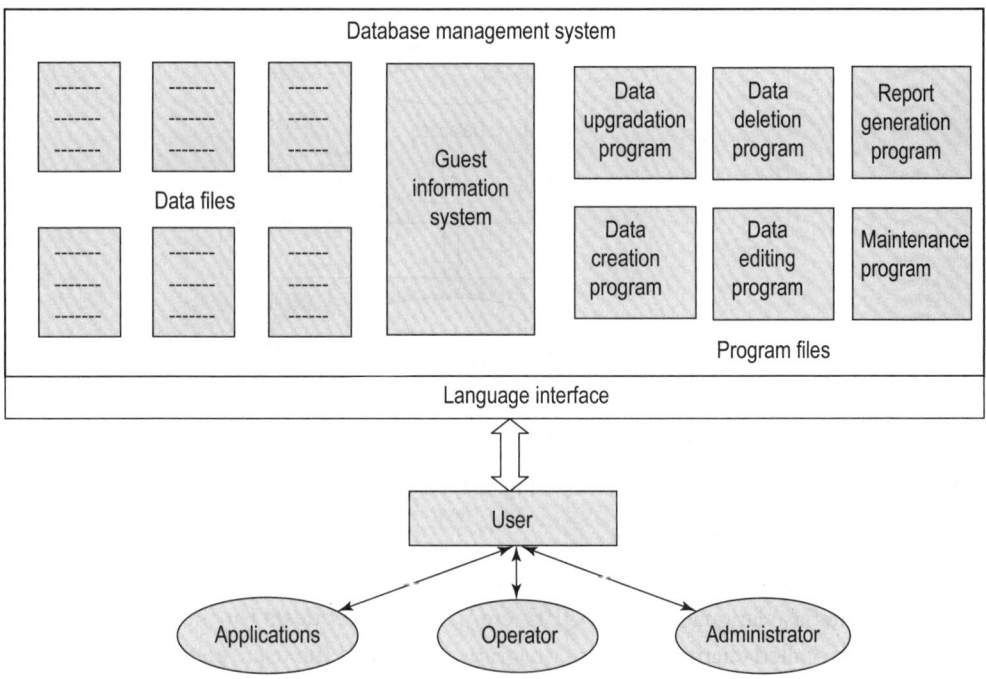

Fig. 5.3   Complex architecture of DBMS

In addition to human users, even programs can interact with the DBMS. Hence, a database is said to be back-end while the user program is front-end.

DATABASE TYPES

The DBMS is the result of continuous research and development over the decades. At present, a large number of databases exist in actual practice. The database types have been briefly described in the following sections.

On Basis of Localization

Stand-alone database   A stand-alone database is used by a single user, is simple, lacks strict security controls, and is mainly used as a desktop database in home computers.

Centralized database   When all the components of a database that is stored at a single client location can be accessed from a different location, it is known as a centralized database.

Distributed database   Databases that are stored on different sites on a network with strict control mechanisms are referred to as distributed databases. Moreover, the DBMS itself may be located at more than one site (as shown in Fig. 5.4).

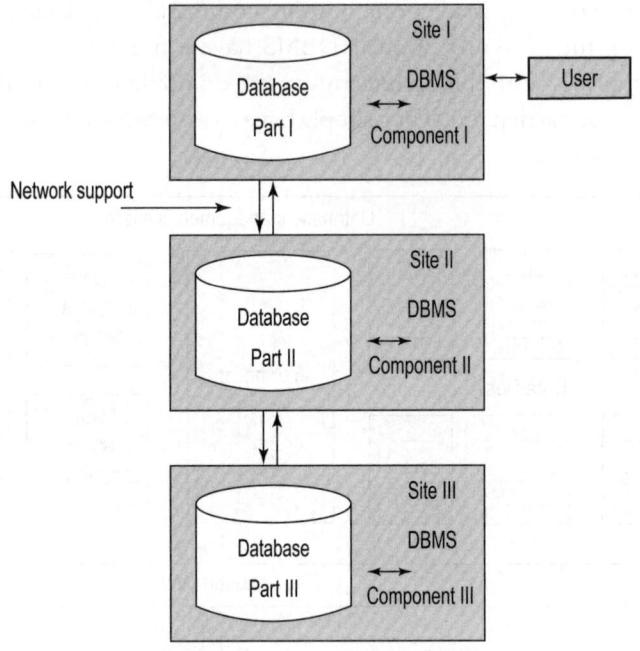

Fig. 5.4   Distributed DBMS

On Basis of Access Methods

Client–Server database   Access to this database is via a request–reply method. With the help of a database language (SQL, QBE, etc.), the user requests for the desired data using a special program called client. The client program then sends the user request to the DBMS (also called a database server, in this arrangement). The database server then retrieves the requested data and transfers it back to the client (as shown in Fig. 5.5).

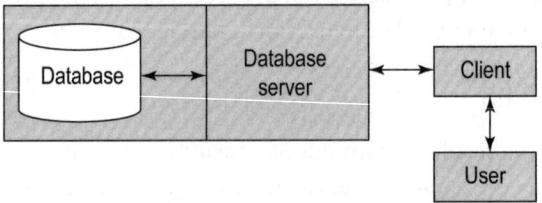

Fig. 5.5   Architecture of client–server DBMS

The client and server might reside on the same computer or on different computers, physically separated but having a network connection between them. Oracle and My-SQL server are two examples of a client–server database.

Web-based database   This database stores data that can be accessed through the Internet. The data can be stored in XML documents on a website (as shown in Fig. 5.6).

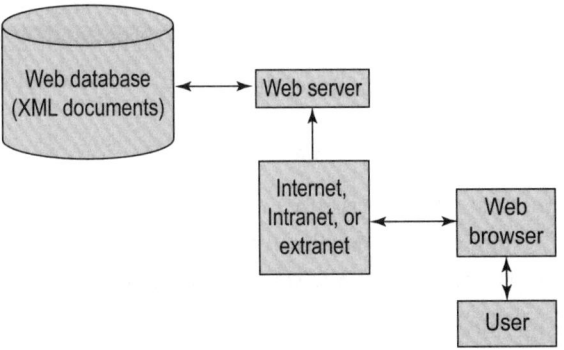

Fig. 5.6   Architecture of web-based DBMS

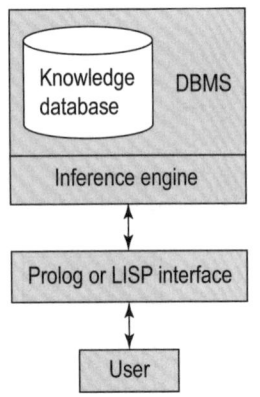

Fig. 5.7   Architecture of deductive DBMS

Users come into contact with a web server on the Internet through a program called a web browser, which uses hypertext markup language (HTML) coding. The web browser sends a request to the indicated web server. The web server retrieves the desired data, compiles it in HTML format, and then sends the page back to the web browser. After receiving the data, the web browser formats the received data and presents it to the users.

Deductive database   It is a type of database that stores a large amount of information in an appropriate form. A deductive DBMS is an intelligent component that is capable of deriving knowledge and making inferences from the stored knowledge. A user interacts with the database using an artificial language such as Prolog or LISP, as shown in Fig. 5.7.

On Basis of Applications

Production database   A database system that is used while developing an application is said to be a production database. It mainly contains sample data items that are used to assist programmers and application developers.

Operational database   A database that participates in the day-to-day business transactions of an organization is known as an operational database. Therefore, strict security and a backup recovery mechanism must be employed in its DBMS.

Enterprise database   An incorporated database that combines all the individual database needs of various divisions and departments of an organization is known as an enterprise database. Presently, big organizations implement this database system. The platforms most common for their implementation are JEEE (an enterprise version of Java) and .NET.

Data warehouse   A very large database that stores a lot of historical data and whose analysis might produce business intelligence is known as a data warehouse. The user interacts with the database using online analytical processing (OLAP) tools to analyse the data, thus helping managements take important strategic decisions.

On Basis of Data Model

Hierarchical database   This database was developed in the 1970s. A hierarchical database is a database model in which data elements are present in the form of an inverted tree structure with the root at the top and the branches below. The data located at different levels within a particular branch is called node. This model of data supports one-to-many relationships.

Let us consider the following facts about the various service outlets of a hotel.

- Assume that a hotel has three restaurants: Sip-n-Bite, Parikrama, and Maxims.
- Each restaurant may have many tables.
- Each table may have made several transactions.

The aforementioned relationship (1:M or one-to-many) could be modelled as a hierarchy as shown in Fig. 5.8.

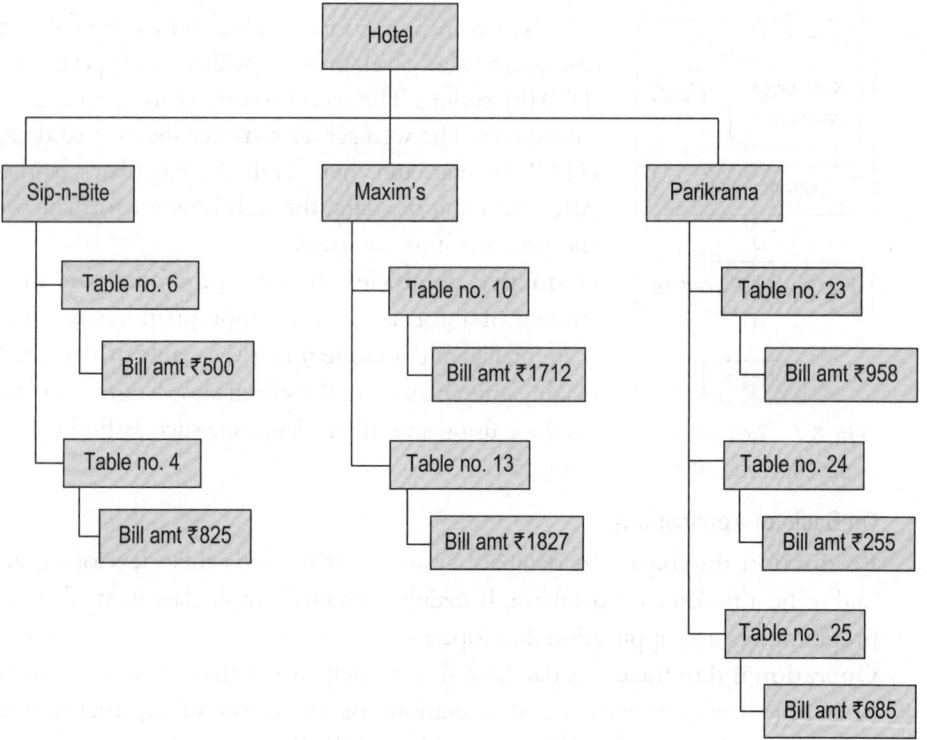

Fig. 5.8   Architecture of hierarchical DBMS

Advantages and disadvantages   The following are the advantages offered by hierarchical databases over the traditional file system of handling data:

- Simplicity: In most situations, data naturally has a hierarchical relationship. Therefore, it is easier to view data arranged in this manner.
- Database integrity: Database integrity is highly promoted in this system because of its inherent parent–child structure.

- Security: Database systems can enforce varying degrees of security features unlike normal flat-file systems.
- Efficiency: These types of databases are very efficient for one-to-many relationships.

Though hierarchical databases were a technical breakthrough, they have the following disadvantages:

- Structural dependence: These databases have a rigidly defined relationship. Therefore, any change in any part of the structure of the database would also require a change in the programs accessing it, thus making maintenance very difficult.
- Difficulty in implementation: The implementation of a hierarchical database depends on the physical storage of data, thus making it more complicated.
- Difficulty in management: The movement of a data section from one location to another might cause the accessing programs to be modified, hence making database management a complex affair.

Network model   Network databases were developed in the early 1980s with a view to eliminate the problems faced by hierarchical databases.

A network database is a collection of records that are connected to each other through links. A link is an association between precisely two records. Figure 5.9 shows a sample network database.

Advantages and disadvantages   The advantages of network databases are as follows:

- Flexibility in data access: The course of action to be followed by the data items can be planned in more than one way, providing the much desired flexibility to data access.
- Availability of standards: Universal standards have been developed and fixed in these types of databases.

The disadvantages of network databases are as follows:

- System complexity: The implementation of this type of database is not simple.
- Structure dependence: Since access depends on the navigational paths that exist in the database at any time, the programs are not independent of the database structures and need to be modified whenever the database structure is modified. Due to this limitation, this type of database lost popularity among users and was replaced by relational databases.

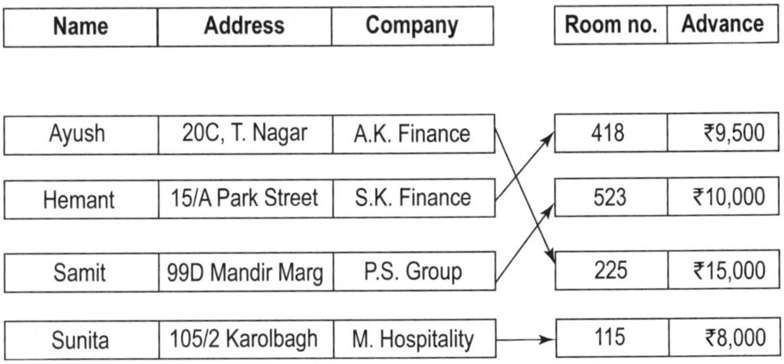

Name	Address	Company		Room no.	Advance
Ayush	20C, T. Nagar	A.K. Finance		418	₹9,500
Hemant	15/A Park Street	S.K. Finance		523	₹10,000
Samit	99D Mandir Marg	P.S. Group		225	₹15,000
Sunita	105/2 Karolbagh	M. Hospitality		115	₹8,000

Fig. 5.9   Architecture of network model database

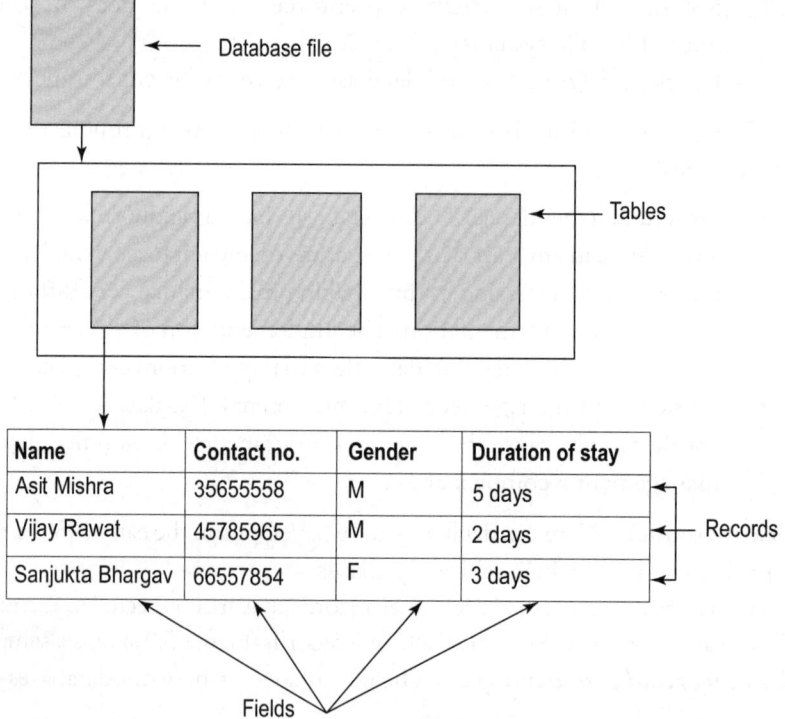

Name	Contact no.	Gender	Duration of stay
Asit Mishra	35655558	M	5 days
Vijay Rawat	45785965	M	2 days
Sanjukta Bhargav	66557854	F	3 days

Fig. 5.10 Architecture of relational model database

Relational model Dr E.F. Codd was the first to introduce the relational database model. This model evolved from a theory based on relation (or tuple) calculus. The relational model allows data to be represented in a simple table (row–column format). An individual data field is considered to be a column and the record is considered to be a row of a table (as shown in Fig. 5.10).

In a relational database, data is arranged in files called database files. A database file may contain one or more tables. A table is made up of a number of rows and columns. A row is said to be a record and a column is said to be a field, as shown in Fig. 5.10.

Relational database management systems (RDBMS) are the most popular database systems. Several varieties and versions of RDBMS are commercially available in the market. Some of these are Microsoft SQL server, Oracle, Sybase, Microsoft FoxPro for Windows, and Microsoft Access.

In Chapter 3, we discussed Microsoft Access. In the later part of this chapter, we will learn to use Microsoft Visual FoxPro 6.0. However, the concepts are quite general and are applicable to almost all RDBMS.

Advantages and disadvantages A relational database model has significant advantages over hierarchical and network databases. They are as follows:

• Simplicity of design and use: A relational database has data and structural independence. Therefore, design and implementation are easier than other types of databases.

• Structural independence: This type of database offers complete structural independence. Therefore, any change in the structure of the database does not affect the DBMS's data access mechanism.

- Advanced query capabilities: The process of querying in the relational database is quite simple, efficient, and powerful. It is supported by structured query language (SQL) because both have their foundation in relational algebra and relational calculus.

The following is the most noted disadvantage of a relational database:

- Increased overhead: Relational database management increases the user's responsibilities leading to more complexity, thereby increasing system overhead.

Object-oriented model   The object-oriented database model has been introduced to overcome the shortcomings of conventional database models. An object-oriented database is a set of objects whose behaviour, state, and relationships are defined in agreement with object oriented concepts (object class, class hierarchy, etc). Figure 5.11 shows an example of an object-oriented database structure. *School* is the root of a class composition hierarchy that includes school *specification*. The class *school* is also the root of a class hierarchy involving *primary, secondary,* and *senior/higher secondary* classes. The students of the school and the teachers have different qualities and characteristics, which are the attributes.

Advantages and disadvantages   The advantages of an object-oriented database are as follows:

- Semantic content handling: The representative object of the data carries more meaning.
- Database integrity: Database integrity has been ensured to the maximum level because of the safe nature of the objects.
- Supports conventional programming languages: It usually ensures compatibility between object-oriented programming languages and database languages. The database application is usually implemented using some conventional programming languages (C++, PASCAL, COBOL, or C) and some database languages (data definition language, data manipulation language, or query language), which are a part of the DBMS.

Object-oriented databases have the following disadvantages:

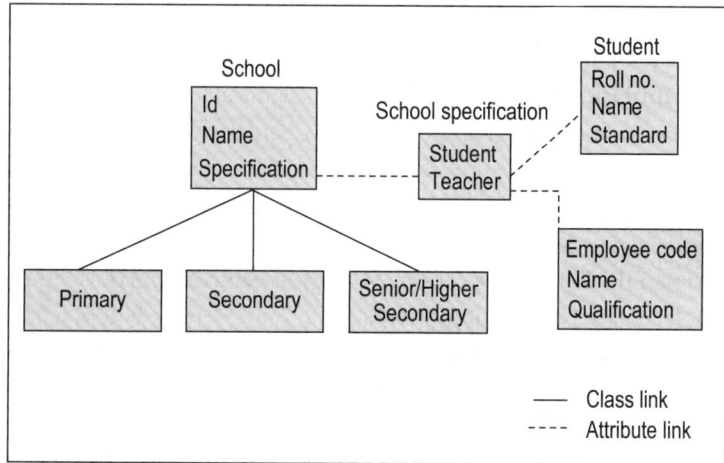

Fig. 5.11   Object-oriented database structure

- Lack of standards: Object data model standards have not evolved as yet. The database is still likely to be vendor specific.
- Increased system overhead: Due to the high system complexity, the system overhead is also proportionally high, which might result in retarded performance.

DATABASE USERS

The users of a DBMS can be categorized as per their role in the database environment. They could belong to any of the categories mentioned here:

End user   A user who utilizes the data stored in the database and the derived processed information is said to be the end user. A manager can be considered to be belonging to this category.

Operator   A user who performs data entry and other related activities is called an operator.

Programmer   Advanced users who develop and create application programs that interact with the database are called database programmers.

Database designer   A user who actually designs the structure of a database that is appropriate for a given application is called a database designer.

Data analyst   A user who recovers data stored in a database, analyses it statistically or otherwise, and comes up with results is called a data analyst.

Database administrator   A user who is in charge of the administrative control over the database system is said to be a database administrator. This individual is responsible for the smooth operation of the database.

MAIN COMPONENTS OF DBMS

A DBMS enables users of the system to organize, process, and retrieve selected data from the database, even if they do not know much about basic database structure (organization and location of the data). The following are the four major components of a DBMS:

Data definition language   The data definition language (DDL) is used for defining the composition of the database. A data dictionary is developed and utilized in a database that is used to document and maintain data definitions. It is automatically created (or updated) by the DDL module of the DBMS as and when the database scheme is defined (or changed).

Data manipulation language   The data manipulation language (DML) is a command that enables users to enter and manipulate data. Users can add various new records to the database, find their way through existing records, view the contents of the fields in the record, modify contents of one or more fields of the record, if necessary delete existing records, and even sort records in the desired sequence.

In some DBMS the DDL and DML have been combined together, whereas in others they are supported as separate components.

Query language   All DBMS provide a query language that enables users to define their requirements as queries for retrieving desired information from the database. Earlier, the DBMS used to have its own query language. In this approach, queries developed for one DBMS could not be used for another DBMS. Later, a query language called SQL became an industry standard. It was originally developed by IBM and was based on an earlier query language called

SEQUEL, an acronym for structured English query language. Today SQL is the standard query language used in many DBMS. A query language can be easily learnt by a non-programmer. This enables normal users to obtain the desired information from the database without the help of any programmer.

Report generator   A report is a presentation of information obtained from a database. Report generators assist users of a database to design the layout of a report in the required format. The user could define the layout of a report and store it for future use.

We have discussed data, information, DBMS, and types of database models. The DBMS is used for programming in FoxPro, which will be discussed in the following section.

MICROSOFT VISUAL FOXPRO 6.0

Bill Gates coined the phrase 'Information at your finger tips' to describe the future of corporate computing, in the 90s. To achieve this goal, computers became easier to use and information easier to access. Microsoft Windows made a major move in this direction. Now we do not have to remember cryptic command strings to format diskettes, copy files, or even manage directories. A graphical interface confirms the maxim, 'A picture is worth a thousand words'. By implementing a powerful RDBMS within a graphical environment, Visual FoxPro for Windows has taken a giant step forward, giving you information at your fingertips.

Visual FoxPro is a DBMS that supports a general-purpose programming language. We can easily create and maintain a list of data in FoxPro tables. The first DBMS, dBase II was developed by Ashton Tate. The advanced versions are dBase III and dBase III Plus. The latest version of FoxPro, Visual FoxPro 6.0, is compatible with DOS, Windows, and UNIX operating systems.

The FoxPro interface and development tools are explained step by step for better understanding.

Features

The following are the features of Visual FoxPro 6.0:

- FoxPro for Windows is a full-fledged programming environment that supports the creation and maintenance of data files.
- The Windows environment makes it easier for the inclusion of user-friendly features such as menu pop-up, pull down, and mouse support.
- The introduction of object linking and embedding (OLE) helps FoxPro share data with other Windows applications.
- The integrated environment allows programmers to code, test, and debug applications without switching packages. It also allows them to switch applications (to word processor, spreadsheet, or any graphical program) and later return to the starting point.
- Visual FoxPro has many information management features. For example, it has basic calculation abilities that use the database fields to create new information.
- FoxPro supports various arithmetic functions such as addition, subtraction, multiplication, and division. It also provides many special date, financial, mathematical, and string functions.
- FoxPro has an important feature known as Rushmore technology, which is very user-friendly to the programmer. This has been discussed later in the chapter.

Starting FoxPro

The steps to start FoxPro are as follows:

Step 1: Turn on the computer.

Step 2: Click the Start button to display the Program menu.

Step 3: Choose the Programs group from the pop-up menu.

Step 4: Choose FoxPro for window from the Program pop-up menu.

Step 5: Choose Visual FoxPro 6.0 for Windows from the program group FoxPro for window.

or

Step1: Double-click the Visual FoxPro 6.0 icon (as shown in Fig. 5.12).

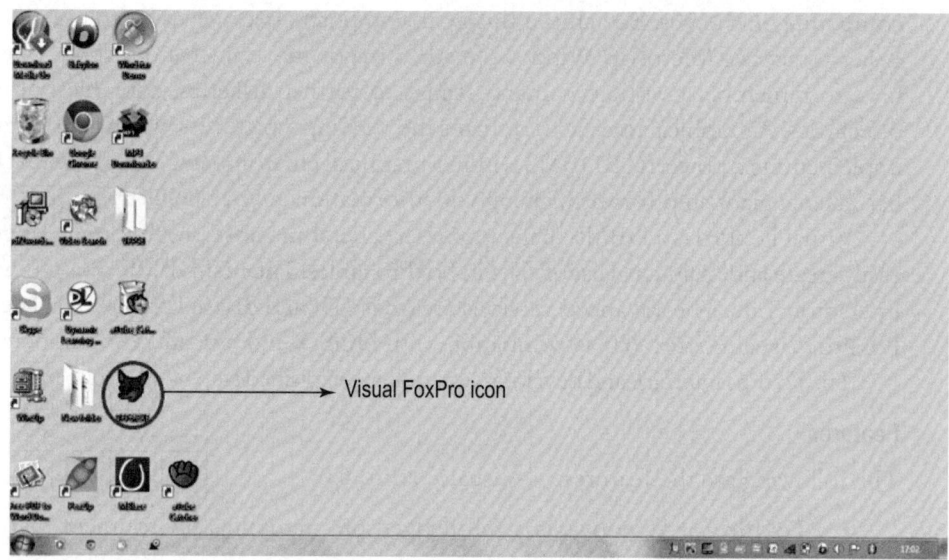

Fig. 5.12   Icon to open MS Visual FoxPro in Windows

Step 2: The welcome screen of Visual FoxPro 6.0 will appear.

Components

Title bar   It is the bar at the top of the window (as shown in Fig. 5.13).

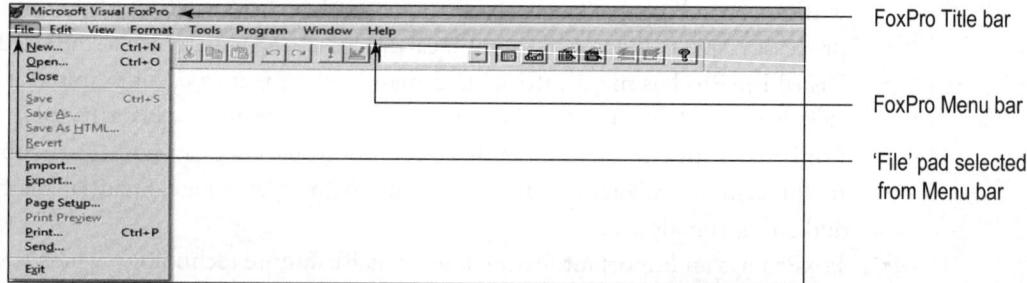

Fig. 5.13   MS Visual FoxPro welcome screen

Menu bar   This bar typically appears at the top of the screen and just below the Title bar (as shown in Fig. 5.13). However, it is possible to place it anywhere. Each menu item is called a pad. To access the Menu bar using the mouse, click the menu name. To access the Menu bar using the keyboard, press the Alt key or F10. To exit the Menu bar, press Esc, Alt, or F10. Else, with the mouse, click anywhere except on the Menu bar or Menu pad.

Menu pad   The Menu pad is horizontally placed on the Menu bar. It is an option on the Menu bar that can be selected and chosen. To select a menu pad with the mouse, click the pad (as shown in Fig. 5.13). For example, click the File menu pad.

To select a Menu pad with the keyboard, click the Menu bar and press the right or left arrow keys. When the desired Menu pad is selected, press Enter. A menu pop-up appears.

Menu pop-up   A menu pop-up consists of a list of related options. To display a menu pop-up with the mouse, click the appropriate Menu pad. To display a menu pop-up with the keyboard, do one of the following:

Step 1: Press the Alt key.

Step 2: Select the hot key on the Menu pad. (The hot key is the underlined letter in the menu name)

or

Step 1: Press F10 from the keyboard (F10 will highlight the File menu by default).

Step 2: Press the hot key in the Menu pad (as shown in Fig. 5.14).

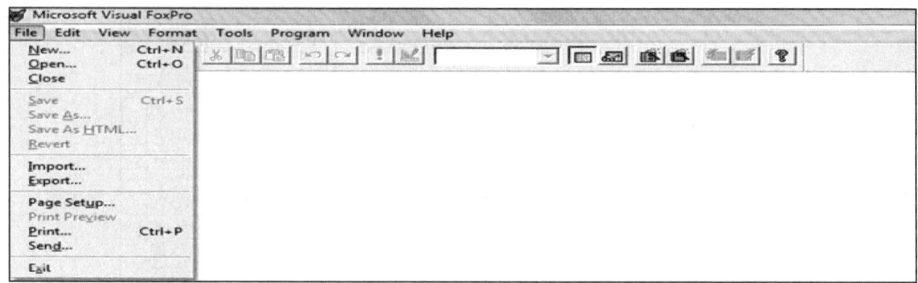

Fig. 5.14   MS Visual FoxPro menu pop-up

To deactivate a menu without choosing an option, perform any of the two given options:

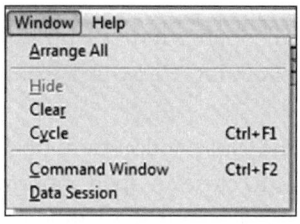

- Press Esc, Alt, or F10.

or

- Using the mouse, click anywhere other than on the menu.

Menu options   Menus contain options that are logically related to the menu name. If a menu option is followed by an ellipsis (...), a dialog box appears when this option is chosen.

Fig. 5.15   MS Visual FoxPro menu options

Choose an option with the mouse (as shown in Fig. 5.15).

or

To choose an option with the keyboard, use the up or down arrow to select the option and press the Enter key. Some menu options have a keyboard short cut (a key or combination of keys to be pressed at the same time) listed next to them. Use this key combination to choose the menu option without displaying the menu pop-up.

Fig. 5.16   MS Visual FoxPro toolbar

Toolbar   The toolbar consists of tools such as New, Open, Save, Undo, Redo, Print, Cut, Copy, Paste, Run, Help. It appears below the Menu bar (as shown in Fig. 5.16). However, it is possible to place it anywhere else.

Command window   We can use this window to enter FoxPro commands. The command window maintains a history of the commands we enter during a FoxPro session. We can re-execute

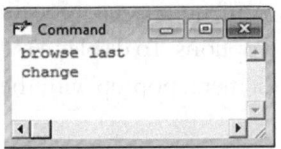

commands by positioning the cursor on the desired command and then pressing the Enter key. We can copy these commands into a text file. Figure 5.17 shows a command window of the FoxPro programming language.

Fig. 5.17   Command window in MS Visual FoxPro

Open   The Command window is automatically opened when we begin a FoxPro session. If we close this window, we can open it again by choosing command from the Window menu or by pressing the Ctrl+F2 keys from the keyboard.

We can open the Browse window by choosing Browse from the View menu. Figure 5.18 shows how the data from a table is displayed in the Browse window.

Dishcode	Dishname	Category	Price
101	DAL MAKHANI	VEG	150.00
112	MALAI KOFTA	VEG	225.00
210	MURG MAKHNI	NON VEG	350.00
125	PANNER TIKKA	VEG	210.00
215	KADHAI MURG	NON VEG	345.00
245	GULAB JAMUN	VEG	100.00
111	TOMATO SHORBA	VEG	80.00
135	JEERA PULAO	VEG	155.00
140	BOONDI RAITA	VEG	75.00
315	MUTTON BIRYANI	NON VEG	375.00
10	ROOMALI ROTI	VEG	50.00

Fig. 5.18   Browse window in MS Visual FoxPro

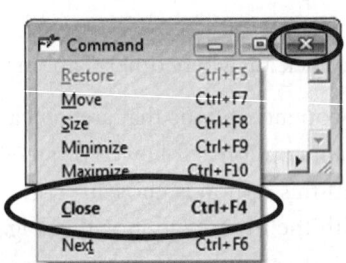

Fig. 5.19   Command window control box in MS Visual FoxPro

Close   We can close all open windows by holding the Shift key down or by choosing Close All from the File menu.

We can also choose the Close option from the File menu or press Ctrl+F4 in order to close an open window. We can close a window by double-clicking the window's Control menu box, which is located in the upper left-hand corner of all system windows (as shown in Fig. 5.19).

Every window has a Close button on the upper right-hand corner that can be selected to close the window and quit the program (as shown in Fig. 5.20).

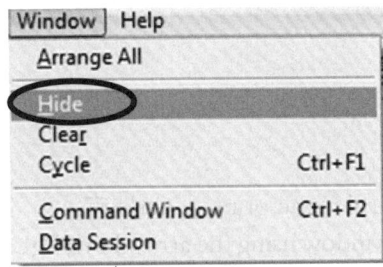

Fig. 5.20   Hide option in
MS Visual FoxPro

Hide   We can hide the window and make it invisible even without closing it. To hide a window, choose Hide from the Window menu. To hide or display all open windows, hold down the Shift key and choose the Window menu. The Hide and Clear options are replaced with the options Hide and Show. Choose the Hide option as shown in Fig. 5.20.

Size   To change the size of the window, move the pointer to the side, bottom, or corner of the window border. When the pointer becomes a double-ended arrow, hold the mouse button down, drag the border to the desired size, and then release it. We can also click the size control on the lower right-hand corner of the window and drag it in any direction.

Maximize   We can expand the current window to the full size of the screen display using the maximize button. To maximize a window, click the maximize button on the upper right-hand corner of the window (as shown in Fig. 5.21). We can also maximize the window by choosing Maximize from the Control menu (as shown in Fig. 5.21) or by pressing Ctrl+F5.

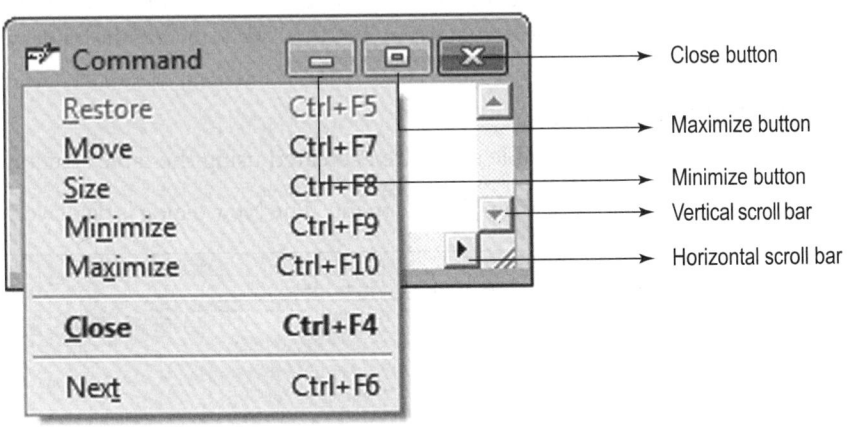

Fig. 5.21   Buttons and the scroll bar on the right-hand side of
the Command Window in MS Visual FoxPro

Minimize   We can reduce the current window to a push button, which is placed on the task bar of the desktop. Although the window appears closed, it is only temporarily removed from view. To minimize a window, click the Minimize button on the left side of the Maximize button, as shown in Fig. 5.22. We can also minimize a window by choosing the Minimize option from the Control menu (as shown in Fig. 5.19) or by pressing Ctrl+F9.

Restore   We can restore only those windows that have been maximized. The Restore button returns the window to the previous size. To restore a window, click the Restore button on the upper right-hand corner of the window (as shown in Fig. 5.21) or choose Restore from the Control menu or press the Ctrl+F5 keys on the keyboard.

Fig. 5.22   Restore button
in MS Visual FoxPro

Move   We can also move a window to a new location by following the steps mentioned here:

Step 1: Choose the window Title bar.

Step 2: Drag the window to its new location.

Step 3: Release the mouse button.

or

Choose the Move option on the windows Control menu or press Ctrl+F7.

Scroll We can move through the contents on the window using the arrow keys, PgUp, PgDn, Home, and End keys. A window contains more information than it can display. Scroll bars are located along the right-hand side and sometimes along the bottom edge of the window (as shown in Fig. 5.21). We can click the arrows on either ends of the scroll bar or move through the contents of a window one line at a time. We can move up or down one page by clicking the regions between the thumb and the arrows at the ends of the scroll bar. We can also drag the thumb to move through the window contents rapidly.

Creating Tables

Earlier in the chapter, we identified the data items we need to store in a table. Each kind of information will be one field in our table. Let us build a table called Menu.dbf. There are several ways of creating a table structure in FoxPro. We can use the table wizard and appropriate commands.

Creating Table Structure

We will first see how a table structure is created using the table structure dialog:

Step1: Choose New from the File menu. The New dialog is displayed (as shown in Fig. 5.23).

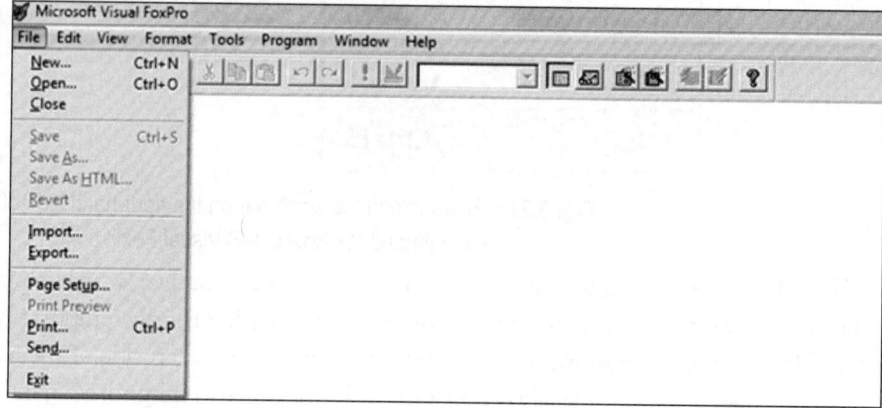

Fig. 5.23   File menu commands

Step 2: Choose the Table radio button from the New dialog.

Step 3: Choose the New file option (as shown in Fig. 5.24).

Step 4: Type a table name (e.g., Menu).

Step 5: Click the Save button and from the Save inbox choose the path, VFP98 folder (as shown in Fig. 5.25).

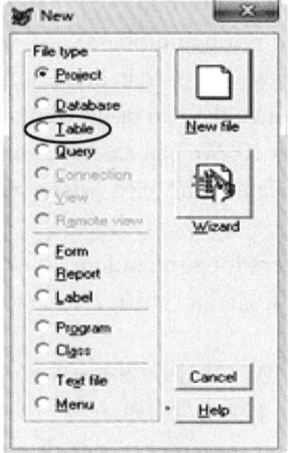

Fig. 5.24 New dialog box in MS Visual FoxPro

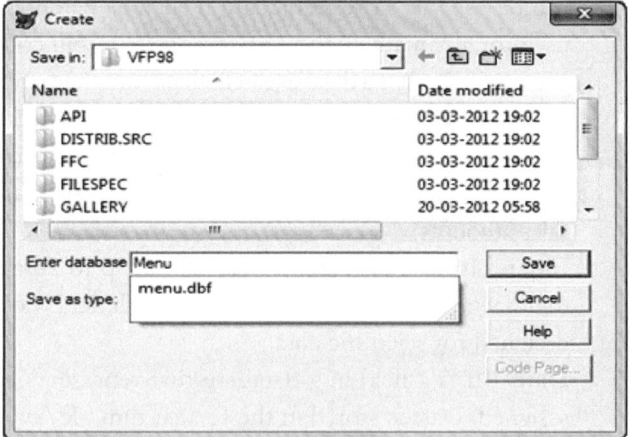

Fig. 5.25 Save dialog box in MS Visual FoxPro

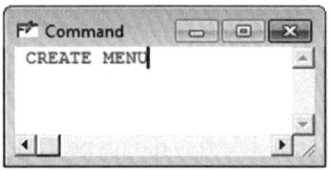

Fig. 5.26 Creating a table using command box in MS Visual FoxPro

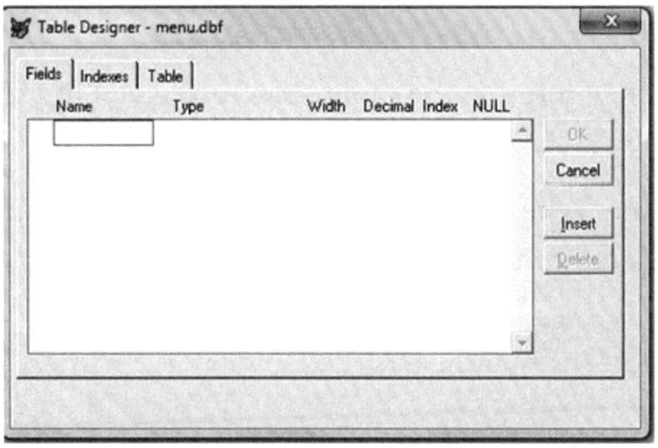

Fig. 5.27 Table structure in MS Visual FoxPro

or

Step 1: Type the following in the command box: CREATE Table name (e.g., 'MENU') as shown in Fig. 5.26 and press Enter.

The table structure dialog opens as shown in Fig. 5.27.

We can see the cursor in the Name textbox (as shown in Fig. 5.27). We can provide the following information for a field:

- A field name under text name
- Data type under text width
- Width of the field under text width
- Decimal precision under text (position of digit to the right of a decimal point), decimal in case of numerical data type
- Create an index for a table when we define a field for a table or after we create the table

We should ensure that the data type selected for a field is of adequate width to accommodate the information. The Table structure dialog has a Status bar at the bottom. As we define the fields, the status line in the Table structure dialog shows the following information: name of the table, number of fields we have defined, and the total length of all defined fields.

Data Types in FoxPro

The following data types are available in FoxPro:

Character It is a type of field in a database file that can be up to 254 letters or symbols long and can easily hold any ASCII character set. The default character width is 10. Character fields can be used to store information such as names, addresses, and numbers that would not be used for any mathematical operations (e.g., pin codes and phone numbers).

Numerical   It is a database field that stores numerical values to be used in numerical expressions. We can enter a field width that is up to 20 characters long, including an optional plus/minus sign and decimal place. The default width for numerical data is 10. We can also use numerical fields to store numbers that would be used for mathematical calculations (e.g., prices, quantities, and sales amounts).

Float   It helps us enter a field that is up to 20 characters long, including an optional plus/minus sign and decimal place. The default field width for float data is 10. This field is especially designed for scientific data.

Date   It is a field in a database that represents a date. A width of 8 bytes is automatically assigned. Date is stored in the format mm/dd/yyyy (e.g., 19/12/2012).

Logical   A character whose length is one is automatically assigned to this data type. Data is stored as T (true) or F (false).We can also enter 'Y' or 'N', which is later converted by FoxPro to 'T' and 'F' respectively.

Memo   A width of 10 characters is automatically allotted to this data type. It usually corresponds to the amount of space used in a table. However, the actual field size of the memo field depends on the amount of data entered into it. The field size is restricted by the amount of memory available to the user.

General   A width of 10 characters is automatically allocated to this data type. It corresponds to the amount of space used in the table. The actual size of the general field however, depends on the amount of data entered into it. The general field size is limited only by the amount of memory available to the user.

Defining Fields

Let us define the fields for menu.dbf. We may refer to the table structure given in Fig. 5.28.

   Step 1: Type the dish code in the Name text box.

   Step 2: Press the Tab button or click the Type pop-up. The Character option appears on the Type pop-up by default. Select Numeric from the Type pop-up.

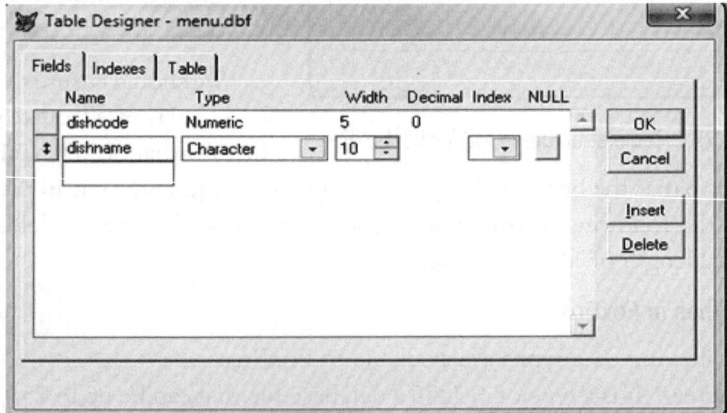

Fig. 5.28   Table structure with type pop-up in MS Visual FoxPro

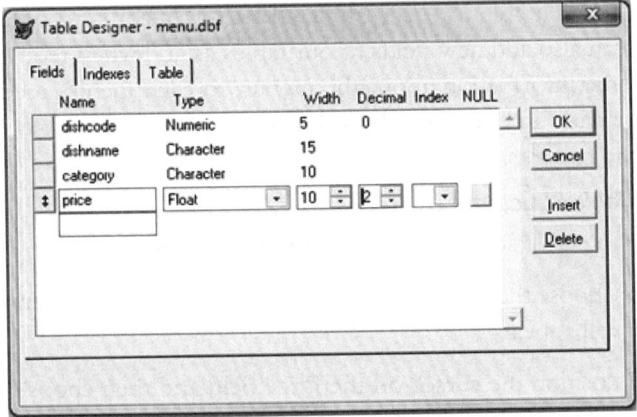

Fig. 5.29   Table structure with fields in MS Visual FoxPro

Step 3: Press the Tab button or click the width spinner. Type 6 or use the spinner control to decrease the width to 6.

Step 4: Using Fig. 5.29 as a guide, define the remaining fields. To move to each field, press the Tab button or click the field. To display the Type pop-up, click the pop-up control.

Step 5: Use the Insert or Delete push buttons to insert a field or delete a field that is not required.

Step 6: Choose OK to save the data (as shown in Fig. 5.29).

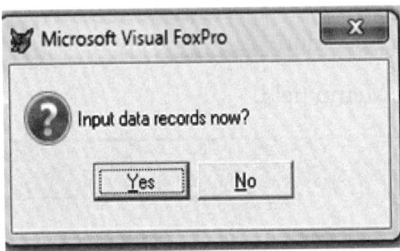

Fig. 5.30   Input records message

Adding Records

FoxPro asks if we want to input data records. Choose Yes (as shown in Fig. 5.30)

A Browse window appears with the cursor in the first field of the first record. Start entering data in the newly created table. Add at least 10 records so that we can view the data using Browse window (as shown in Fig. 5.31).

Fig. 5.31   Entering data in a table in MS Visual FoxPro

We have been able to successfully add a few records to our table by typing data in the fields. We can also add new fields to our tables as and when required. We may add a note that gives us some information about the menus, to each menu's records. We can store pictures, remarks, or dates in our table. We will need to modify the table structure to add a field that will hold the remarks.

Modifying Table Structure

To modify the table structure, carry out the given procedure:

Step 1: Choose the Table Designer option under the View menu. The Table Designer dialog will appear.

Step 2: Position the cursor on the Price field and then choose Insert. A new field appears above the Price field.

Step 3: Change this field to Remarks.

Step 4: Choose Memo from Type pop-up.

Step 5: Repeat Steps 2–4. Add another general type field name for photo, logical type field name for availability, and a data field name for prepared date.

Step 6: Choose OK. When FoxPro asks us if we want to make the change permanent, choose 'Yes' (as shown in Fig. 5.32).

Storing Memo Text

To store memo text in a table, carry out the given procedure:

Step 1: In the Browse window, double-click the Memo field.

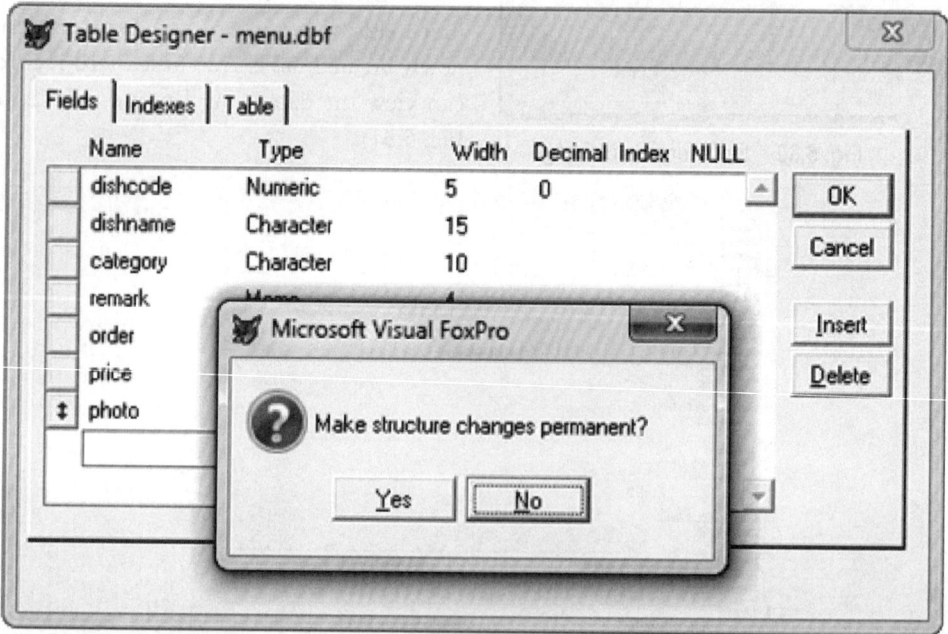

Fig. 5.32   Modifying table structure

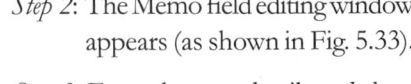

Fig. 5.33 Memo field with details in MS Visual FoxPro

Fig. 5.34 General field edit or window with
picture in MS Visual FoxPro

Step 2: The Memo field editing window
appears (as shown in Fig. 5.33).

Step 3: Enter the note details and close
the edit window.

We will notice that M of the Memo
field is in capitals as a remark has been
inserted in this field.

Storing Graphics

To store a graphic in a general field, we need to
collect a photo of the dish in a format acceptable
to FoxPro. For this example, a photo of the dish
named *Dal Makhani* has been scanned and stored
as Dalmakhani.bmp.

Step 1: Activate MS Paint Brush.

Step 2: Open the .bmp file.

Step 3: First select the picture by using the
Select All option from the Edit menu.

Step 4: Copy it onto the clipboard using the Copy option from the Edit menu.

Step 5: Open Menu.dbf using the Open dialog.

Step 6: Choose the Browse option from the View menu.

Step 7: Locate a particular record in the table.

Step 8: Double-click the photo field. The general field editing window opens.

Step 9: Choose Paste from the Edit menu. The field editing window will appear (as shown
in Fig. 5.34).

We will notice that G of the general field is in capitals after a picture has been pasted in it
(as shown in Fig. 5.35).

Double-click the photo field. The general field editing window appears.

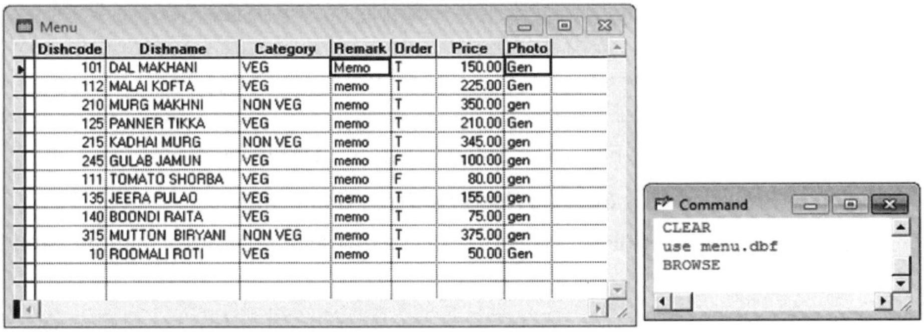

Dishcode	Dishname	Category	Remark	Order	Price	Photo
101	DAL MAKHANI	VEG	Memo	T	150.00	Gen
112	MALAI KOFTA	VEG	memo	T	225.00	Gen
210	MURG MAKHNI	NON VEG	memo	T	350.00	gen
125	PANNER TIKKA	VEG	memo	T	210.00	Gen
215	KADHAI MURG	NON VEG	memo	T	345.00	gen
245	GULAB JAMUN	VEG	memo	F	100.00	gen
111	TOMATO SHORBA	VEG	memo	F	80.00	gen
135	JEERA PULAO	VEG	memo	T	155.00	gen
140	BOONDI RAITA	VEG	memo	T	75.00	gen
315	MUTTON BIRYANI	NON VEG	memo	T	375.00	gen
10	ROOMALI ROTI	VEG	memo	T	50.00	Gen

```
CLEAR
use menu.dbf
BROWSE
```

Fig. 5.35 Browse window after modification of the structure in MS Visual FoxPro

Rushmore Technology

Rushmore technology is a data access technique that permits a set of records to be accessed very efficiently. Rushmore is an exclusive technology for rapidly selecting a set of records from a table. It reduces query response times. This technology automatically uses all the index files, if they are available and open. Thus to enable this technology, all the index files related to a file are kept open. The following commands make use of this technology:

List   This command is used to display records of the current database file, irrespective of the record pointer position (as shown in Fig. 5.36).

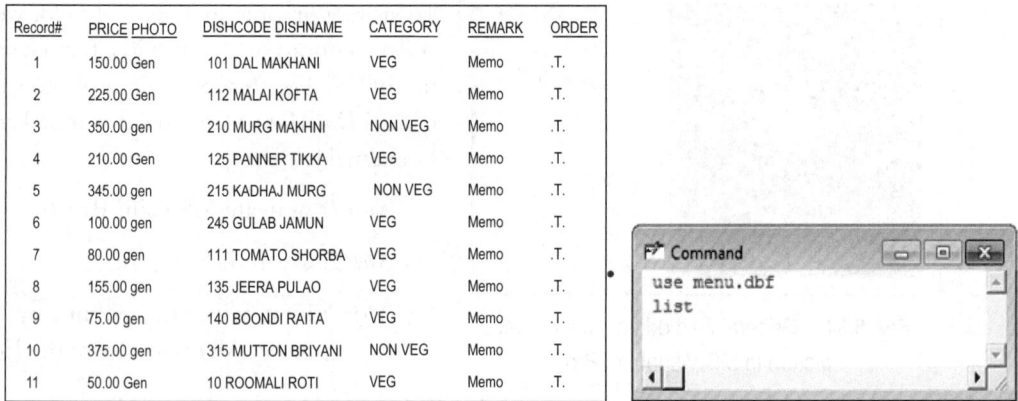

Fig. 5.36   Listing of records in MS Visual FoxPro

Browse   It is used to modify records of the currently opened database file. It displays multiple records together with respect to the record position pointer (as shown in Fig. 5.37).

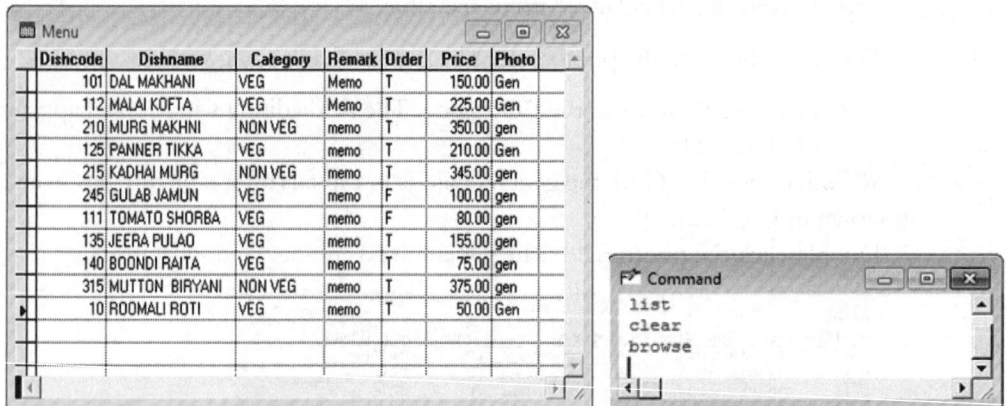

Fig. 5.37   Browsing of records in MS Visual FoxPro

Listing or Browsing for a Specific Records Using Commands

Using Rushmore technology, it becomes easier to search for information using the 'for' clause in the search criteria.

Step 1: Type in the command box, use <database file name> e.g., use Menu.dbf.

Step 2: Type the LIST and BROWSE commands along with the specific field.

In the given example, if we want to search for those records whose dish price is greater than ₹300, we will type LIST for fieldname>=value and then press the Enter key. For example, LIST for PRICE>=300 and then press the Enter key. These items are listed as shown in Fig. 5.38.

Record#	PRICE PHOTO	DISHCODE DISHNAME	CATEGORY	REMARK	ORDER
3	350.00 gen	210 DAL MAKHANI	NON VEG	memo	.T.
5	345.00 gen	215 KADHAI MURG	NON VEG	memo	.T.
10	375.00 gen	315 MUTTON BRIYANI	NON VEG	memo	.T.

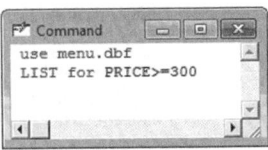

```
Command
use menu.dbf
LIST for PRICE>=300
```

Fig. 5.38 Listing of specific records in MS Visual FoxPro

When we use the LIST command here, we will only be able to see those records whose prices are greater than ₹300. However, in this format, a layman could face problems in viewing the records due to an unjustified database. Therefore, if we want to see the justified database, we have to use the BROWSE command. Suppose we want to view those records that come under the 'veg' category, we will type the following:

BROWSE for fieldname="specified" and then press the Enter key. For example,
BROWSE for CATEGORY="VEG" and then press the Enter key (as shown in Fig. 5.39).

Dishcode	Dishname	Category	Remark	Order	Price	Photo
101	DAL MAKHANI	VEG	memo	T	150.00	Gen
112	MALAI KOFTA	VEG	memo	T	225.00	Gen
125	PANNER TIKKA	VEG	memo	T	210.00	Gen
245	GULAB JAMUN	VEG	memo	F	100.00	gen
111	TOMATO SHORBA	VEG	memo	F	80.00	gen
135	JEERA PULAO	VEG	memo	T	155.00	gen
140	BOONDI RAITA	VEG	memo	T	75.00	gen
10	ROOMALI ROTI	VEG	memo	T	50.00	Gen

```
Command
use menu.dbf
LIST for PRICE>=300
BROWSE for CATEGORY="VEG"
```

Fig. 5.39 Browsing for specific records in MS Visual FoxPro

Marking Records for Deletion

To mark a record for deletion, follow the steps given here:

Step 1: Go to the View menu and choose the Browse option.

The browse window of menu.dbf will appear. Scroll and select a record by clicking any field of the record we want to delete. We can delete one record or a set of records at a time.

Step 2: Select the Table option and choose Delete Records as shown in Fig. 5.40.

Step 3: The delete dialog box will appear (as shown in Fig. 5.41). After that, select the Delete option.

The deletion marker for the selected record is highlighted (as shown in Fig. 5.42).

The record is only marked for deletion. We can now remove a record from the table by permanent deletion.

We can remove all marked records permanently from a table by packing the table. This process is irreversible. Hence, we must be sure that only those records that we really want to erase are marked for deletion.

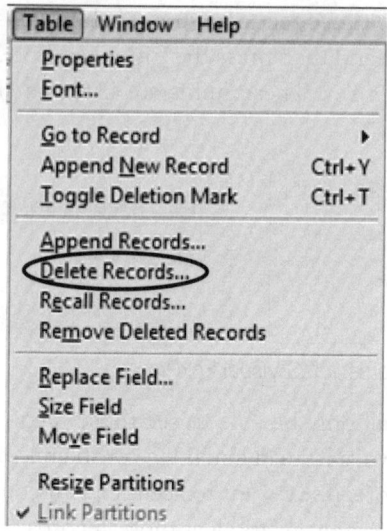

Fig. 5.40   Table pop-up menu in MS
Visual FoxPro

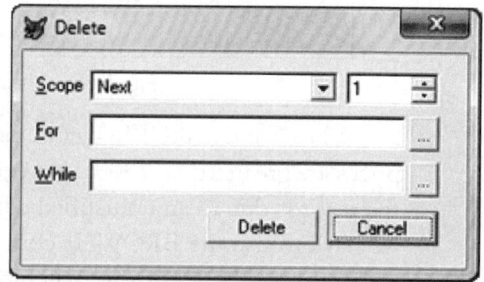

Fig. 5.41   Delete dialog in MS Visual FoxPro

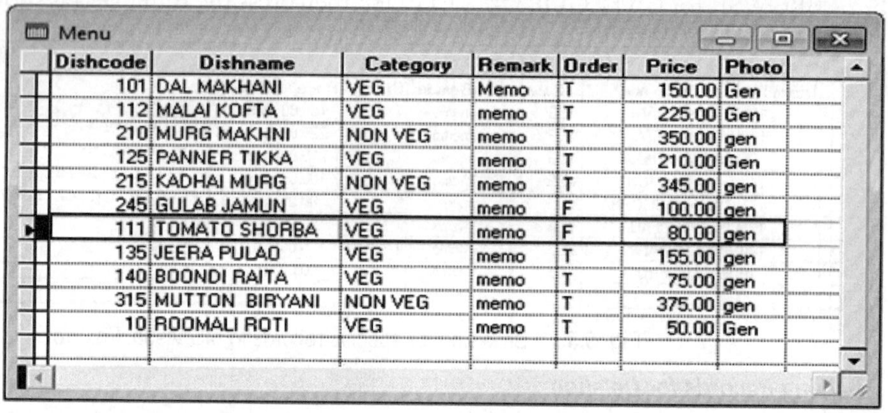

Dishcode	Dishname	Category	Remark	Order	Price	Photo
101	DAL MAKHANI	VEG	Memo	T	150.00	Gen
112	MALAI KOFTA	VEG	memo	T	225.00	Gen
210	MURG MAKHNI	NON VEG	memo	T	350.00	gen
125	PANNER TIKKA	VEG	memo	T	210.00	Gen
215	KADHAI MURG	NON VEG	memo	T	345.00	gen
245	GULAB JAMUN	VEG	memo	F	100.00	gen
111	TOMATO SHORBA	VEG	memo	F	80.00	gen
135	JEERA PULAO	VEG	memo	T	155.00	gen
140	BOONDI RAITA	VEG	memo	T	75.00	gen
315	MUTTON BIRYANI	NON VEG	memo	T	375.00	gen
10	ROOMALI ROTI	VEG	memo	T	50.00	Gen

Fig. 5.42   Record marked for deletion in a Table in MS Visual FoxPro

Recalling Records Marked for Deletion

To recall a record marked for deletion, follow the steps given here:

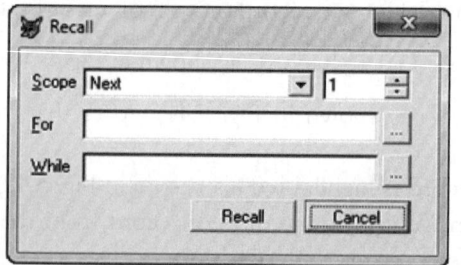

Fig. 5.43   Recall dialog in MS Visual FoxPro

Step 1: Click the deletion marker before Dishcode 111. The deletion marker will no longer be highlighted. Choose Recall records from the Table Menu (shown in Fig. 5.40). A Recall dialog box will appear as shown in Fig. 5.43.

Step 2: Select the Recall button.

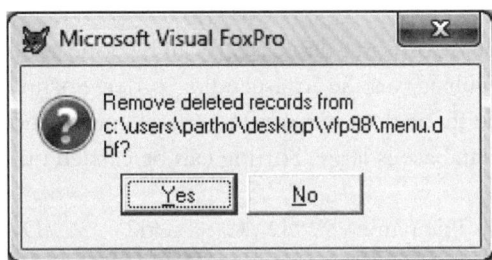

Fig. 5.44   Removing deleted records

Removing Deleted Records

If a record that has been marked for deletion has to be permanently removed, go to the Table menu option (as shown in Fig. 5.40) and choose Remove Deleted Records. A remove delete option dialog box appears (as shown in Fig. 5.44).

We can choose Yes to completely delete or choose No (as shown in Fig. 5.44).

Indexing Records

In a hotel there might be around 20,000 guests arriving and reserving rooms every year. It would be a difficult task to locate the room reservation form of a specific guest. We would have to examine every name from the file/files of reservation forms to locate the desired form. Hence, it is necessary to create an order in which we can access and search for data.

In FoxPro, we can create an order to access data by indexing. Indexing a file is used to provide the order in which records can be accessed. For example, we can access the records in alphabetical order by using the guest's name, prepare a mailing list using postal codes, or organize them in a way so as to speed up searching.

Indexing is done to arrange the records in alphabetical order. The records are not physically arranged in the database. The records within a table may not be sorted out physically in any order but the table appears to be sorted in the form of an index. The index file contains pointers that point to the corresponding value in the indexed file.

Using Index Command

The procedure to using the Index command is as follows:

Step 1: Open the dbf file (e.g., Menu.dbf).

Step 2: Type the proper command for indexing, that is, INDEX on <field> to <index file name>

For example INDEX on DISHCODE to MENU1 (as shown in Fig. 5.45).

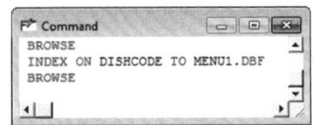

Fig. 5.45   Indexed records in MS Visual FoxPro

Step 3: To see the result of indexing, use the Browse menu command (as shown in Fig. 5.45).

Sorting Records

Sorting is the physical rearrangement of records in a database. It can be done in both ascending and descending order, though by default it is done in ascending order. Sorting can also be done on fields such as character, numeric, or float but not in general and memo. Sorting can be done on one or more fields if the database is large. Sorting can be carried out by using the Sort command.

Syntax: SORT ON <Field1> TO <File name> A /D /C, <Field2> /A /D /C <Scope> [FOR <exp>]

An example, SORT ON DISHNAME TO MENU2.DBF, is shown in Fig. 5.46, where

/A : Ascending order

/D : Descending order

/C : Ignore the case of character field

To see the records in the sorted format, open the newly created sorted file, then type the Browse command, (as shown in Fig. 5.46).

Dishcode	Dishname	Category	Remark	Order	Price	Photo
140	BOONDI RAITA	VEG	memo	T	75.00	gen
101	DAL MAKHANI	VEG	Memo	T	150.00	Gen
245	GULAB JAMUN	VEG	memo	F	100.00	gen
135	JEERA PULAO	VEG	memo	T	155.00	gen
215	KADHAI MURG	NON VEG	memo	T	345.00	gen
112	MALAI KOFTA	VEG	Memo	T	225.00	Gen
210	MURG MAKHNI	NON VEG	memo	T	350.00	gen
315	MUTTON BIRYANI	NON VEG	memo	T	375.00	gen
125	PANNER TIKKA	VEG	memo	T	210.00	Gen
10	ROOMALI ROTI	VEG	memo	T	50.00	Gen
111	TOMATO SHORBA	VEG	memo	F	80.00	gen

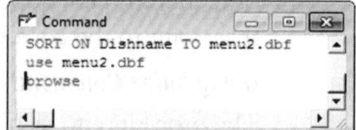

```
SORT ON Dishname TO menu2.dbf
use menu2.dbf
browse
```

Fig. 5.46   Sorted records in MS Visual FoxPro

FoxPro Commands

We will discuss some of the FoxPro commands that are used for working with data. They are as follows:

- CLOSE ALL : To close all opened files
- CLOSE<File name> : To close a specified file
- USE : To close recently opened files
- CLEAR : To clear the FoxPro screen (where the output of executed commands are displayed)
- DISPLAY : To display the current record (where the record pointer is currently placed) of a recently opened database file
- DISPLAY ALL : To display all the records irrespective of the current record pointer
- EDIT : To modify records of a database file
- APPEND : To add records in the currently opened .dbf file; using this command, the record will be added at the end of the file, that is, after the last record.

- LIST OFF   : To display all the records excluding their record number
- DELETE   : To mark the current record for deletion
- DELETE ALL   : To mark all records of the database for deletion
- QUIT   : To exit the FoxPro window

FoxPro Programming

We will learn how programs are created and executed using the FoxPro programming language. We will also learn about memory variables, FoxPro functions, and procedures used in the creation of programs. FoxPro executes commands from the command window in an interactive mode. FoxPro programs are a list of commands that are written in the FoxPro programming language and saved in a text file. We can start creating programs by saving a series of commands that we would often type in a command window.

FoxPro Language

The FoxPro language consists of two elements:

- Commands that perform actions (e.g., the question mark (?) is a command that sends the output to the screen)
- Functions that return a value; the functions contain a pair of parentheses that distinguishes it from the commands.

Commands and functions can be combined to create a FoxPro statement. Functions are always combined with FoxPro commands.

Variables   Variables represent data or information that can be manipulated in a program. The FoxPro language provides the following two types of variables:

- Memory variables: They are also referred to as memvars.
- Table fields: They are the building blocks of a table.

Variable attributes   Every data object, whether a field, a table, or memvar, has certain attributes. These are as follows:

- Name of the object (e.g., room no. and guest ID)
- Data type (e.g., numeric and character)
- A value that refers to the current contents of the data object
- Width or size (in case of numeric or float, decimal places are included)

Planning FoxPro Programs

A programmer would not be able to write instructions for the computer unless he/she knows how to solve it manually. An individual might know how to go about solving a problem, but if the person forgets a step in between or does not write it sequentially, then it would lead to a wrong answer. Keeping this in mind, a programmer must be careful of the sequence of steps while writing a program; otherwise the result will not be correct. The steps involved in writing a program are complex. Thus, we have to plan a program before writing it down.

Algorithm

The term algorithm refers to the logic of a program. It is a step by step process to reach the solution of a problem. The processes should have the following characteristics:

- The instruction should be precise and unambiguous.
- The instruction should be carried out within a fixed time.
- An instruction should not be repeated as this could lead to termination of the algorithm.
- After the instructions are carried out, the algorithm should be terminated to achieve the results.

Sample algorithm   To understand algorithms, let us consider the following example for the calculation of the total bill amount of five items in the menu. Tot is a variable symbolizing total amount.

Algorithm

Step 1: Initialize five dishes (a, b, c, d, e) and tot to 0.00.

Step 2: Take the total amount of the five types of dishes.

Step 3: Add the values of a–e to tot.

Step 4: Print tot.

Step 5: Stop.

Representation of an algorithm   The various ways by which an algorithm can be represented are as follows:

- Flow charts
- Pseudocodes
- Programs

Flow chart   A flow chart is a pictorial representation of an algorithm. Programmers sometimes use it for visualizing the steps. Boxes of different shapes are used to denote the various instructions, which are written in the box. Lines with arrow marks, which determine the flow of operation, are used to connect the boxes. Symbols in a flow chart are standardized, thus helping in communicating the program logic. The symbols that are used in a flow chart are shown in Fig. 5.47.

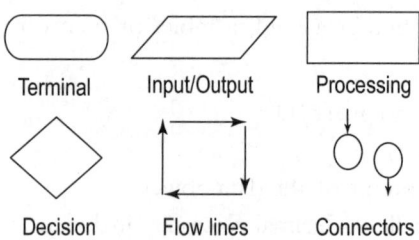

Fig. 5.47   Flow chart symbols

Terminal: This symbol indicates either the beginning or end of the program. At times, it denotes a pause.

Input/Output: This symbol denotes any input–output function in a program.

Processing: This symbol represents movement of arithmetic and other kinds of data instructions. The logical process of moving data from one location in the main memory to another is also denoted by it.

Decision: This symbol indicates a decision point, where a branch to one or more alternatives is possible.

Flow line: This symbol indicates the flow of operation and shows the sequence in which the instructions are carried out. Usually, the sequence of instructions in a flow chart is from top to bottom or left to right.

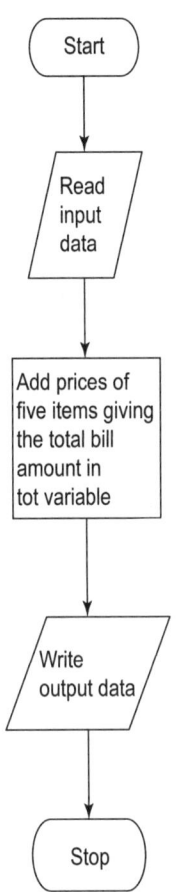

Start

Read
input
data

Add prices of
five items giving
the total bill
amount in
tot variable

Write
output data

Stop

Fig. 5.48 Sample
flow chart

Connector. Sometimes a flow chart can become complex if there are too many flow lines. If a flow chart spreads to more than one page, it is preferable to use a connector instead of flow lines. A sample flow chart is shown with steps describing how billing is done in Fig. 5.48. The billing programme is shown in Fig 5.51.

Pseudocode A pseudocode is a program planning tool for planning the program logic. *Pseudo* means false and *code* refers to the process of writing instructions in a programming language. Pseudocode is a replication of instructions and sentences written in an ordinary language, which the computer does not understand. When a programmer uses pseudocode, he/she can concentrate on the development of the program without being worried about the syntax for writing the program instructions, as these can be easily converted into a programming language.

Creating Programs

FoxPro programs are text files containing a series of commands. The steps to create a program in FoxPro are as follows:

Step 1: From the File menu, choose the New option.

Step 2: In the New dialog box, click the radio button Program.

Step 3: Choose the New file push button (as shown in Fig. 5.49).

or

In the command window, type the following: modify command filename (as shown in Fig. 5.50).

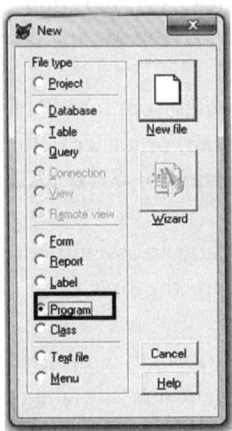

Fig. 5.49 New dialog box in MS Visual
FoxPro

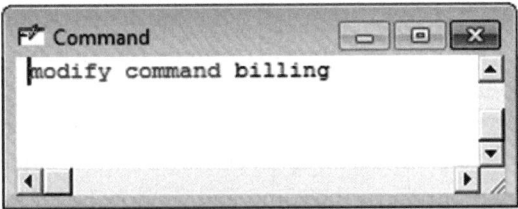

Fig. 5.50 Command window in MS Visual FoxPro

FoxPro opens a new program window named billing.prg. We can now type our program in this window (as shown in Fig. 5.51).

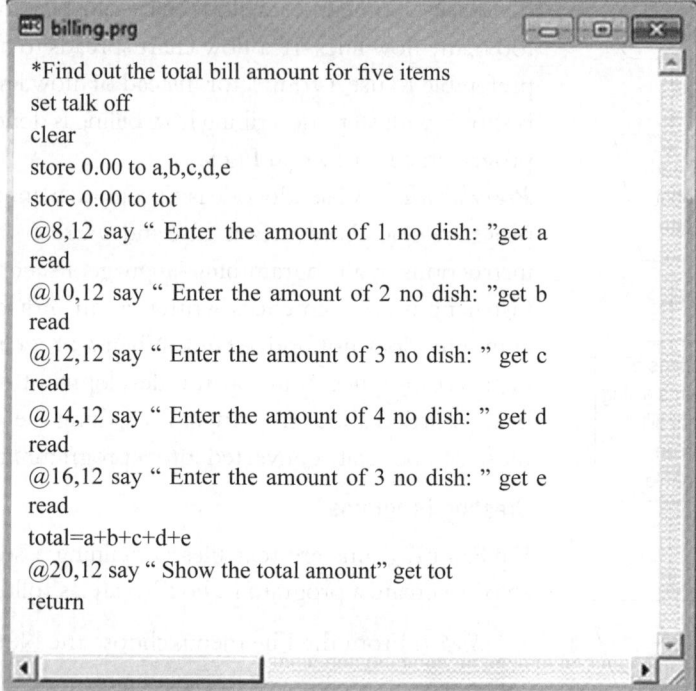

```
billing.prg

*Find out the total bill amount for five items
set talk off
clear
store 0.00 to a,b,c,d,e
store 0.00 to tot
@8,12 say " Enter the amount of 1 no dish: "get a
read
@10,12 say " Enter the amount of 2 no dish: "get b
read
@12,12 say " Enter the amount of 3 no dish: " get c
read
@14,12 say " Enter the amount of 4 no dish: " get d
read
@16,12 say " Enter the amount of 3 no dish: " get e
read
total=a+b+c+d+e
@20,12 say " Show the total amount" get tot
return
```

Fig. 5.51   Program window in MS Visual FoxPro

Problem 1: Find out the total bill amount for five items.

After typing the program, press CTRL+W to save it. Then to execute the program, type the following in the command window box: DO <program file name>, for example, DO BILLING. PRG and press the Enter key. On entering the values, the result is displayed as shown in Fig. 5.52.

SET TALK ON command: It allows talk to be sent to the main Visual FoxPro window, the system message window, the graphical status bar, or a user-defined window. If SET TALK is set to OFF and then changed to ON, the talk is directed to the same location it was sent to before SET TALK OFF command was issued.

SET TALK OFF command: It prevents talk from being sent to the main Visual FoxPro window, the system message window, the graphical status bar, or a user-defined window. Note that for in-process .dll automation servers, the default setting of SET TALK is OFF. These are shown in Fig. 5.51 and Fig. 5.53.

GET command: We can assign default values by adding an assignment statement before each get. A more structured approach puts them all before the first GET. This is shown in Fig. 5.51 and Fig. 5.53.

RETURN command: RETURN terminates execution of a program, procedure, or functions and returns control to the calling program, the highest-level calling program, another program, or the command window (as shown in Fig. 5.51 and Fig. 5.53). Visual FoxPro releases PRIVATE variables when RETURN is executed. RETURN is usually placed at the end of a program, procedure, or function to return control to a higher-level program. However, an implicit RETURN is executed if we omit the command.

The process of solving the program is shown in Fig. 5.52.

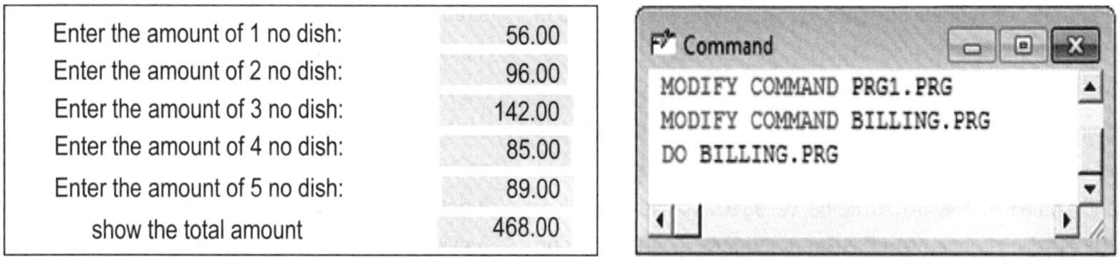

Enter the amount of 1 no dish:	56.00
Enter the amount of 2 no dish:	96.00
Enter the amount of 3 no dish:	142.00
Enter the amount of 4 no dish:	85.00
Enter the amount of 5 no dish:	89.00
show the total amount	468.00

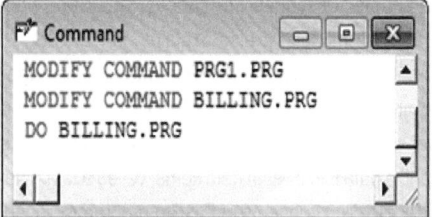

Fig. 5.52   Executing the program in MS Visual FoxPro

Problem 2: Find out the average sale amount of five rooms.

To execute the program, follow the procedure mentioned in Fig. 5.51. Figure 5.53 shows the execution of *Problem 2*. Through similar procedures, we can make simple programs.

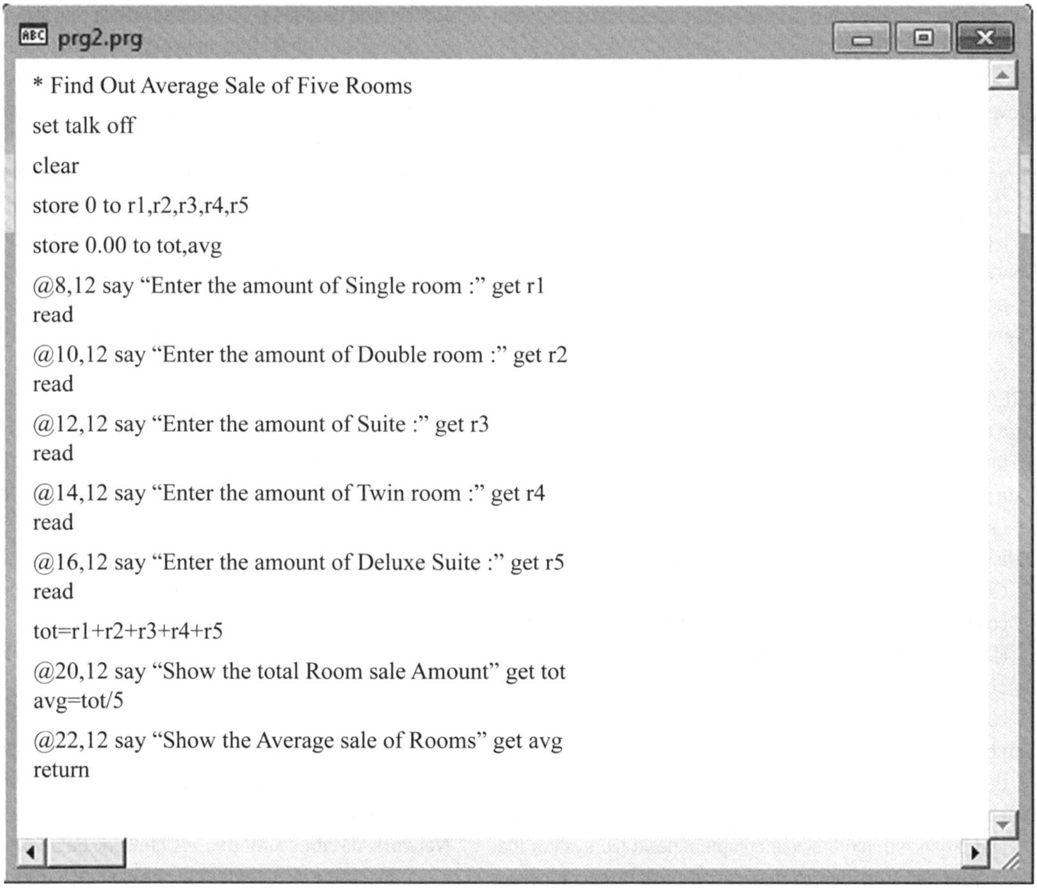

```
prg2.prg

* Find Out Average Sale of Five Rooms
set talk off
clear
store 0 to r1,r2,r3,r4,r5
store 0.00 to tot,avg
@8,12 say "Enter the amount of Single room :" get r1
read
@10,12 say "Enter the amount of Double room :" get r2
read
@12,12 say "Enter the amount of Suite :" get r3
read
@14,12 say "Enter the amount of Twin room :" get r4
read
@16,12 say "Enter the amount of Deluxe Suite :" get r5
read
tot=r1+r2+r3+r4+r5
@20,12 say "Show the total Room sale Amount" get tot
avg=tot/5
@22,12 say "Show the Average sale of Rooms" get avg
return
```

Fig. 5.53   Program window of Problem 2 in MS Visual FoxPro

We have familiarized ourselves with some of the basic commands of Visual FoxPro 6.0 (adding, deleting, and indexing of records, search and display of those records). We were also able to create an algorithm and run a simple program using Visual FoxPro 6.0.

SUMMARY

This chapter describes the concept of DBMS with the types of database models. It also explains an RDBMS. This chapter introduces the theory of FoxPro interface, navigating in the interface, and menu systems. It gives us a brief idea of the options available in FoxPro. We also discussed how easy it is to organize data into a list of data items, called a table, database file, or DBF file. The concepts of FoxPro programming language and structured programming are also discussed. The chapter explains the concepts of variables, commands, and functions. The chapter also talked about the expressions used in FoxPro programs and control flow commands in the FoxPro language. The uses of commands have been explained with an example of a complete functional program to get the required output. The chapter also provides an overview of frequently used commands and functions.

KEY TERMS

.Net   The .NET (pronounced *dot net*) is a software framework that runs primarily on Microsoft Windows. It includes a large library and provides language interoperability (each language can use codes written in other languages) across several programming languages.

Bit   This is an acronym for binary digit. It can have values of 0 or 1.

Browse window   This window displays information about a table. We can modify data and append records in the browse window.

Centralized database   Database components that are stored on a single location and which can be accessed from a different location is called a centralized database.

Character   A character is the basic unit of human perception and is a combination of multiple related bits.

Client–Server database   Access to this type of database is in a request–reply manner.

CREATE command   This command creates a table.

Data definition language   It is a language that is used to define the structure of a database.

Data manipulation language   This refers to a language that is used to enter and manipulate data in a database.

Data warehouse   A database that the user interacts with using OLAP tools to analyse data so as to arrive at a conclusion that could help managements take important strategic decisions, is known as a data warehouse.

Database file   In a relational database, data is arranged in files called database files.

Database model   This is the manner in which various files of a database are linked together.

Database   It is a collection of data files that are integrated and organized into a single comprehensive file system that is arranged to minimize duplication of data and provides convenient access to information within that system to satisfy a wide variety of users' needs.

Data   It is a collection of facts in raw form, which could become information after processing.

Deductive database   It includes an intelligent component that is capable of deriving knowledge and inferential reasoning from the stored knowledge.

Delete dialog   This box give us the option of deleting records from the table.

Distributed database   It is stored on different sites on a network and is equipped with network support and strict control mechanisms.

End user   A user using information generated by a computer system is an end user.

Enterprise database   An integrated database that combines all the individual database needs of all the divisions and departments of an organization is known as an enterprise database.

Field   In a record, a meaningful collection of one or more related characters that is treated as a unit is called a field.

File   It is a collection of multiple related records that are treated as a unit.

Get command   This command is used to assign default values by adding an assignment statement before each GET.

Hierarchical database model   In these types of databases, data is organized in the form of child–parent relationships, thus forming a tree structure.

Information   It refers to processed data that is obtained as an output of data processing. Users use it to achieve specific purposes.

Key field   It is a unique field in a record that is used to distinguish one record from another.

List processing (LISP)   It is a family of computer programming languages that have a distinctive, fully parenthesized notation. It is the favoured programming language for artificial intelligence (AI) research.

Network database model   A network database consists of a collection of records that are connected to one another through links.

New dialog   It creates a new file, table, index, report, menu, screen, etc.

Object-oriented database   It is a database model that captures object-oriented concepts such as class hierarchy, inheritance, and method.

Online analytical processing (OLAP) It is a category of software tools that provides analysis of data stored in a database.

Operational database It refers to the actual database that participates in the day-to-day business transactions of an organization.

Production database A database system that is used while developing an application is referred to as production database.

Programmer A person who designs, writes, tests, and maintains computer programs is called a programmer.

Prolog This is a general-purpose logic programming language associated with artificial intelligence and computational linguistics. Modern Prolog environments support creating graphical user interfaces and also administrative and networked applications.

Query by example (QBE) It is a software tool that allows database users to specify the search criteria by providing them with front-end for the query language, which collects the facts about a query from the user and composes query language statements internally.

Relational model The relational model allows data to be represented as a simple table in the row–column format.

Report generator This software enables users of a database to design the layout of a report, which can be presented in the desired format.

Return command Return terminates execution of a program, procedure, or function and returns control to the calling program.

Set talk off command It prevents talk from being sent to the main Visual FoxPro window, the system message window, the graphical status bar, or a user-defined window.

Setup dialog We can perform several operations in the Setup dialog that is associated with a table in the current work area.

Standard query language (SQL) SQL is the abbreviation of standard query language for relational databases. A query language helps us to obtain information from a database as it is very tedious to search a record from a database possessing many records and files.

Web-based database It is a type of database that can be accessed through the Internet. Data can be stored in XML documents on a website.

REFERENCES

Antonovich, Michael P., *FoxPro 2.5/2.6 for Windows Programming Guide*, Galgotia Publications, New Delhi, 1999.

Gupta, S. and S. Gupta, *Computer Aided Management*, Excel Books, New Delhi, 2004.

Mukkhopadhay, A. K and A. Das, *Elements of Computer Science*, Vol II, Kalimata Pustakalaya, Kolkata, 2002.

Sharma, S.R., *Text Book of Computers for Hotel Management*, Aman Publications, New Delhi, 2008.

Singh, Ranbir, *Understanding FoxPro*, Excel Books, New Delhi, 2000.

Sinha, P.K. and P. Sinha, *Computer Fundamentals*, Fourth edition, BPB Publications, New Delhi, 2007.

Web References
http://en.wikipedia.org/wiki/Lisp_(programming_language), last accessed on 29 March 2012.

http://en.wikipedia.org/wiki/Online_analytical_processing, last accessed on 29 March 2012.

http://en.wikipedia.org/wiki/Prolog, last accessed on 29 March 2012.

http://en.wikipedia.org/wiki/Query_by_Example, last accessed on 29 March 2012.

http://madhurasrecipe.com/index.php?p=1_154_Dal-Makhani, last accessed on 26 March 2012.

http://msdn.microsoft.com/cn-us/library/33a5zy93(v=vs.80).aspx, last accessed on 28 March 2012.

http://msdn.microsoft.com/en-us/library/b660264t(v=vs.80).aspx, last accessed on 28 March 2012.

EXERCISES

Concept Review Questions

1. Explain the concepts associated with databases. What are the advantages and disadvantages of DBMS?
2. What is RDBMS? What is the role of a database administrator?
3. Explain how DBMS can be of help in the hotel industry.
4. What are the different methods of data organization?
5. What are the features of Visual FoxPro 6.0?
6. What is a data warehouse?
7. Write down the main components of DBMS.
8. What is the object-oriented database model? What are the advantages and disadvantages of this model?
9. Write down the various types of data in Visual FoxPro 6.0.
10. What is database mining?
11. What is the difference between List and Browse commands?
12. Write down the major differences between hierarchical and network model databases.
13. Discuss the railway reservation DBMS.
14. What does back-end and front-end refer to?

Multiple Choice Questions

1. Information is
   - (a) raw data
   - (b) processed data
   - (c) both (a) and (b)
   - (d) none of these
2. The command that is used to add a new field to an active database file is the
   - (a) modify label
   - (b) modify report
   - (c) modify command
   - (d) modify structure
3. A suite of programs that handles an organization's database responsibilities is called
   - (a) DBMS
   - (b) Database processing system
   - (c) Both (a) and (b)
   - (d) All of these
4. A float field is specially designed for
   - (a) character data
   - (b) photographs
   - (c) scientific data
   - (d) optional data
5. The default width of numeric data is _____ characters.
   - (a) 10
   - (b) 8
   - (c) 254
   - (d) 4
6. Which of the following is an example of artificial intelligence language?
   - (a) SQL
   - (c) HTML
   - (b) LISP
   - (d) None of these
7. Which of the following models of data support one-to-many relationships?
   - (a) Network model
   - (b) Object-oriented database model
   - (c) Hierarchical model
   - (d) All of these
8. A patented data access technique that extracts data quickly is
   - (a) Rushmore technology
   - (b) Information technology
   - (c) Database technology
   - (d) Access technology
9. Which of the followings commands is correct?
   - (a) Listing record
   - (b) List to see
   - (c) List off
   - (d) None of these
10. Which of the following is used to add records to the currently opened .dbf file?
    - (a) Add record
    - (b) Append
    - (c) Entry
    - (d) Skip

Project Work

1. Create a database of guests visiting a hotel for a span of three months (one quarter). A guest who visits six or more times in a quarter will be given a loyalty card and will get a 10 per cent discount on room rent for the next quarter.

2. If the bill amount for a loyal guest is greater or equal to ₹1,00,000, the guest would be considered for a Gold card. A guest with this card will get an additional discount of 5 per cent on room rent and 10 per cent on food.

CASE STUDY

Hotel Landmark has been doing business for over two decades and is the best business hotel in the city. The hotel has around 500 employees working at various levels in the organization. A new human resource manager, Sonia Sachdeva, joined the organization. She was amazed to know that the hotel didn't have an employee state insurance facility (ESI) for the employees. Since the government has made it mandatory that ESI should be provided to all employees drawing a gross salary of less than ₹15,000, she wanted to have a list of employees who were eligible for ESI.

The general manager of the hotel, Srikant Bhalla, also instructed Sonia to give an incentive to employees, since the sales achieved by the hotel were 30 per cent more than the target sale. The incentive would be given to only those employees whose monthly gross salary was less than ₹40,000.

(a) Using FoxPro 6.0, how will you find out the employees who were drawing a salary that was less than ₹15,000?
(b) Using FoxPro 6.0, how will you find out employees who were drawing a salary that was less than ₹40,000?
(c) How will you take a printout of the list of employees selected in questions 1 and 2?

Answers to Multiple Choice Questions

1. (b)	2. (d)	3. (a)	4. (c)	5. (a)	6. (b)	7. (c)	8. (a)
9. (c)	10. (b)						

PART II

Computer Applications in the Hotel Industry

- Introduction to Hotel Information System

- Computerized Reservation System and Room Management

- Account Management Module

- Food and Beverage Management

- Property Management System Interface

- Management Information System

CHAPTER 6

Introduction to Hotel Information System

LEARNING OBJECTIVES

After reading this chapter, you will be able to understand the following:

- Departments of a hotel and their functions
- Revenue and non-revenue generating departments
- Use of property management system
- Some property management systems used in hotels

Exhibit 6.1   Need for Computerization in Hotels

Hotel Tripti, a hundred-room hotel, is partly computerized. The front office, Food and Beverage (F&B), and stores departments had computers but they were only used for their respective departmental work. The front office and the F&B service departments issued guest bills with the help of computers. The food production department stored standardized recipes of all their dishes on computers. This was convenient when it came to briefing new staff about the hotel's dishes.

However, when a guest checked out of the hotel, the front-office staff had to call telephone outlets, such as room service and coffee shop, to determine if all the bills of the guest had been settled. At the end of each day, bills were sent to the front office before the night audit. Many a time, the front-office assistant would forget to call one of the outlets, resulting in loss of revenue to the hotel. There was lack of coordination between departments such as the front office and housekeeping, about room allocation, clean-up, those that were under repair, etc. There was also disorganization in

the stores department regarding the purchase, quantity, and specification of items ordered and received, and in the F&B department regarding kitchen order ticket (KOT) ordering time, etc. Since all operations were manual, it lead to a lot of human error, which not only translated to revenue loss but also misunderstanding among employees and hence dissatisfaction at the workplace.

Keeping these problems in mind, a property management system (PMS) was suggested. This could not only connect various departments but also provide speedy and efficient services to guests. When a guest checks out of the hotel, the PMS automatically updates bills from outlets (speciality restaurant, gift shop, lounge, etc.) that the guest may have availed services from. Activities within a department were also streamlined. For example, the housekeeping department had a clear idea about the number of rooms that had to be readied within a particular time frame, and as and when they were ready, they could immediately be allotted to the waiting guests.

A hotel is an organization that comprises many departments, which work together and coordinate with each other for better efficiency. Some departments in the hotel are revenue generating while some are not. Nevertheless, each of them has an important role in the overall operation of the hotel. We will discuss revenue and non-revenue generating departments in the later part of this chapter. We will first talk about the different departments of a hotel in brief.

DEPARTMENTS IN A HOTEL

The following are the main departments of a hotel:

- Front office
- Housekeeping
- Food and beverage
- Sales and marketing
- Accounts
- Maintenance and engineering
- Safety and security

These departments are discussed in detail in the following sections.

Front Office

The front-office area in a hotel is where the front-office staff assists arriving and departing guests. Front-office employees are very critical to a hotel's success as they may sometimes be the only people that the guests actually see when they arrive or depart from the hotel.

The main function of the front office is to make guest reservations, an essential and complex job. A room not sold in a day can never be sold on that day and the revenue loss cannot be recovered. Hence, it is very important that a hotel does the best it can by matching room availability with room demand. The room inventory of a hotel is fixed since it is not possible to increase or decrease the number of rooms available for sale each day. Therefore, it requires an efficiently managed and talented front-office staff to optimize total revenue.

Another important task of the front office is the collection of money from the guests. Hence, the front-office manager must formulate and administer revenue management systems that charge guests the right amount for the services they use. Nowadays, hotels use revenue or yield management software to increase revenue.

Housekeeping

Guests, quite naturally, always want clean rooms. It is the role of the housekeeping department to provide clean rooms as well as maintain other areas of the hotel. The job of the housekeeping department is complex and is becoming even more so every day. Proper cleaning requires knowledge of equipment and chemicals that make the cleaning job easier.

Other than cleaning rooms and public areas such as lobbies and corridors, the housekeeping department in the hotel is also responsible for cleaning sheets, towels, and other items in the hotel's laundry. Though some hotels may not have their own laundry on-site (off-premise laundry), in many hotels the on-premise laundry (OPL) plays a significant part of the housekeeping department's daily activities. In some hotels, even horticulture comes under the housekeeping department.

Food and Beverage

A hotel offers both food and beverages to guests staying at the hotel or visiting the restaurant. The F&B department consists of the F&B production and F&B service departments. A small establishment may have a food and beverage manager who may manage employees responsible for both food production and service. In a large organization, the kitchen brigade takes care of the food production outlets (e.g., the garde manger for cold preparations, patisserie for bakery items and confectionery, and specialized chefs for speciality restaurants) of a hotel.

The food production department's preparation begins with menu planning, focusing on guest needs and what the department could prepare profitably. After a menu or dish is planned, the ingredients for the menu have to be noted, purchased, received, stored, issued, and used. The dishes are then finally served to the guests by the staff.

The F&B service department is an essential link between the customer and the menu, beverages, and other services on offer in an establishment. This unit is the bridge between the food and beverage production, and the customers. As trends are changing, the department has become more customer-oriented. Nowadays, since both the husband and wife earn well, more families are eating out. To meet this demand, diverse food and beverages are on offer. Food service operations are enjoying improvement and development with considerable advances in the quality of food offered. The food served by the F&B department includes a wide range of styles and cuisines, which can be classified country-wise. Alcoholic and non-alcoholic drinks are served. Alcoholic beverages include wines, cocktails, beers, spirits, and liqueurs. Non-alcoholic beverages include mineral water, juices, aerated water, tea, coffee, chocolate, milk, and milk drinks.

Sales and Marketing

A well-run hotel must attract, maintain, and expand a strong guest base so as to increase room sales. The goal of the hotel's sales and marketing team is to increase sales and thus increase revenue. Though every employee in a hotel may have an impact on a guest's experience, it is the job of the sales and marketing personnel to attract potential guests.

The sales and marketing department along with the personnel working in other concerned departments, create marketing plans, set aside budgets, and implement proposals. A well chalked out marketing plan will help in maximizing hotel sales. The hotel professionals responsible for sales and marketing must identify and develop new clients, plan the hotel's marketing efforts, provide inputs on appropriate room rates, and negotiate sale contracts.

Accounts

A hotel generates financial data obtained from the sale of rooms, food and beverages, and other products and services. The money received from guests has to be collected and safeguarded, and the hotel's operating expense carefully recorded, monitored, and managed. Accounting involves the maintenance of accurate records of all the financial activities of a hotel.

The accounting function usually begins with budget allocation. A hotel must be able to estimate the income and expenditure involved in its operations.

The money collected from guests for renting out rooms and other services must be correctly accounted for. The accounting process is done on a daily basis as a guest's stay in a hotel might be just for a day. In addition to rooms, a hotel also operates retail outlets such as restaurants, lounges, banquet facilities, pastry shops, and gift shops, all of which account for daily sales.

Maintenance and Engineering

In some hotels, the maintenance department is also known as the maintenance and engineering department. Usually, the head of the maintenance department is referred to as the chief engineer. The chief engineer, along with the department staff, is responsible for proper maintenance of a hotel's building and surroundings.

A well-run maintenance department boosts a hotel's sales by providing the guest with the best experience possible. The maintenance department works not only for the guests but also for the employees who expect the tools and equipment they need for their job to be safe and in proper condition. The hotel owner expects that the building and all that it houses within it, is correctly repaired and maintained to protect the value of investments. The maintenance department must develop an effective preventive maintenance programme so as to prolong the life of the hotel's equipments to ensure maximum operating efficiency.

Safety and Security

A guest who checks in counts on a safe and comfortable environment in the hotel. Guests and employees wish to be at a place that is free from risks. The hotel industry is committed to the safety and security of their guests. In order to achieve this, the hotel industry relies upon a variety of security tools such as recodeable locks, alarm systems, surveillance systems, and emergency plans that could be used to increase safety and reduce security risks.

Depending on the location and the services that they offer, hotels have safety and security issues to deal with. Adequate safety measures should be provided to guests in the swimming pool area, spa, gym, and in a hotel's parking lot. Though a hotel cannot completely guarantee a guest's safety, it is their responsibility to exercise reasonable care in protecting the well-being of their guests.

A threat to the security of a hotel's assets could be from both internal and external sources and therefore to protect these assets adequately, programmes should be in place to guard against such threats. The possibility of a safety or security crisis is prevalent even in the best managed hotels and therefore all employees must be prepared to respond appropriately, if the situation so demands.

REVENUE AND NON-REVENUE GENERATING DEPARTMENTS

Based on whether they directly earn money for the hotel or not, the departments of a hotel can be divided into revenue generating and non-revenue generating. The types of departments that fall under these two categories are discussed in the following sections.

Revenue Generating Departments

The revenue generating departments of a hotel are as follows:

Laundry   A guest's personal linen is laundered and charged by the hotel.

Telephone   A guest is charged for local and long-distance calls such as STD and ISD.

Room division   The front office and housekeeping departments together make room division. The front office is associated with the sale of rooms and comes into direct contact with the guest. Housekeeping, on the other hand, keeps the guest rooms clean and in a position to be rented out.

Food and beverage   The food production department prepares food in its respective sections (e.g., bakery, pantry, grill) and also works behind the scenes for the guest.

The F&B service department has a number of outlets where the food prepared by the F&B production is sold to the guest.

The following are some common F&B service outlets:

Restaurant   They could be speciality cuisine (single cuisine) or multi-cuisine restaurants that usually have fixed hours of operation during lunch and dinner.

Coffee shop   Open round-the-clock, they mainly serve snacks and beverages. They also serve meals during lunch and dinner.

Bar   This is an outlet, which serves a wide array of alcoholic and non-alcoholic beverage along with limited snacks.

Banquet   It is a major revenue producing department among all F&B service outlets. It caters to a group of people at a time. Usually banquet functions are held for special events such as birthday parties, weddings, anniversaries, conferences, and meetings. The menu for a banquet function is usually elaborate, though it depends upon the host.

Room service   From this outlet, guests are served food and beverages round-the-clock within the comfort of their rooms. The menu for room service is limited and not as elaborate as it is in the restaurants.

Other outlets   The other revenue generating outlets are travel desk, gift shop, florist, spa, beauty parlour, and cake shop.

Non-revenue Generating Departments

The following are some of the non-revenue generating departments of a hotel:

Human resource This department deals with recruitment and training of employees and trainees. The department is also responsible for staff induction, promotion, and appraisals.

Security   It deals with the safety and security of a hotel, its employees, and guests and also deals with any unusual event in a hotel.

Accounts   This department is responsible for all the financial activities of the hotel. It maintains financial records, and helps in budgeting and evaluating the total operations of the hotel.

Sales and marketing   It is concerned with getting and maintaining a clientele for the hotel and also improving the goodwill of the hotel.

Engineering   It is responsible for hotel maintenance and upkeep of the rooms and public area.

Need for Computerization

The departments described until now, both revenue and non-revenue generating, work with a lot of information that has to be processed and retrieved at very short notice. If a hotel does not have an automation facility, then each department has to have files, filing cabinets, and racks for storage. Apart from occupying a lot of space, files require manpower for the maintenance and upkeep of the documentation process. If a query is put forth, an individual will have to look through a number of files before coming to a conclusion. On the other hand, a computer can search through all the files and display the results on the monitor within a few seconds.

Computerization in a hotel not only increases the efficiency of departments, but also of various sub-departments which have their own functions, as well as those that are related to each other. Thus each sub-department has separate modules, which perform a particular or a set of functions. A hotel with departments that take care of various guest needs requires an arrangement by which information can be passed from one system at a particular location to another at a different location in the hotel. To meet this purpose, a PMS has been put in place.

In modern hotels, computers are used in front-office operations. They are also used in housekeeping, F&B, marketing and sales, engineering, accounting, and human resource departments of a hotel. All departments in a hotel play an important role in serving and satisfying guest needs, before, during and, even after a guest's stay.

HOTEL PROPERTY MANAGEMENT SYSTEMS

Although the components of a PMS may vary, the term is generally used to describe a set of computer programs that directly relate to front-office and back-office activities.

Front-office Applications

Computerized front-office applications consist of a series of software programs (or modules) that include hotel reservations, room management, and guest accounting functions. A variety of stand-alone applications may also be interfaced with a PMS. Popular interfaces include microcomputers, point-of-sale (POS) systems, call accounting systems, electronic locking systems, energy management systems, auxiliary guest service devices, and guest-operated devices. PMS packages contain modules covering accounting and internal control functions. Hospitality accounting applications are discussed in Chapter 8.

While hotel PMSs differ, many of them offer front-office applications software in relation to reservations, room management, guest accounting, and revenue management.

Reservation Module

A reservation module enables a hotel to quickly process room requests, generate timely and accurate room revenues, and forecast reports. Reservations received at a central reservation location can be processed, confirmed, and communicated to the appropriate branch of the hotel before the reservationist actually ends the call with the guest on the telephone. When the destination hotel uses a PMS, the reservation module receives the data forecasts, which are immediately updated. In addition, the reservations data can be automatically reformatted into preregistration material and an updated 'expected arrivals' list could be directly generated from the central reservation system. This is explained in detail in Chapter 7.

Room Management Module

A room management module maintains up-to-date information regarding the status of rooms, assists in the assignment of rooms during registration, and helps coordinate guest services. Since this module replaces most traditional front-office equipment, it often becomes a major determinant in the selection of one PMS over another. This module alerts front desk employees on each room's status, similar to what 'room information' racks do in non-automated systems.

In a computerized system, the front desk employee simply enters a room number using a keyboard and the current status of the room is immediately displayed on the screen. Once the room becomes clean and ready for occupancy, housekeeping changes the room's status through a terminal in the housekeeping desk, and the information is immediately communicated to the front desk. This is explained in detail in Chapter 7.

Guest Accounting Module

A guest accounting module increases a hotel's control over guest accounts and also adjusts the night audit routine. Guest accounts are nowadays maintained electronically, thereby removing the need for folio cards, trays, or posting machines, which were prevalent in non-automated systems. The guest accounting module monitors the predetermined guest credit limits and provides flexibility through multiple folio formats. Since the revenue centres are connected to the PMS, remote electronic cash registers or POS terminals communicate with the front desk, and the guest charges are automatically posted to the appropriate guest folios. At checkout, outstanding account balances are transferred automatically to the city ledger (accounts receivable) for collection. This is explained in detail in Chapter 8.

Revenue Management Module

Revenue management or yield management is a set of demand-forecasting techniques used to determine whether prices should be raised or lowered and whether a reservation request should be accepted or rejected in order to maximize revenue. Yield management is based on supply and demand. Prices tend to rise when demand exceeds supply; prices tend to fall when supply exceeds demand. One of the principal computations involved in yield management is the yield, which is the ratio of actual revenue to potential revenue. Actual revenue is the revenue generated by the number of rooms sold. Potential revenue is the amount of money that the property would receive if all of its rooms were sold at full rack rates.

There are many formulae used to implement yield management strategies. Although individual computations involving yield management can be performed manually, doing so is very difficult and time-consuming. The most efficient means of handling data and generating yield statistics is by using a computer. Sophisticated yield management software can integrate room demand and room price statistics, and hence project the highest revenue-generating combination.

Property Management System Interfaces

PMS interface applications are stand-alone computer packages that may be linked to a hotel computer system. Although the number and kinds of software packages that can be linked to a hotel system are growing, the most popular interfaces have been discussed in the following sections. The PMS interface is explained in detail in Chapter 10.

Microcomputer Interfaces

Microcomputer interfaces to larger hotel computer systems have become a popular means of expanding data processing capabilities. Downloading (transferring) data from the hotel systems to the microcomputer enables the management to use data contained in the hotel system's software with software applications designed for the microcomputer. A lodging system might maintain all of the hotel's accounting data while designing the following year's budget; the management might wish to support the projections on the actual transactions of the current accounting period. If the management is able to access the necessary accounting data from the PMS to the microcomputer, the data could be used by software applications that may include word processing, electronic spreadsheets, database management, and communication programs.

Point-of-sale System

A POS system is a network of electronic cash registers that is capable of taking data at POS locations and transfer them to the system's guest accounting and financial modules. The ability to communicate the data to both front and back-office components could result in numerous benefits derived through comprehensive reporting. This is explained in detail in Chapter 10.

Telephone Call Accounting System

A telephone call accounting system (CAS) enables a hotel to take control over both local and long-distance telephone services. A call accounting system can also place and charge outgoing calls. When a CAS is interfaced with a front-office guest accounting module, telephone charges can be immediately posted to the appropriate folios. This concept is explained in Chapter 10.

Electronic Locking Systems

Many types of electronic locking systems are available today. Often, these systems interface with a front-office computer system, thereby enabling the management to exercise important control measures. This is explained in Chapter 10.

Energy Management Systems

Interfacing energy management systems with a hotel computer system links energy controls of a room with the front-office room management package. An energy management system monitors room temperatures with the help of a computer. This may lead to significant reduction in energy consumption and lower energy costs. This is explained in Chapter 10.

Auxiliary Guest Service Devices

Automation has made easy many auxiliary guest services, such as the placement of wake-up calls and the delivery of messages to guests. These functions are often performed by devices (electronic message-waiting systems and voice mailbox systems) that are marketed as stand-alone systems.

Guest-operated Devices

Guest-operated devices can be located in a public area of the hotel or in the rooms. The devices located within a room are designed to be user-friendly systems. Variations of such devices within the rooms can provide concierge-level services to guests.

Housekeeping

Computerization helps in keeping a check on rooms as and when they are cleaned by the housekeeping department. As soon as the rooms are ready, the personnel in the housekeeping desk alter the room status so that the room is available for letting out. After cleaning each room, the housekeeping staff can update the status to the housekeeping supervisor, who after checking the room can immediately inform the housekeeping desk, rather than waiting for a block of rooms to become free. The housekeeping department can also assign rooms to their staff to analyse their productivity. This is explained in detail in Chapter 7.

Food and Beverage Module

Computerization in the F&B module reduces lot of paper work as well as telephone calls from the restaurants and other outlets to the front desk. It also aids in the accounting process thus verifying the integrity of the POS system. The various features of the F&B module include inventory listing, recipe management, sales analysis, and report generation from each outlet at the end of the day. This is explained in detail in Chapters 8 and 9.

Sales and Catering Applications

The sales staff and food service catering managers spend a great part of the day processing paperwork related to information collected from prospecting, selling, booking, and finally reporting. In most establishments these days, much of this time-consuming and costly activity is handled by computers.

In a fully automated sales office, every salesperson having a computer terminal has immediate access to room information. Bookings and cancellations can be processed quickly even as the salesperson is on the phone with the client. This ensures that every salesperson has access to exactly the same information and that 'definite' and 'tentative' bookings are clearly identified to prevent errors. In addition, an automated sales office system can produce reports that provide information on accounts, bookings, market segments, sales staff productivity, average room rates, occupancy, revenue, service history, lost business, and important marketing data. If done manually, many of these reports would have taken several hours to produce.

Accounting Applications

The number of accounting software modules provided by a back-office PMS may vary widely. A typical back-office system contains application software designed to monitor and process 'accounts receivable' and 'accounts payable' transactions, payroll accounting, and financial reporting. Other back-office programs streamline inventory control, purchasing, and budgeting. This is explained in detail in Chapter 8.

Engineering

Property management system in engineering streamlines work orders that have to be processed. Repair orders entered by various departments are prioritized as per importance, cost, and availability of equipments and parts.

Human Resource

The maintenance of employee files is enhanced by using a PMS. Information pertaining to an employee (designation, date of joining, salary structure, deductions, taxes, and performance and promotions) can be obtained very easily. A PMS also reduces the paperwork involved in maintaining records and easily generates reports on employees.

A variety of PMSs are available in the market, both in India and abroad. We will discuss a few in detail in this chapter and some others in the following chapters. Micros, IDS Fortune, and Shawman will be briefly discussed in this chapter. PMSs, such as Hotelogix, Datamannet, and hotel management systems, will be discussed in Chapter 7.

MICROS

MICROS Systems, Inc. (NASDAQ: MCRS), headquartered in Columbia, Maryland, is the world's leading developer of enterprise applications serving the hospitality and speciality retail industries. MICROS serves table service and quick service restaurants, hotels, the leisure and entertainment industry, and speciality retail stores, with complete information management solutions including software, hardware, enterprise systems integration, consulting, and support.

OPERA ENTERPRISE SOLUTION

The OPERA Enterprise Solution (OES) is a fully integrated suite of products that can be easily combined for deployment in organizations ranging from single-property hotels to global, multi-branded hotel chains. Hotels can choose the products and features they need. OPERA ES is modular and scalable. OPERA modules include property management, sales and catering, quality management, gaming and computerized accounting, and condo/hotel room management. In addition, the OES offers central management products including the OPERA Reservation System for both guestrooms and sale of functional space; the OPERA Customer Information System, a customer relationship management (CRM) package specifically designed for the hospitality industry; and Sales Force Administration, which provides centralized lead management and sales support for regional and national sales teams.

Central systems:

- OPERA Business Intelligence
- OPERA Customer Information System
- OPERA GDS Interface
- OPERA Sales Force Automation (SFA)

- myfidelio.net
- OPERA Reservation System
- OPERA Web Suite
- OPERA Revenue Management
- OPERA Comp Accounting and Gaming

Property systems:

- OPERA Property Management System
- OPERA Xpress
- OPERA Lite
- Operetta Hotel Software Solution

- OPERA Vacation Ownership System (OVOS)
- OPERA Sales and Catering
- OPERA Kiosk
- OPERA Activity Scheduler

We will now discuss a few systems in detail.

OPERA Reservation System

The OPERA Reservation System (ORS), a true centrally managed central reservation system, is at the heart of the hotel industry's first enterprise-wide room inventory management system. It is the most advanced central reservation system. The ORS hotel reservation system is designed for seamless integration with the entire OPERA product family: the OPERA Customer Information System (OCIS), the OPERA Property Management System, OPERA Sales and Catering, and OPERA Sales Force Automation. This comprehensively managed central reservation system offers reservation agents and global sales staff the tools to maximize bookings and increase revenue in a chain or multi-property environment. ORS easily handles all types of reservations such as individual, group and party, company, travel agent, multi-legged, multi-rate, and wait-listed.

Key Features

The following are the key features of ORS:

Global perspective   ORS supports multi-currency and multi-language features to meet the hotel reservation system requirements of global operations. Rates and revenues can be dynamically converted from the local currency to any other currency. The appropriate language for guest correspondence can be automatically determined from the guest's profile language. During the reservation process, key information such as the property, room, and rate descriptions can be displayed in multiple languages. Country-specific address formats are supported.

Automatic rate and inventory controls   The ORS central reservation system lets you set-up rate structures for individual properties, groups of properties, or chains. Multi-level rate and inventory controls make inventory management easier and increase profitability. Rates and room types can be automatically restricted based on percentage occupancy, minimum stay, arrival date, etc. Agents can use the hotel reservation system to easily determine the best available rates for any duration of stay. To maximize property revenues, these hotel reservation computer systems also support interfacing to major yield management systems.

Full reservation functionality   Agents making reservations with the ORS central reservation system can easily handle complex operations such as routing instructions, splitting charges, sharing reservations, and frequent flyer and loyalty programme memberships, as well as negotiate rates and rate discounts (percentage or fixed amount). Multiple advance deposit requirements and cancellation penalties may be applied to reservations with ORS. This system also deposits transactions automatically to the establishment.

Group and block features   Creating and managing group and block reservations is as easy in the ORS central reservation system environment as it is using a stand-alone hotel PMS. Room blocking, room lists, room sharing, deposits, tour series, and other group booking features are handled by the ORS.

Sales flow control   This hotel reservation system allows you to customize the flow of the sales process by the chain or property. ORS operations can be optimized for any operational environment. Dynamic global messaging, scripts selling, and scripts closing guide the agent through the selling process. Agents can easily change or cancel existing bookings and move bookings to different properties or dates any time prior to check-in.

Multi-property rate display   The ORS central reservation system can display rates, room types, and packages for one property or for multiple properties. Agents can easily re-query for alternative dates when the requested dates are not available. Colour-coding throughout the sales screen lets the agent see at a glance why a property, room, or rate might be unavailable.

Efficient searching   The ORS hotel reservation system can be used to conveniently search for room availability across properties and chains. By entering just a few criteria, agents can narrow a property search. Searching can be controlled by one or more criteria, such as property name, city, region, property features, property type (e.g., three-star, four-star), package elements, attractions, and rate range. The ORS central reservation system's property information displays are comprehensive, with details on transportation services, restaurants, amenities, etc. Area maps and images of the property can also be provided. ORS automatically logs the original search criteria and registers whether the call resulted in a booked reservation or a turn away.

Reports and logs   The ORS hotel reservations computer systems offer dozens of standard reports that provide extensive data for analysis.

Channel management   Hotels and chains can also use the central reservation system to review business volume within a channel, and open or close channels based on the channel, property, rate, or room type. Much of the channel management set-up information is pre-configured to make it easy to get started.

OPERA Property Management System

OPERA PMS is fully integrated with all the OPERA modules and offers the most extensive list of certified interfaces in the industry.

Key Features

The following are the key features of OPERA PMS:

Reservations   The features are integrated with other functionalities such as profiles, cashiering, and deposits. This property management software module provides a complete set of features for creating and updating individual, group, and business reservations, including deposit handling, cancellations, confirmations, wait-listing, room blocking, and sharing.

Rate management   This system has an extensive set of features for setting and automatically controlling rates, rate quotation, and revenue forecasting and analysis to create the most comprehensive rate management system in the industry. OPERA's PMS interfaces with OPERA Revenue Management Systems and other major yield management applications.

Profiles   Complete demographic records of guests, business accounts, contacts, groups, agents, and sources can be stored. Profiles include addresses, phone numbers, membership enrolments, stay and revenue details, guest preferences, and additional data that make reservations handling and many other activities faster and more accurate.

Front desk   This OPERA system handles individual guests, groups, and walk-ins and has features for room blocking, managing guest messages and wake-up calls, and creating and following up on inter-department advisories.

Back-office interface   Revenue transfers, market statistics transfers, daily statistics transfers, and city ledger transfers can be easily carried out from the OPERA PMS to a back-office system.

Rooms management   All facets of room supervision including availability, housekeeping, and maintenance and facility management, can be handled. The 'queue rooms' feature of the property management software coordinates front office and housekeeping efforts when guests are waiting for rooms, which are not immediately available for assignment.

Cashiering   Posting guest and non-guest charges (including taxes and other generates), making posting adjustments, managing advance deposits, settlements, checkout, and folio printing are a few of the many activities handled by OPERA cashiering. Cashiering accommodates multiple payment methods per reservation including cash, cheque, credit cards, and direct bills. In multi-property environments, guest charges can be cross-posted from any property in the hotel complex.

Accounts receivable   Fully integrated with the OPERA PMS database, it includes direct billing, invoicing, account aging, bill payments, reminder and statement generation, and account research. Old balances from external accounting systems may be entered.

Commissions   This system calculates, processes, and follows up on travel agents and other types of commission payments, either by cheque or via electronic funds transfer (EFT).

Reporting   Over 360 separate standard reports can be prepared. Reports can be customized for each hotel and new reports may be created as per need using OPERA's in-built report writer.

Fully configurable   The choice of OPERA features, system behaviours and priorities, and system-wide defaults are controlled by the property. User permissions determine which property management software features may be accessed by each user and user group. Many OPERA screens may be customized by the property.

Global perspective   This system supports multi-currency and multi-language features to meet the requirements of global operations. Rates and revenues can be dynamically converted from the local currency to any other currency. The appropriate language for guest correspondence can be automatically determined from the guest's profile language; country-specific address formats are also supported.

Hospitality system interfaces   OPERA PMS includes interfaces to hundreds of third-party hospitality systems including yield management, telephone and electronic switching, TV and video entertainment, key lock, restaurant POS, activities scheduling, minibar, and wake-up call systems.

OPERA Xpress

OPERA Xpress offers a scaled-down edition of their property management systems for smaller properties or properties offering limited services. Based on the core OPERA property management software products, properties may choose the features they want from a menu of product options.

Key Features

The following are some of the prominent features of OPERA Xpress:

- Resides on a robust Oracle platform
- Uses one system across multiple properties
- Configures screens according to business processes, reducing costs by streamlining workflow
- Facilitates check-in with a single swipe of a credit card
- Manages guest messages efficiently and on time
- Simplifies night audit, eliminating down time
- Contains a user dashboard for simple navigation
- Flexibility for branding

Key Benefits

The following are some of the prominent benefits of OPERA Xpress:

- Tailored to fit a hotel's operational business needs
- Scalable to suit the size of the hotel
- Helps to become more productive, profitable, and professional
- Delivers fast, accurate, and online information about the property

OPERA Lite

MICROS Systems, Inc. offers products with a pre-configured solution for a hotel with simple technology needs. A majority of limited service, limited facility hotels prefer to not complicate a guest's experience with unnecessary features in their software. What these hotels need is an economical, timely solution with a straight-forward training approach and easy to understand concepts.

Key Features

The following are some of the prominent features of OPERA Lite:

- A dashboard that serves as a central navigation point for all basic functions such as reservations, check-in, and checkout
- Simplified accounts receivable (optional) quick business blocks and family groups (optional)
- Charge routing between folios
- Tax-exempt guests
- Guest message functionality
- Credit card and call accounting interfaces
- Simple base rates for easy rate management
- Multiple payment methods
- Accompanying guests
- Add-on reservations for quick duplicate reservations
- Provides alerts and a front desk logbook
- Tracks basic housekeeping and out of order rooms
- Clean reports, colourful screens, and easy to navigate menus

MICROS offers fully integrated POS solutions for restaurant clientele across the globe whether it is a stand-alone store or a large chain with hundreds of locations. The POS systems provided by MICROS are not just a cash register, but a whole enterprise solution with options for front-of-house management, back-office applications, restaurant and enterprise operations, and provides the best possible customer experience.

Tools have been created and are continuously being developed to cater towards the changing needs of MICROS' customers. With easy upgrade options and scalable solutions, MICROS provides products that are perfect for shifting an individual restaurant to a central enterprise. Quotations are provided by an experienced MICROS specialist after a detailed consultation including an overview of the potential customer's business needs, operations, and logistics.

MICROS RES: Back-office, Guest Services, and Restaurant POS Software

MICROS RES is a complete restaurant POS software solution that adapts to the way a hotel runs its business, with tools for back office, restaurant operations, and guest services. This fully-integrated restaurant POS system is both powerful and flexible and helps operations within the property flow more cohesively, efficiently, and profitably. The MICROS RES POS system offers restaurants multiple solutions that are designed specifically for table and quick service operations, as well as fast casual concepts.

This comprehensive restaurant POS system also allows restaurants to manage complex employee environments, monitor inventory levels, streamline ordering and receiving processes, enhance the profitability of operation with customizable reporting and analysis tools, compare operational conditions against established standards, create wait-lists and reservations, and implement gift card and loyalty programmes for guests.

Complete Solution for Restaurant POS System Needs

For owners of multiple restaurant locations, this restaurant POS software offers enterprise management. MICROS RES provides end-to-end communications throughout operations to increase efficiency, speed service, and ultimately provide a better guest experience for customers. The end result of the MICROS RES restaurant POS system is to increase customer satisfaction, maximizing their experiences at the restaurant.

When security, data integrity, easy software deployment, and real-time business intelligence top the list of requirements, MICROS RES restaurant POS software is the solution.

MICROS 3700 POS

The ruggedized and powerful MICROS 3700 POS system is designed for optimized speed of service and enhanced guest experiences. Beyond all standard POS functions, MICROS 3700 POS offers a vast range of functionalities for process optimization, increased level of service, and guest satisfaction.

Its intuitive user interface leads even untrained staff through the selling process, driving up revenue by increasing the average guest ticket (by suggesting suitable promotional offers).

The standard configuration includes the following functions:

- Case management
- Table management
- Scheduled price management
- Scheduled (de-)activation of articles
- Freely configurable automatic discounts (e.g., on quantity, price, a combination of quantity and price, articles, and article groups)
- Connection to drink dispensers
- Free definition of article groups/report groups
- Simple configuration of articles (SLU groups)
- Freely configurable tracking down to article level for highly detailed controlling
- Create menus including automatic information display of ingredients
- Waste module

- Conversational ordering module (COM), an innovative module that has been developed with the aid of psychological research. The ordering process at the POS is completely aligned to the ordering process of the customer. COM allows hotel staff to easily change menu items without going through time-consuming cancellations. COM displays alternatives or menu items that have not been ordered as yet.
- Smart keys for clearly laid out displays
- Integrated system control for cancellations
- Comprehensive reporting (more than 150 default reports)
- Hand-held support with full POS functionality
- Control of digital menu boards

Optional functionalities   The following are some of the other functions of MICROS 3700 POS:

- Control of kitchen display systems (KDS) to optimize the order and food preparation processes
- Time and attendance
- Alert managers to inform employees regarding events in the restaurant

Kitchen Display System

The MICROS KDS is like having a second expeditor in your kitchen. It provides highly visible, real-time information to manage and control kitchen efficiency, thus driving customer satisfaction. Mounted conveniently in the kitchen or food preparation area, this seamlessly integrated, intuitive, graphical software application displays food orders for preparation and monitors the timing of orders to check 'speed of service'. MICROS KDS also provides feedback about the status of each table and captures service times for management reporting.

Fully integrated with the MICROS 3700 POS system, KDS runs on standard PC hardware using colour touch screen monitors or colour monitors and bump bars. By managing food preparation, KDS provides a higher level of management control and customer service.

Key Features

The following are some of the prominent features of KDS:

- Timely order preparation
- Highlights orders in yellow or red to indicate an order that has exceeded expected preparation time
- Displays each order in either list mode or chit mode and monitors time to prepare
- Allows users to define preparation times for both appetizers and entrees
- Intuitive icons display rush orders, VIP, and void statuses
- Displays features, such as all day, order done, and order recall, to make information readily available
- Displays speed of service

- Helps view the status of each table in the restaurant at a glance
- Change in the colour of table buttons as an indication
- Records table vacancy
- Records guest seating
- Keeps track of the kitchen working on the order
- Entree that has been served
- Order that is late
- Reporting and statistics
- Captures service times for different courses at the various preparation stations
- Generates real-time reports on kitchen performance

Guest Service Solution

As a module of the complete MICROS Restaurant Enterprise Solution (RES), the Guest Services Solution (GSS) provides restaurants with a single-source, all-in-one guest marketing system that helps a hotel build loyal and repeat customers. Implementing a frequent-diner programme and tracking guest preferences help build a loyal following that comes back more often. Gift certificates or gift cards can be sold and redemption can be easily tracked. MICROS GSS can be used to bring customers in during off-hours by establishing promotions at certain times of the day or certain days of the week.

GSS is seamlessly integrated into the MICROS 3700 POS system and all other modules of the RES suite.

Key Benefits

The following are some of the prominent advantages of this solution:

- Build a guest information database of loyal customers
- Set up multiple promotional offers to keep guests coming back
- Establish times of the day or day of the week promotions to fill 'softer hour' seats with guests
- Create marketing programs in-house
- Import guest information from most other programs
- Configure POS printers for coupon printing

This solution also offers an optional delivery and carry-out solution to streamline the ordering process for take-away businesses as listed here.

- Print addresses and delivery information on the POS guest cheque
- Store order details
- Retrieve delivery information for repeat customers
- Enter additional 'notes'
- Recall the previous order
- Target delivery time
- Caller ID functionality

Table Management Solution

MICROS Table Management is a simple, easy-to-use software that seamlessly integrates customer preferences, seating capacity, and available staff, while effortlessly managing the customer's dining experience. Capturing time-sensitive guest demands, MICROS Table Management puts the hotel staff in complete control from the moment a guest is greeted until the next diner is seated.

Customer Seating

The following are some of the solutions offered:

Wait-list capabilities

- Adds a customer to the wait list manually or through an integrated MICROS GSS database
- Provides and records estimated seating times
- Records preferences for table requests such as window view or handicap accessibility
- Manages wait-time based on preconfigured table turn times or course timing provided by an integrated MICROS KDS, or on customer preference
- Provides customer-viewable waiting list

Customer management

- Adds customer information in customer database
- Links customer name to guest cheque when seated
- Pages customer when table is ready via the JTECH GuestAlert Pager

Table Management

The following are some of the solutions offered:

- Pages server when customer is seated via JTECH ServAlert Pager
- Supports multiple table layouts
- Facilitates large party management (combine tables)
- Manages next available table based on server sections and wait times

Reservations

The following are some of the solutions offered:

- Manages table inventory by time period
- Creates guest records in the customer database when a reservation is made
- Attaches special requests to a reservation
- Integrates API for third-party web reservations
- Provides user-friendly interface

Reporting

The following are some of the solutions offered:

- Waiting times for customers
- Abandonment rate, time duration between seating and greeting by a waiter, and order time

- Combined with KDS for end-to-end guest experience reporting
  - Captures greet time and promise time
  - Time gap between bill settlement and re-laying table
- Roll-up to mymicros.net
- Table reporting (determine optimum table seating, if more tables of a certain size are necessary)

Fig. 6.1   Example of a digital signboard

Source: http://www.micros.com/Solutions/ProductsAM/DigitalSignageAndMenuBoards/

One of the latest tools to emerge in the quick-service market is the digital menu board. A fully integrated feature of the MICROS POS system, this novel technology displays menus while simultaneously drawing customers' attention to other information like current store promotions. While the quick-service market becomes increasingly aware of the benefits of digital versus traditional menu boards, restaurant owners seek a solid solution that will improve operations and boost customer experiences. Typically sold as an added module or an interfaced product, MICROS includes the digital menu board functionality as a core feature in its POS solutions. So when a hotel is ready to deploy this innovative technology for business, MICROS is ready with digital signage and menu boards. Figure 6.1 shows a digital signboard.

IDS

Established in 1987, IDS Softwares Pvt. Ltd is Asia's largest dedicated provider of integrated, full-service enterprise property management software for the hospitality and leisure industry. Powered by 24-plus years of experience and enabled by a vibrant mix of domain experts from the hospitality and technology spheres, IDS designs, develops, markets, and maintains a comprehensive range of information management systems for various hospitality businesses including hotels, restaurants, clubs, and resorts.

IDS software is an ISO 9001:2008 certified company that has established its presence in 40 countries worldwide. IDS has a PMS with back office, one of which is Fortune NEXT. Figure 6.2 shows a flow diagram of Fortune NEXT.

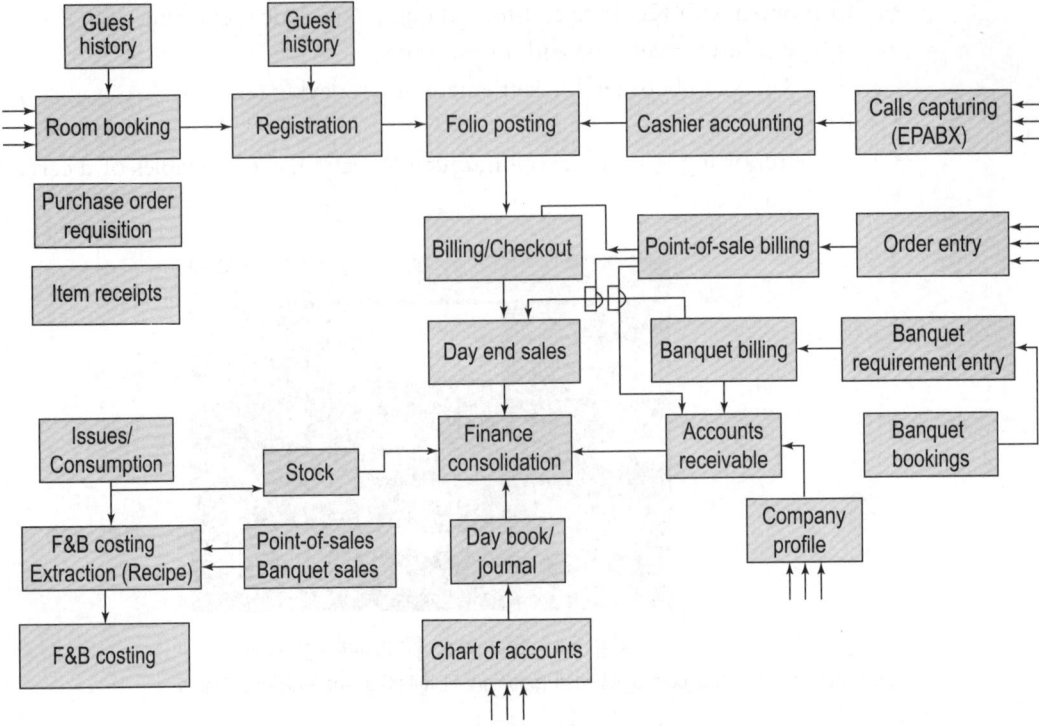

Fig. 6.2   Fortune NEXT flow chart

Source: http://idsnext.com/

The opening screenshot of the Fortune NEXT main screen is shown in Fig. 6.3.

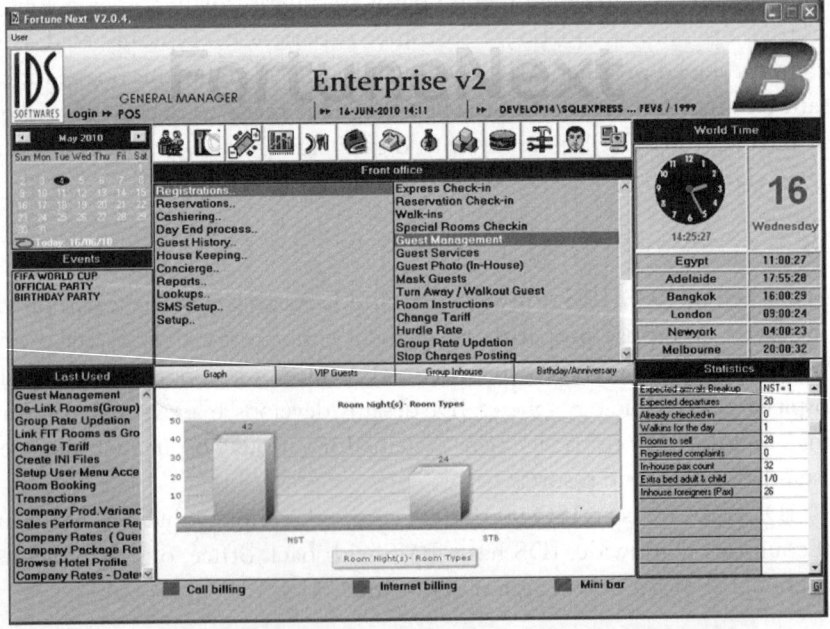

Fig. 6.3   Screenshot of Fortune NEXT (*Courtesy*: IDS)

The modules and sub-modules of Fortune NEXT V2 are shown in Table 6.1.

Table 6.1   Modules and sub-modules of Fortune Next V2

Modules	Sub-modules	Modules	Sub-modules
Front-office management	Reservation; Registration; Group management; Guest management; Billing or cashiering; Day end process`; Quick look ups; Demand calendar; Housekeeping; Concierge; Reports	Point-of-sale	Multi-level menu definition; Table layout; Table reservation; Order generation and billing; Advance order and billing; Delivery management; Kitchen display system/ KOT printing; Online table status; Session statistics; Sales promotions; Happy hours
Account receivable	Manual sales receipt and journal entries; Payment follow-up console; Bill matching for ad hoc receipts; Online account balance; Agent commission; Ageing report; Invoice and reminder printing; Standard debtors statement; Look ups and MIS	Sales and marketing	Manage company travel agent rates; Defining company-wise discount policy; Daily sales calls; Entertainment entry; Competitors–Occupancy entry; Business lost entry; Exhaustive company profile; Allocation, forecasts, and budgets; Executive planners; Sales executive performance; Watch list companies
Banquets	Banquet booking; Requirement entry/FP; Advance receipts; Banquet billing; Event calendar; Block/ release room; MIS/Reports	Telephone management	Call accounting; Wake up calls; Minibar billing; Clear rooms; Message display; Auto check-in/checkout; Operator console
F&B costing	Automatic sales and consumption transfer; Manual sales and consumption entries; Costing methodologies; Recipe and sub-recipe definition; Kitchen stock entry; Inter-cost and inter-kitchen transfer; Sales and cost budget; Kitchen/bar stock analysis; Standard vs Actual reports; Profitability analysis; Cost report consumption; Cost report recipe; Missing recipe	HR and payroll	All statutory reports; PF/ESI challans; Flexible user definable computation; Leave management; Bonus management; Loan management; Interface with IDS fortune; Personnel database available; Recruitment process; Interface with attendance recording devices; User tool to design own reports; Multi-unit/Multi-department processing with different payroll execution period
Material management	Multi-store/Sub-store accounting; Indent management; Requisition management; Auto requisition based on minimum stock; Purchase order/ Standing purchase order; Multiple authorization levels; Receipts (direct purchase order/cash); Physical stock update; Stock variance report; ABC/ FSN analysis; VAT compliant; Cost centre wise consumption; Priced	Maintenance management	Equipment masters; Preventive maintenance scheduling; Complaint registering; Job order generation; Job order printing; Equipment reading; Maintenance staff roistering; Spares and cost analysis; Action taken; Complaint register; Complaint status report; Cost spares report; Resolution time analysis; Duty chart; Employee-wise action taken report; Interface with energy management system

Table 6.1   *(Contd)*

Modules	Sub-modules	Modules	Sub-modules
	stock status; Inter-store transfer; Vendor analysis; Quotation requesting; Quotation updating		
Financial management	Uniform system of accounting; Departmental cost centre wise P&L; Integrated with all modules; Multi-currency handling; Provisional entries posting; Bank reconciliation; Cash flow statements; Budgeting; Voucher printing; Detailed vendor master; Payment advice generation; Provisional PJV booking; Cheque printing; TDS system; Interactive payment match; PDC transaction; Expense allocation; Voucher authorization; Audit reports; Payable outstanding; Drill down option up to transaction entry level; User define tool for designing own report; Audit block	Fortune care	Call centre concepts; Guest feedback; Improve guest service
SMS iAlert			

SHAWMAN

It started with the vision of a man, the Late Pesi M. Shaw, who identified the need to develop end-to-end software that could manage his business, a growing chain of leading hotels in India. Till 1997, the software remained more of a passion and a personal hobby of the Shaws and was neither commercially marketed nor sold and remained in use at more than a dozen hotels, restaurants, and hospitality institutes, at a premier standard marketed and serviced by a dedicated seven-member team.

The ShawMan brand name got large-scale recognition in the late nineties for its decade-long spirit in sustaining cutting-edge developmental efforts, and is today considered a leader in providing hospitality solutions in India and overseas.

Key Features of ShawMan PMS

The following are some of the prominent features of ShawMan PMS:

- PMS reservations are linked to web-based central reservation system with payment gateway feature.
- Hot key helps in quick access to information.
- The desktop can be customized individually for every user.
- Room blocking feature is available for smooth billing and housekeeping operations.

- In guest billing folio, posting instructions can be set at the time of reservation.
- A built-in utility enables external programs to be automatically called before the day ends. Users can predetermine which reports need to be generated by the end of the day.
- PMS has been developed in three-tier architecture with defined business rules that can be easily distributed with the support of intelligent patch control logic. It also has inbuilt system maintenance tools for repairing databases and purging audit.

Key Features of ShawMan POS

The following are some of the prominent features of ShawMan POS:

- Enables control of remotely located outlets from a central head office with rule-based set ups, process definition, and consolidation of data, which is automated at the end of the day
- Supports multi-currency
- *A la carte* orders can be placed
- Choice of different style sheets for F&B operations such as fine dining, regular, counter sale, around-the-clock, and room service as well as non F&B operations such as direct biller and leisure activities
- Provision for franchisor–franchisee style of operation
- Can constantly update the chef's speciality depending on seasonal availability and non-availability of items, happy hour, and on the house specials
- Descriptive free flow text/coded modifiers with or without a charge
- Supports multiple F&B and non F&B outlets at hotels, clubs, and service apartments
- Automatic prompts to communicate requests and parcel orders with the kitchen
- Multiple settlement types possible
- Reservation diary for guest bookings
- Remote printing of KOTs to 20 pre-defined distant locations per outlet
- Member entry management system using magnetic cards
- Debit card system
- Canteen management system
- Supports smart cards for CRM applications
- PortaPOS, the pocket PC application for complete Wi-Fi operations
- Digidiner gives a virtual view of the restaurant
- Kitchen display system (KDS)
- Supports touch screen as well as keyboard operations

ShawMan's contributions to the hospitality and food service industry in India over the years include the following:

- Intelligent remote printing of multiple orders in remote food processing zones
- Integrated software kitchen display system (KDS)
- Touch screen-based POS also compatible with keyboard
- Only freeware comprehensive PMS software
- Integrated debit card solution
- Pocket PC based applications using Wi-Fi networks

- Introducing magnetic smart card and RFID technology
- Integration to EDC terminals to transport secure data over banking networks
- Interfaced to banks for instant credit card authorization from within billing systems
- Integrated SMS gateways using CDMA and GSM technology
- Integrated data warehouse solutions for the food service, hospitality, and retail industry

SUMMARY

The chapter talks about the various departments in a hotel, their requirements, and importance. We also discussed revenue generating and non-revenue generating departments of a hotel.

A hotel requires a property management system (PMS) for smooth and easy operations, and fast retrieval of reports and information. The chapter also focuses on PMSs such as Micros, IDS Next, and Shawman with their features.

KEY TERMS

Call accounting system   It is a system that enables a hotel to take control over local and long-distance telephone services. A call accounting system can place and price outgoing calls.

Customer relationship management (CRM)   It is an information industry term for methodologies, software, and usually Internet capabilities that help an enterprise manage customer relationships in an organized way.

Electronic data capture (EDC) terminal   It is a point-of-sale device that reads information encoded in the bank card's magnetic strips, performs authorization functions, and transmits data for processing.

Electronic locking system   It is a locking system that uses plastic keys instead of metal keys for guests to open and close their rooms.

Energy management system   It is a system that helps to minimize power costs of a hotel without affecting the comfort of guests or employees.

Guest accounting module   This module gives the management considerable control over the financial aspects of a hotel. This front-office module is responsible for making postings online, automatically update (audit) and maintain files, and display/print folios on demand.

Human resource information system (HRIS)   It is a method by which an organization collects, analyses, and prepares reports about people and their roles.

Kitchen display system (KDS)   It is a highly visible, real-time information system to manage and control a kitchen's efficiency.

Non-revenue generating departments   It refers to departments in a hotel that operate without earning any revenue for the hotel.

Point-of-sale system   A point-of-sale system with a network of electronic cash registers is capable of capturing data and transferring them to the system's guest accounting and financial tracking modules. The ability to communicate such data to both front and back-office components can result in numerous benefits derived through comprehensive reporting.

Property management system   It is a system which manages guest interactions and at the same time acts as an information 'hub' for other computer systems.

Reservations module   This module enables a hotel to rapidly process room requests and generate timely and accurate rooms, revenue, and forecasting reports.

Revenue generating departments   These departments of a hotel help it to earn revenue.

Revenue management module   Revenue management is a set of demand-forecasting techniques that are used to determine if room rates should be raised or lowered and whether a reservation request should be accepted or rejected in order to maximize revenue.

Room management module   This module maintains up-to-date information regarding the status of rooms, assists in assignment of rooms during registration, and helps coordinate guest services.

Yield management   This technique involves planning to achieve maximum room rates and attract the most profitable guests.

REFERENCES

Bardi, J.A., *Hotel Front Office Management*, Second edition, John Wiley & Sons, Canada, 1996.

Bhakta, Anutosh, *The Professional Front Office Management*, B. Sukla, Kolkata, 2006.

Bhatnagar, S.K., *Front Office Management*, Second edition, Frank Bros & Co., New Delhi, 2008.

Hayes, D.K., J.D Ninnemeier, and A.A. Miller, *Foundations of Lodging Management*, Second edition, Pearson Education, New Delhi, 2012.

Lillicrap, Dennis, and John Cousins, *Food & Beverage Service*, Seventh edition, Book Power, London, 2006.

Web References

http://2.bp.blogspot.com/-Cr-FiRuSKn0/Tjt-OTrzWoI/AAAAA AAAEPY/4ARUcztkucM/s1600/Housekeeping-could-you-do-it.jpg&ir, last accessed on 23 March 2012.

http://idsnext.com/products/by-industry/hotels/, last accessed on 14 April 2012.

http://magicsweepsu.blogspot.com/2011/06/what-happen-between-housekeeping-and.html&docid, last accessed on 23 March 2012.

http://searchcrm.techtarget.com/definition/CRM, last accessed on 11 April 2012.

http://www.hospitality-school.com/must-follow-food-beverage-service-rules&docid=vtAtVvdyhFHhpM&imgurl1, last accessed on 23 March 2012.

http://www.hospitality-school.com/wp-content/uploads/2009/12/food-beverage-production-cooking.jpg, last accessed on 23 March 2012.

http://www.micros.com/AboutUs/CompanyProfile/Default.htm, last accessed on 14 April 2012.

http://www.micros.com/Products/HotelSolutions/CentralSystems/OPERAReservationSystem, last accessed on 14 April 2012.

http://www.micros.com/Solutions/ProductsAM/3700POS/, last accessed on 14 April 2012.

http://www.micros.com/Solutions/ProductsAM/Digital SignageAndMenuBoards/, last accessed on 14 April 2012.

http://www.micros.com/Solutions/ProductsAM/GuestService SolutionGSS/, last accessed on 14 April 2012.

http://www.micros.com/Solutions/ProductsAM/KitchenDisplay Systems/, last accessed on 14 April 2012.

http://www.micros.com/Solutions/ProductsNZ/OPERALite/, last accessed on 14 April 2012.

http://www.micros.com/Solutions/ProductsNZ/OPERAXpress/, last accessed on 14 April 2012.

http://www.micros.com/Solutions/ProductsNZ/Restaurant EnterpriseSeriesRES/, last accessed on 14 April 2012.

http://www.micros.com/Solutions/ProductsNZ/Simphony/, last accessed on 14 April 2012.

http://www.micros.com/Solutions/ProductsNZ/TableManagement Solution/, last accessed on 14 April 2012.

http://www.micros.com/Solutions/RestaurantsAndFoodService/, last accessed on 14 April 2012.

http://www.sathiyams.com/about.html&docid, last accessed on 23 March 2012.

http://www.scribd.com/doc/29919138/Hotel-Property-Management-Systems, last accessed on 4 October 2011.

http://www.shawmansoftware.com/Hospitality_PMS.htm, last accessed on 12 April 2012.

http://www.shawmansoftware.com/Hospitality_POS.htm, last accessed on 12 April 2012.

https://www.directpos.com/documents/brochures/micros-kitchen-display-system.pdf, last accessed on 12 April 2012.

www.merchantglossary-com/page_1d=83, last accessed on 11 April 2012.

EXERCISES

Concept Review Questions

1. What are the most common front-office components of a property management system?
2. What are the various revenue generating departments of a hotel?
3. How has computerization helped both revenue and non-revenue generating departments of a hotel?
4. What are the computer systems that can interface with a hotel's property management system (PMS)?
5. What is a PMS?
6. List some PMSs with their features.
7. How does the yield management software help in improving revenue generation for a hotel?
8. Property management systems in all departments help in comparing sales with budgeted sales. Justify your answer.

Multiple Choice Questions

1. The front-office staff members assist in
   - (a) reservation
   - (b) security
   - (c) housekeeping
   - (d) none of these
2. The housekeeping department is responsible for cleaning of the
   - (a) public area
   - (b) laundry
   - (c) rooms
   - (d) all of these
3. The food production department's role is to plan menus with the focus on
   - (a) guest needs
   - (b) management needs
   - (c) owner's needs
   - (d) none of these

4. Which of the following is a revenue generating department?
   (a) F&B   (c) Safety and security
   (b) Sales and marketing   (d) Accounts
5. The maintenance department of a hotel is responsible for
   (a) upkeep of the hotel
   (b) preventive maintenance
   (c) hotel equipment
   (d) all of these
6. Staff induction is taken care by
   (a) Housekeeping
   (b) Sales and marketing

   (c) Human resource department
   (d) Security and safety
7. Property management system (PMS) is
   (a) a set of computer programs
   (b) an interfaced and integrated system
   (c) hardware
   (d) none of these
8. Which of the following is a famous PMS available for hotel and restaurants?
   (a) Micros   (c) ShawMan
   (b) IDS Fortune   (d) All of these

Project Work

1. You have attended a seminar on property management system and its uses. What points will you consider when making a presentation to your general manager?

CASE STUDY 1

Customer Service and Productivity Gains for Theme Dining Pioneers

Since pioneering the concept of themed bar–diners in New York in 1965, TGI Friday's has been associated with a string of industry-leading innovations in menu development (from 'happy hours' to Long Island iced tea), customer service, and staff training. Its iconic red and white stripes rank among the world's most recognized restaurant trademarks and, with 1000 restaurant in 55 countries, the chain continues to set the pace in casual dining.

The international business has tripled in size since 1997 and a system-wide revitalization programme initiated by brand owners Carlson Hospitality Worldwide in 2004 has helped refocus the TGIF experience. This has gone hand-in-hand with significant improvement in back-of-the-house efficiency and customer service speed, aided by the latest MICROS POS restaurant management technology.

The 45-restaurant UK estate operated under Whitbread Group's license, is the world's largest TGIF business outside North America and has applied a process of continuous innovation since it started in the mid-1980s. In the past two years, the touch screen POS terminals used by the chain since 1994 have been upgraded to the state-of-the-art 3700 system as part of the comprehensive Restaurant Enterprise Series management solution.

Key changes include a switch to hand-held Mobile MICROS server terminals for tableside ordering. The new hand-held terminals communicate directly with a highly functional kitchen display system(KDS). The KDS solution breaks down each order and prioritizes preparation tasks based on the dining course and ingredients' cooking times. This high powered combination of new front-of-house and back-of-house technology has had a major impact on productivity and customer service, helping to reduce service delays, improving meal quality, and increasing table turnover.

The Mobile MICROS hand-held terminals are built for harsh food service environments with rugged, spill-proof, and drop resistant specifications. However, changing over was a major decision. While TGIF UK has in the past year reduced its total number of menu items from 134 to 74, menu ordering is still a potentially complex process, both in the range of dishes and in the way customers can modify choices to tailor meals exactly to their taste. With the tight dimensions of a hand-held terminal, this can pose problems. However, this problem is not faced in Mobile MICROS units, thanks to their easy-to-navigate full colour touch screens.

Staff guest interaction and the 'fun' tableside manner of the service staff is very much part of the TGIF dining experience and it was vital that the hand-held technology did not interfere with that. According to James Jackson, Outlet Systems Director, Whitbread groups PLC, there have been no problems at all in accommodating the menu and the hand-held has enhanced the order-taking process. 'It's like having a full MICROS keyboard on a hand-held,' he comments.

Loss of valuable eye contact between the server and guest has not been an issue, either. 'In training staff tend to look down rather than up, with either a pad or hand-held,' James comments but once trained, they find it quicker to enter orders on hand-helds because they can use short cuts. Writing orders down is slower and more open to mistakes.'

The hand-helds also make it easier for staff to repeat orders back to the guest tableside, which provides valuable up-selling opportunities, such as prompting guests to upgrade to ultimate versions and add-on side items. As for staff-customer interaction, the hand-helds can make a positive contribution. 'People are fascinated by gadgets and tend to be curious about the hand-helds, which works in our favour. Basically, it's a win-win situation.'

Was there any initial resistance in switching staff to the new technology? 'It was basically cultural issue,' James observes. 'Generally, young people today are happier using electronic devices than writing their own names but people also need to get used to things. The longer an employee had worked for TGIF, the more attached they were to pen and paper. But if you went into any of our stores now, those same employees would be very resistant about giving back their hand-helds.'

The hand-helds communicate directly via a wireless network to MICROS KDS screens in the kitchen. This link enables fully defined orders to reach line chefs the moment guests place them. MICROS software provides dynamic ordering screen flow, with special MICROS software to breakdown each order to its component parts and feeding this information in a prioritized fashion to relevant members of the kitchen team.

One KDS unit is allocated to each of the four cook stations in each TGIF kitchen. 'Basically each chef receives a cook list arranged by how long each item takes to cook,' comments James. One result is that satisfaction scores are 5%, as recorded by mystery diners, thanks, in a large part to improved meal quality. Complaints about food not being hot enough have been reduced by 70%.

The combination of hand-held terminals and KDS screens in the kitchen has contributed to productivity gains in both front-of-house and back-of-house ends. A window-man is no longer needed to deal with orders entering the kitchen and there has been a total savings

of up to four employees per store. In a comparison with TGIF with a similar volume in the USA (where KDS are employed but without a switch to hand-held order terminals), the UK store outperformed the USA store by 350 man hours per week.

'We found that by upgrading our POS system to a combination of hand-held and KDS, we have slashed at least 20 minutes off the total guest experience time,' James observes. That means less of a wait between starter and the main course and greater flexibility for the guests in managing their time. In time trials at one TGIF branch, average time between a guest entering and leaving dropped from 105 minutes to 72 minutes.

Speeding the total experience has three important outcomes:

- The customers get the freshest possible meals hot from the kitchen.
- The total order execution time is significantly accelerated.
- Seat availability increases significantly by up to 50% during busy times.

While the TGIF meal experience epitomizes casual dining, the company has found that speed is not incompatible with a relaxed meal experience. Flexibility is more of an issue with today's generation of customers. In a recent interview in the international trade magazine Food Service Europe and Middle East, Richard Snead, President–CEO of Carlson Hospitality Worldwide, commented: 'People want to relax and be taken care of. However, they also want to be in control of their time, which may sound like contradiction. When guests are in hurry the restaurant staff needs to be able to establish exactly how much time the guest has and adapts the service accordingly.'

'Technology for instant transmission of orders to the kitchen plus scheduling system which achieve best utilization of kitchen equipment, in line with order flow, can contribute to this process,' he said.

TGIF UK managers also have password-controlled access to mymicros.net, a content-rich Internet portal, which enables them to monitor real-time reports on branch sales, promotions, and overhead costs from a web browser anywhere in the world, at any time. This enables them to react immediately to any emerging issues, develop customized reports, and generally be more proactive in their day-to-day activities. James Jackson is enthusiastic about the benefits: 'You get the information more quickly and more accurately.'

The system also has inbuilt functionality for highly strategic activities like data warehousing, enabling management to drill down through huge amounts of data to make strategic decisions about all aspects of business performance, from staff utilization and costs to customer loyalty programmes.

The MICROS 3700 electronic POS system with the hand-held interface gives TGIF UK management up-to-the-minute access to the company, with data, including sales performance and staff costs. This helps managers run their operations more effectively and has helped eliminate many paper-based processes. This has been particularly relevant to menu planning. TGIF is notable for its highly responsive approach, adding new items quickly in a strategic reaction to new opportunities and pruning under-performers whenever necessary.

The MICROS Enterprise Management (EM) solution used in conjunction with the 3700 system distributes new menus, price adjustments, special offers, and other changes to all restaurants for immediate introduction or at a predetermined time. This powerful tool not only ensures chain-wide database consistency, but also eliminates the labour intensive effort of manually updating store databases one at a time.

What about return on investment? Transmitting orders straight to the kitchen from the table side as opposed to having to walk to a terminal to put in an order has been a big time saver. 'With the upgrade to the MICROS 3700 with hand-helds, plus the added benefit of being able to access mymicros.net, the activity story has just gotten better and better.'

James comments, 'The new system costs less to maintain because it is more robust and also gives us a lot more flexibility, both now and in the longer term. We are not yet using all of the built-in functionality, such as Chip and Pin (to be introduced soon) and data warehousing, but when we do, that will also contribute to return on investment.'

TGIF's association with MICROS and its restaurant system dates back to 1994 when the MICROS 2700 system was introduced across all UK stores operated by Whitbread, to help the chain achieve higher levels of operational control. The recent upgrade to the 3700 system with hand-held technology followed detailed testing. Key factors included the more flexible server interface, enabling speedier retraining for managers and staff, as well as more effective control of IT costs.

The Whitbread Group has been a major user of the MICROS systems for the past decade at its Beefeater, Brewers Fayre, Brewsters pub–restaurant chains, TGIF, Pizza Hut chains, and also at its Marriott hotels. MICROS Systems, Inc. provides enterprise applications for hospitality and retail industries worldwide. Over 2,20,000 MICROS systems are currently installed in table and quick service restaurants, hotels, motels, casinos, leisure and entertainment, and retail operations in more than 130 countries, and on all seven continents.

Source: http://www.micros.com/NR/rdonlyres/E387 A766-6CED-4D8D-8FC4-1C172F8400B3/0/TGIFridays.pdf

(a) What is KDS?
(b) What is Mobile MICROS? How has it changed the overall employee experience in TGIF?
(c) Why didn't waiters opt for MICROS hand-held devices for taking orders over the conventional pen and paper?
(d) How has the overall guest experience changed after using KDS and MICROS hand-held device?

CASE STUDY 2

Green Park Group of Hotels Leverage iAlert to Improve Internal Communication

IDS NEXT assisted Green Park Hotels deliver enhanced guest experiences.

Customer Brief

Green Park Group of Hotels is a successful four-star hotel chain in South India with three four-star properties at Hyderabad, Vishakhapatnam, and Chennai. In operation since 1991, the hotel chain endeavours to achieve positive guest experience by offering world-class services and comforts for business travellers.

Business Challenge

Green Park Group of Hotels is in the process of expanding operations across other cities of South India and one of the challenges they face is the time taken to work through all the data, which had been gathered as reports.

Another challenge faced by employees was that email communication would sometimes get a delayed response as staff were constantly on the move. A more feasible communications tool was required to ensure that information reached recipients in a timely manner.

Solution

Fortune Enterprise is IDS NEXT's all-in-one solution for large-scale hotels and its various facets including iAlert proved to be a good fit for Green Park Group of Hotels. While the hotels previously used a Unix Fortune Enterprise, the change to Fortune NEXT Enterprise helped them automate every facet of their business through user-friendly interfaces.

iAlert, Bridges Internal Communication Gap

iAlert, a module in Fortune NEXT Enterprise helps create cohesive internal communication through SMS alerts that are generated for various operational duties, ongoing hotel activities, and other in-house changes. It enables operational heads to manage necessary activities, even when they are on the move, offering them the advantage of mobility.

The following are the key modules of IDS Fortune NEXT Enterprise deployed at Green Park Group of Hotels:

- Front-office management
- Accounts receivable
- Banquets
- Material management
- Financial management
- Maintenance management
- SMS alert
- Point of sale
- Sales and marketing
- Telephone management
- F&B costing
- HR and payroll
- Quality management

Cross-linking of hotel related activity across all modules of the software, enabled management at Green Park Group of Hotels access a comprehensive view of all ongoing hotel activities. Since the introduction of Fortune NEXT Enterprise, all operations have moved online, which has helped the hotel increase operational efficiency and avoid redundancy.

Business impact With mobility no longer an issue, managers at Green Park Group of Hotels could effectively dedicate their time towards enhancing customer satisfaction.

Benefits of iAlert Alerts are sent to guests to confirm booking or notify amendments, and to managers reagrding special guest requests. Information about events (e.g., birthday alerts) can also be sent. iAlert informs the financial department on high-billing rooms and the need to obtain approvals or advances from companies before a guest checks out.

In addition to these, Green Park Group of Hotels innovatively implemented iAlerts to increase customer satisfaction and drive their vision. iAlert SMSes were used to notify front-office managers or housekeepers about VIP arrivals and departures, thus enabling the manager to be present at the reception to personally greet the guests. This human connect was crucial to ensuring guest satisfaction.

'IDS NEXT's expertise helped us apply latest technology trends like web interfaces and centralized reporting systems, to efficiently manage our hotels. iAlert in particular contributed significantly to improving customer delight with SMS updates that helped us make managerial decisions on the go,' Mr K. Mohan Krishna, vice-president–Operations, Green Park Group of Hotels.

Source: http://idsnext.com/wp-content/uploads/2011/06/Casestudy-Template-Green-Park1.pdf

(a) What were the problems faced by the employees of hotel Green Park?

(b) What are the various modules of iAlert.

(c) How has iAlert helped the employees to provide customer satisfaction?

CASE STUDY 3

Pallav Modi, the front-office manager of The Retreat Plaza was comfortable working with computers. He was given a promotion and joined The Retreat Resort Inn, a riverside resort, as General Manager. The Retreat Resort, with just 30 rooms and around 80 per cent occupancy, was an establishment run by The Retreat Group.

Pallav, who had always worked with computers and property management systems, had difficulties working manually with records and files. He found it tedious to go through the daily reports, some of which ran into pages. Some of the departments had computers but there was no property management system

(PMS). The reason for the resort not having a PMS was the size of the hotel and the increase in costs. Non-computerizations lead to human errors and delays in guest services, from the F&B department to housekeeping services. Comments from guests suggested that maximum customers to Retreat Resort were from the Retreat Plaza. The guests loved the services of Retreat Plaza, hence were under the impression that the same would be offered at the Retreat Resort. As the expectations of the guests were not met, they were dissatisfied.

(a) What went wrong in this case?

(b) If you were Pallav Modi, what would you have done? Justify your action.

(c) Is property management system an expensive proposal for a 30-room property? Justify your answer.

(d) How does a PMS help an organization run more efficiently?

CHAPTER 7

Computerized Reservation System and Room Management

LEARNING OBJECTIVES

After reading this chapter, you will be able to understand the following:

- Computerized room reservation system
- Sources of reservation
- Making reservation enquiries and checking availability of rooms
- Creation and maintenance of reservation records
- Room management module: room status, type, and rates
- Housekeeping functions
- Generation of reports on reservations and room management

EXHIBIT 7.1   Guest History

Hotel Coral Beach, a seaside resort, is usually busy with customers. The hotel has a computerized reservation system for better efficiency. Pankaj Gupta, a walk-in guest, came to the resort requesting for a two-day stay. The resort sales had dipped a little and so Sunitha Verma, the receptionist, hurriedly agreed to accommodate him, and filled up the guest details and registration card.

While Pankaj was waiting for the room allocation to be done, Deepak Sharma, the Assistant Front Office Manager, came to the reception desk. Deepak recognized Pankaj as a blacklisted guest, which was further confirmed after verification. Deepak asked Sunitha how she had assigned a room without checking the records. For the sake of meeting target sales, Sunitha had not cross-checked with the guest

history before assigning the room, a prerequisite for room allocation.

When a guest checks out of a hotel, the guest details, preference of food and beverage, and remarks about the guest are stored in a file known as the guest history. If a guest is a skipper or has caused some damages to the property, he/she may be blacklisted. Whenever a guest makes a reservation, the reservation assistant checks the guest history to see if he/she had previously stayed at the hotel. In a chain property, the guest history is stored in the server and can be retrieved by all the owned properties and franchises. Guest history helps hotels give their guests a better and personalized service as well as spot blacklisted guests.

RESERVATION SYSTEMS

A reservation system is a system that is used to display room availability, make individual and group reservations, track guest deposits and travel agent commissions, and generate reports such as arrival lists, reservation forecasts, and preregistration cards.

Sources of Reservation

Hotels can draw reservations from market sources within the hospitality industry. The most common sources of reservation are as follows:

- Central reservation system
- Global distribution system
- Intersell agencies
- Cluster reservation office
- Property direct reservation system
- Internet distribution system

By supporting a variety of reservation sources, hotels can handle a large volume of reservation transactions.

COMPUTERIZED/CENTRAL RESERVATION SYSTEM

In the 1970s, the airline industry began modifying and enhancing their internal reservation systems to make the sale of airline tickets through travel agents more efficient. The central reservation system gave travel agents access to information about flight schedules, fares, and seat availability. It also enabled them to make reservations and issue tickets automatically. Although the central reservation system was initially introduced in airlines, it was later used by hotels, railways, and other means of transport. Statistics show that around 50 per cent of all reservations in US hotels are made through the central reservation system. A majority of the hotel groups belong to one or more than one central reservation system.

Definition

A central (or computerized) reservation system controls and maintains reservations for several hotels from one location and automatically redirects the reservation to the required hotel. A

Fig. 7.1   Oberoi central reservation system

Source: http://www.oberoihotels.com/

central reservation system (CRS) is composed of a central reservation office and member hotels that share room availability information with the central reservation office.

Figure 7.1 shows how reservation is carried out through a central reservation system in the Oberoi Group of Hotels.

A person can choose a country, city, and hotel using the Find hotel option. On clicking the city option, the screen lists the cities where the Oberoi hotels are situated (as shown in Fig. 7.2).

Fig. 7.2   Cities in India where the Oberoi has its presence

Source: http://www.oberoihotels.com/

On selecting the city Udaipur, the hotel is chosen. After this, the availability, check-in and checkout dates, and the number of persons are selected (as shown in Fig. 7.3).

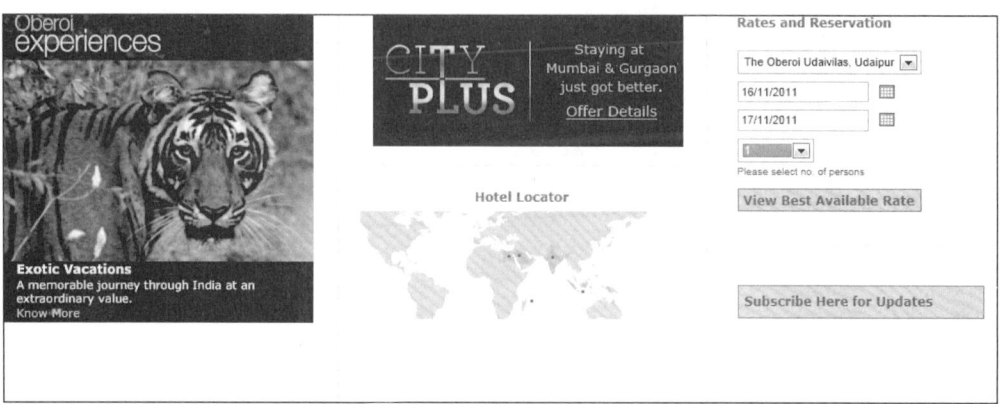

Fig. 7.3   Selection of hotel and stay dates

Source: http://www.oberoihotels.com/

After the dates are chosen and all requirements have been met, select continue. The best available rates and the location of the hotel, its address, and policy of the hotel can be read (as shown in Fig. 7.4).

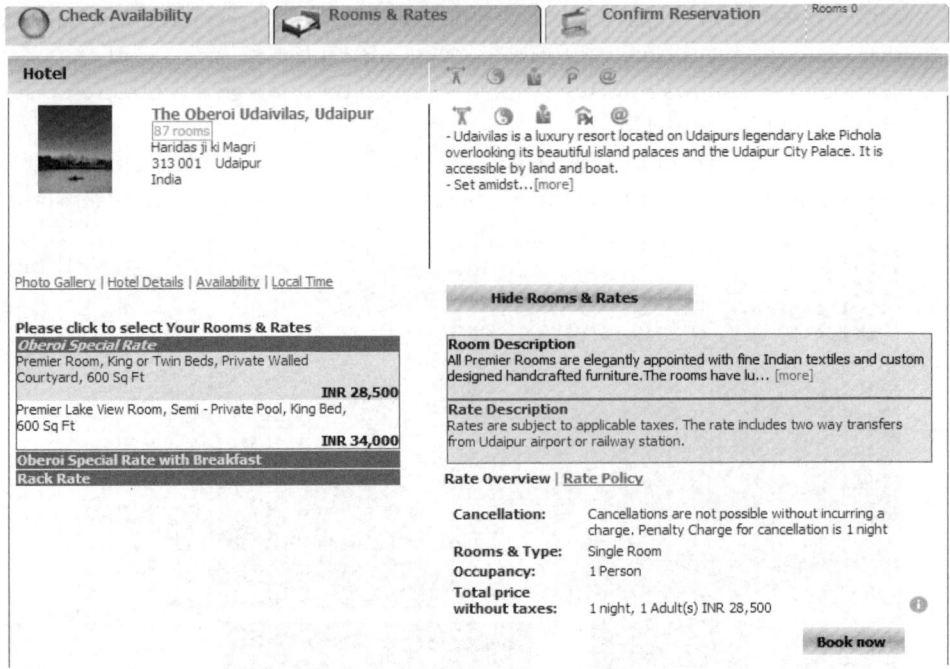

Fig. 7.4   Rates and a brief desription of Hotel Udaivilas

Source: http://www.oberoihotels.com/

After viewing the rates, a person can go to the Book now option, where the personal details of an individual are filled (as shown in Fig. 7.5).

Fig. 7.5   Form to fill in personal details for reservation

Source: http://www.oberoihotels.com/

After the room has been finalized, guest information details are entered, payment is made, and a confirmation is generated. (This is discussed later in the chapter).

Traditionally, the central reservation system was operated independently of the reservation systems in individual hotels. The reservations made through the central reservation had to be manually transferred to the property-based system periodically. Hence, the two systems would often not be in sync with each other. Nowadays, however, automatic real-time postings of reservations that are directly fed into an individual property's reservation system are available.

The central reservation office (CRO) offers its services via 24-hour toll-free telephone numbers. These offices charge a fee for using their services. Depending upon the agreement, a flat fee or variable fee is charged. A percentage of the potential room revenue, actual room revenue, and/or a gross profit of the rooms division may also be charged.

For a hotel group, a central reservation system has marketing benefits as it acts as a valuable channel of distribution. Customers—direct clients or travel agents—have access to room availability and rates from one source. Some companies also use this facility to cross-sell their hotels. If the hotel requested for is sold out, the central reservation system will automatically suggest the nearest hotel that has rooms available to try and keep the sale within the group. Valuable information that is of use to the management is also provided by the central reservation system (for e.g., a detailed breakdown of the number of rooms sold, the percentage of cancellations, and no-shows). The average rate at which bookings are made by each travel agent can also be found out and thus, with whom it is preferable to do more business.

A central reservation system might also serve as an inter-property communications network, an accounting transfer system, or a destination information centre. For instance, a central reservation system is used as an accounting transfer system when the hotel chain communicates operating data with the company headquarters for processing. When a central reservation system communicates reports on local weather, special events, and seasonal room rates, it serves as a destination information centre.

There are two types of central reservation systems—affiliate and non-affiliate.

Affiliate System

It refers to a hotel chain's reservation system where all the participating properties are contractually related. Hotel chains link their reservation operations to streamline the processing of reservations and reduce overall system costs. Another intended outcome is that one property will attract or promote business to another property of the same chain. The main advantages of an affiliate reservation system are as follows:

- Streamlines the process of reservation
- Reduces overall system costs
- Attracts business to another property of the chain

Affiliate reservation networks might serve, in addition to their main function, other duties as mentioned here:

- Inter-property communication network
- Accounting transfer tool

- Destination information centre
- Connection with a global distribution system (GDS), a system that includes several central reservation offices connected to each other

Reservations are often passed from one property of a chain to another property through a reservation network. If one property is booked, the reservation agent handling the caller's transaction might suggest accommodation at another property of the chain in the same geographical area. Some properties might even acknowledge such reservations through a specially prepared note.

Referrals may also be made to properties whose locations might appear more convenient or suitable to the guest's needs. Affiliate reservation networks, which even allow non-chain properties to join the system, are able to represent a broader market. These non-chain properties are referred to as overflow facilities. Reservation requests are routed to overflow facilities only after all available rooms within the properties of a chain in a specific geographical area have been booked.

Non-affiliate System

It is a system designed to connect independent or non-chain properties. Non-affiliate reservation networks enable independent hotel operators to enjoy many of the same benefits that chain affiliated operators do. A non-affiliate reservation network is composed of a central reservation office, potential guests, and independent member hotels. Like an affiliate reservation network, a non-affiliate network usually takes up the responsibility of advertising its services.

GLOBAL DISTRIBUTION SYSTEM

A central reservation system may be linked to other large reservation systems forming a GDS. The GDS 'Sabre' was founded in the mid-1960s by four major North American airlines, most prominently, the American airlines. Later, in the eighties and nineties, several other GDS came into existence. These are generally based on the various airline reservation systems.

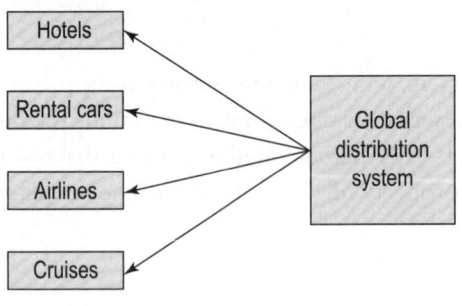

Fig. 7.6 Types of reservation in a global distribution system

It is a computerized system in which reservation-related information is stored and retrieved for multiple organizations. Selling of hotel rooms is accomplished by connecting the hotel reservation system with a GDS.

Through a GDS, a person can not only book hotel rooms but also cruise liners, airlines, and even car rentals (as shown in Fig. 7.6).

Listing a hotel on an airline system means that it can immediately be booked by travel agents worldwide. As travel agents are also suffering from shrinking commissions because of the fall in airfares, they are keen to find a replacement source of income. Since people like to travel, travel agents target hotel bookings also. The ability to view room availability and room rates electronically are attractive features, as telephoning each individual hotel is expensive in terms of both, time spent and telephone charges. Using a GDS, a travel agent with a client flying to London can easily check availability of hotels on the required dates, find out what room rate each hotel is offering, and make the booking electronically in few seconds and at minimum cost. Carrying out the same booking by telephone would be expensive and long-distance phone calls might have to be followed by a fax to confirm the details, which together would make the transaction unprofitable for the travel agent. As a result, many agencies opt for booking rooms only electronically.

The type of reservation system that a hotel should use depends on its booking profile. If most of its bookings come from local sources or from personal recommendations, a property-based reservation system would be sufficient. However, if most reservations are made by travel agents or other international sources, the hotel would consider using electronic distribution through a central reservation and global distribution system. The most popular GDS are Sabre, Galileo International, Amadeus, and World Span. An example of the use of GDS is explained here.

Venkatesh, a resident of Chennai, wants to study in either London or Paris. He had appeared for the entrance examinations and has been shortlisted by a few universities at both locations. Venkatesh does not have any idea about hotels in the UK or Paris. He wants a single-point solution to his travel to both locations.

GDS is a solution for students like Venkatesh, any traveller, or tourist. With the help of a GDS, Venkatesh can book his airline tickets from Chennai to London, and also choose from various airlines, their schedules, and tariffs. He can also opt for hotels based on their locations, proximity to universities he wishes to visit, their facilities and tariffs, and book rooms accordingly. He can also book a ticket on an Euro-rail from London to Paris from Chennai itself. This would make Venkatesh's tour comfortable and free him from the hassle of finding rooms at unknown destinations.

INTERSELL AGENCIES

An intersell agency is a central reservation system that gives out contracts to handle more than one product line. Intersell agencies typically handle reservation services for airline companies, car rental companies, and also hotel properties in a 'one call does it all' approach. Although intersell agencies typically channel their room reservation requests through a central reservation system, they might also communicate directly with hotels.

CLUSTER RESERVATIONS OFFICE

It is a single reservations office serving several geographically connected hotel chains. All the direct calls to each hotel are channelled to the cluster office from where the reservations are made and again redirected to the concerned hotel. A cluster reservation office saves labour as all reservations are made by a particular office and then redistributed. Cross-selling of rooms is possible. Availability of rooms can be matched with their rates.

PROPERTY-DIRECT RESERVATION SYSTEM

Hotels usually handle around 40 per cent of their reservation transactions directly. Depending upon the volume of direct customer contacts, a hotel may have a reservation department apart from the front desk. A reservation department handles direct requests for accommodations, monitors any communication links with central reservation systems and intersell agencies, and maintains an updated room availability status report.

Property-direct reservation requests could reach a hotel in any of the following ways:

Telephone   A guest may directly telephone the hotel for room reservation. This used to be the most common method of making reservations.

Mail   Written requests for reservations were common for group, tour, and conventions. Generally, mail requests were directly sent to the reservations department of the property.

Property-to-property   Hotel chains typically encourage guests to plan their stays ahead of time by offering direct communication between affiliated properties, thereby increasing the overall number of reservations.

Fax   It is a method of communication that accounts for a small proportion of the total reservation.

E-mail   It is the modern way of sending mails for faster and easier communication with hotels.

Website   Hotels have their own websites and individuals wanting reservations can check room availability, rates, packages, and special events with the click of a mouse.

INTERNET DISTRIBUTION SYSTEM

The Internet distribution system (IDS) is a collection of more than 2000 Internet reservation systems, travel websites, online reservation systems, and travel portals, which specialize in Internet marketing of travel and related services directly with the consumers. These online systems have distinctive features that could be used to drive potential travellers to a given destination and/or travel company. Unlike booking through a travel agent or tour operator, consumers with access to the world wide web have the ability to book and travel on their own.

IDS offers technology that allows customers to build complete trips that combine flight and hotel (and other lodging) bookings, transportation, and tourist activities. Travellers have the ability to research, plan, and book their travel from a broad selection of agents. Technology allows travel providers to quickly change offers so that consumers can find great last minute deals to purchase. Examples of IDS include Expedia, Hotwire, Travelocity, and Priceline.

RESERVATION MODULE

The most significant outcome of the reservation process is having the room ready when the guest arrives. To achieve this, the hospitality department must have an efficient reservation procedure in place. Proper methods allow reservation agents to attend to accommodation details, market hotel services, and ensure an effective reservation system. In this chapter, we will discuss typical activities associated with the reservation process. These are as follows:

- Conducting a reservation inquiry
- Determining room availability and rates
- Creating a reservation record
- Confirming a reservation record
- Maintaining a reservation record
- Generating reports

Reservation Inquiry

A hotel receives reservations in many ways. Reservation requests may be made in person, over the telephone, via mail, using a central reservation system, or through an intersell agency. A reservation agent will collect information about the guest's stay through a process known as reservation enquiry. The agent will ask for information such as guest name, address, telephone number, company or travels name, date of arrival and departure, and the type and number of rooms needed.

Details gathered through the enquiry process can be used to create a reservation record. The reservation agent enters the information on a reservation form or into a computer terminal according to defined procedures.

Reservations can be made for individuals, groups, tours, or conventions. In this way, reservations for groups can be differentiated from that of individuals.

Reservation Availability

When a hotel receives a reservation enquiry, it is important to compare the data with previously processed reservations. A hotel can process the reservation request in any of the following ways:

- Accept the reservation as requested
- Suggest alternative room types, dates, and rates
- Suggest alternative hotel properties

In any reservation system, it is necessary to closely monitor the number of reservations in order to avoid overbooking. A hotel should be careful while accepting reservations, after all its rooms have been occupied or reserved. A hotel may wish to book every room to achieve 100 per cent occupancy. Overbooking is usually done so that a hotel can fill all its existing rooms. Experienced reservation managers can forecast cancellations and no-shows (guests with confirmed reservations who do not turn up). Hence, hotels book rooms slightly beyond their actual capacity to ensure that as many rooms as possible are occupied. Reservation availability can be checked across room categories (as shown in Fig. 7.7).

Start Date 30/Jun/2005							Show Detail			Previous Days			Next Days				Close		
Room Type	Thu	Fri	Sat	Sun	Mon	Tue	Wed	Thu	Fri	Sat	Sun	Mon	Tue	Wed	Thu	Fri	Sat	Sun	Mon
	30/06	01/07	02/07	03/07	04/07	05/07	06/07	07/07	08/07	09/07	10/07	11/07	12/07	13/07	14/07	15/07	16/07	17/07	18/07
<<Total>>	21	21	21	21	21	21	21	21	21	21	21	21	21	21	21	21	21	21	21
DELUXE	6	6	6	6	6	6	6	6	6	6	6	6	6	6	6	6	6	6	6
ECONOMY 0.00	1	1	1	1	1	1	1	1	1	1	1	1	1	1	1	1	1	1	1
FAMILY	1	1	1	1	1	1	1	1	1	1	1	1	1	1	1	1	1	1	1
STANDARD	8	8	8	8	8	8	8	8	8	8	8	8	8	8	8	8	8	8	8
SUPER DELUX	6	6	6	6	6	6	6	6	6	6	6	6	6	6	6	6	6	6	6

Fig. 7.7   Room availability category-wise (Datamannet copyright protected)

Source: www.datamannet.com/pps/presentation-aatithya.pps

Overbooking should be tackled carefully. If a reservation manager books too many rooms, guests with confirmed reservations may have to be turned away. To control overbooking, hotels must monitor room availability by coordinating reservations. A computerized reservation system helps in maintaining and monitoring reservations in a better way.

Computerized Room Reservation System

A computerized reservation system keeps a close track of all reservations. A computer system can efficiently control room availability data and automatically generate many reservation-related reports. In addition, this report projects estimated revenue based on reported reservation, information about the number of rooms sold, and rates. Computerized systems can also generate reports summarizing reservations by room type, guest profile, and other characteristics. The biggest advantage of a computerized reservation system is the accuracy of room availability information. Reservation agents feed in reservations, modifications, and cancellations to the system so that the record of available rooms is immediately updated. In addition, any front desk transaction involving no-shows or walk-ins can be immediately entered into the computer and again, room availability is instantly updated.

Once all the rooms in a specific category are sold, the computer can be programmed to refuse any further reservations in that category. Some systems will automatically suggest alternative room types, rates, or even other nearby hotel properties. Computers can also be programmed to list room availability for future periods. Systems might also display open, closed, and special event dates. *Open dates* refer to available room days while *closed dates* show lack of availability of rooms. *Special event dates* can also be programmed to alert reservation agents that a convention or large group is expected to occupy the hotel before, during, or immediately following the caller's requested day of arrival. A hotel's computer system stores reservation records electronically, thereby allowing the creation of waiting lists for high-demand periods.

Reservation Record

Reservation records identify guests along with their occupancy requirements before they actually arrive. These records enable the hotel to personalize guest services and schedule staff accordingly. These records also contain data that hotel personnel can use to generate important management reports.

Reservation agents create reservation records based on interactions with the guests. These records initiate a guest cycle. However, agents create these records only after determining that the request for reservation has been met. To create a reservation record, reservation agents collect and enter data as mentioned here:

- Guest name (and group name, if applicable)
- Home or billing address
- Telephone number, including area code
- Name, address, and telephone number of the guest's company, if applicable
- Company profile (special instructions such as discounts and special rates)
- Name of the person making the reservation, if not the guest
- Number of people in the party and perhaps the age of children, if any
- Expected date and time of arrival
- Number of nights required
- Expected date of departure
- Reservation type (guaranteed, non-guaranteed)
- Special requirements (accommodation for the differently-abled, non-smokers, etc.)
- Additional information as needed (airport pickup, flight details, room preference)

Figure 7.8 shows a screenshot of a reservation form.

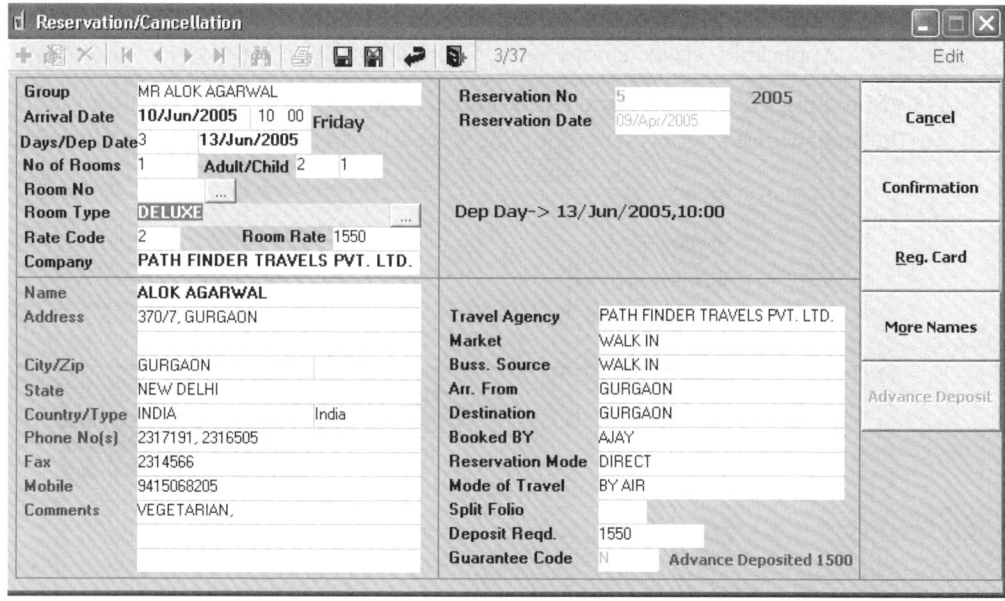

Fig. 7.8   Sample computerized reservation form (Datamannet copyright protected)

Source: www.datamannet.com/pps/presentation-aatithya.pps

Reservation agents need to obtain additional information for guaranteed reservations. For guaranteeing reservations, the following information is required:

Credit card information   It refers to information related to credit card type, number, expiration date, and card holder's name. Nowadays to verify credit cards, transactions are usually carried out electronically.

Prepayment or deposit information   In this mode of payment, an agreement from a guest or client states that a required amount will be deposited to the hotel before a specified date. If the amount is not deposited, the reservation will be considered either cancelled or non-guaranteed.

Corporate or travel agency account information   It includes the name and address of the booking company, the name of the person making the reservation, and the client's corporate or travel agency account number (if assigned by the hotel).

The reservation agents should discuss the aspects of guaranteeing reservations with the guests. Guests should also be aware that the room will be reserved only till the checkout time, following their scheduled arrival. If the guest fails to cancel the reservation before a specified time, he/she may have to either pay the advance deposit or a charge for not honouring the guarantee.

Individual properties and chains might differ in their policies on quoting and confirming room rates during the creation of a reservation record. However, rates once confirmed during the reservation are usually honoured. Reservation agents should be aware of several factors when quoting rates during the reservation recording process. These are mentioned here:

- Supplementary charges for extra services or amenities
- Minimum stay requirements for the dates requested, if any
- Special promotions in effect for the dates requested, if any
- Applicable foreign currency exchange rates, if quoting rates to a foreigner

- Applicable room tax percentages
- Applicable service charge

Sometimes reservations and check-ins are simultaneously carried out for guests who come to the hotel without prior reservation. The screenshot in Fig. 7.9 shows how this process is done.

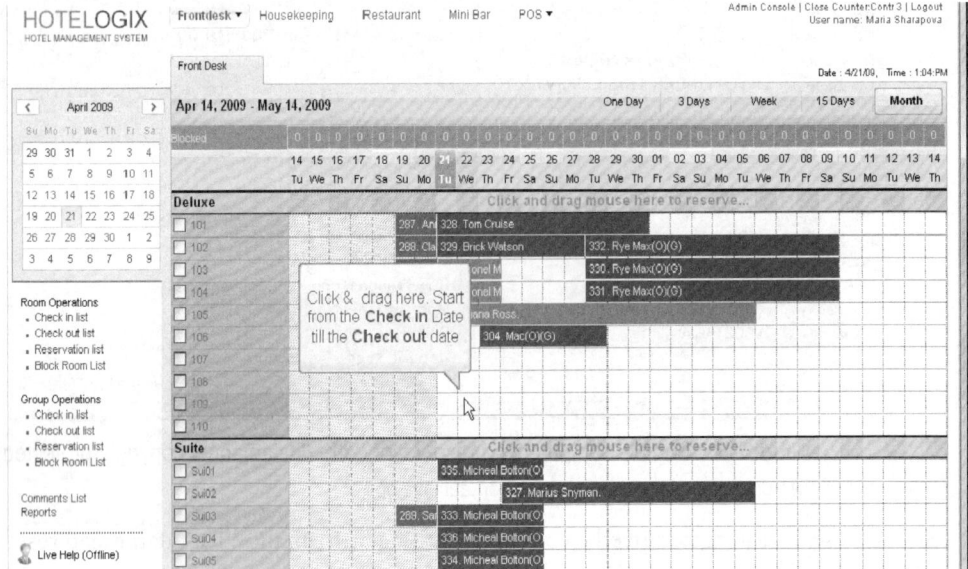

Fig. 7.9   Check-in and checkout dates, and room number defined for a reservation

Source: http://www.hotelogix.com/popup_helpvideo/single-reservation/how_to_make_a_new_reservation.html

After the dates are entered, details such as rates, packages, and names and number of guests (as shown in Fig. 7.10) are fed in.

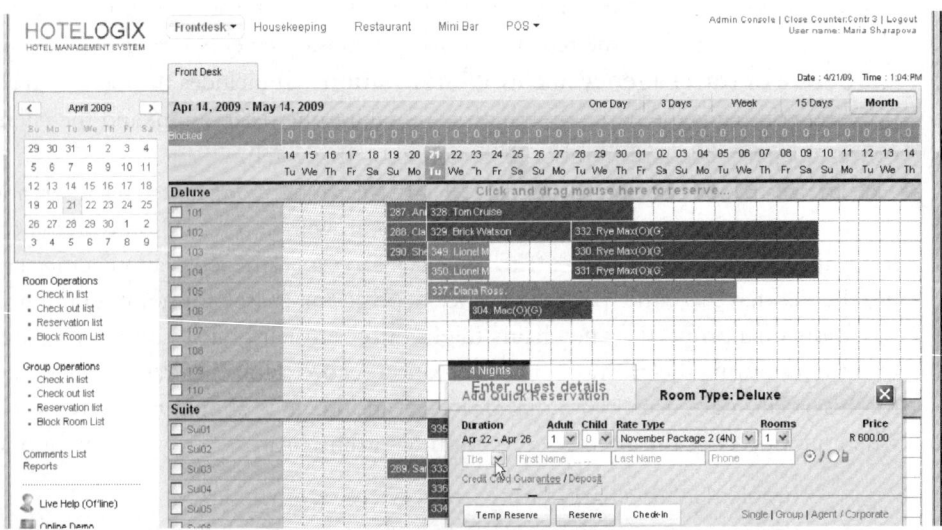

Fig. 7.10   Documentation of reservation details

Source: http://www.hotelogix.com/popup_helpvideo/single-reservation/how_to_make_a_new_reservation.html

After the reservation details are recorded, the Check in button is clicked (as shown in Fig. 7.11).

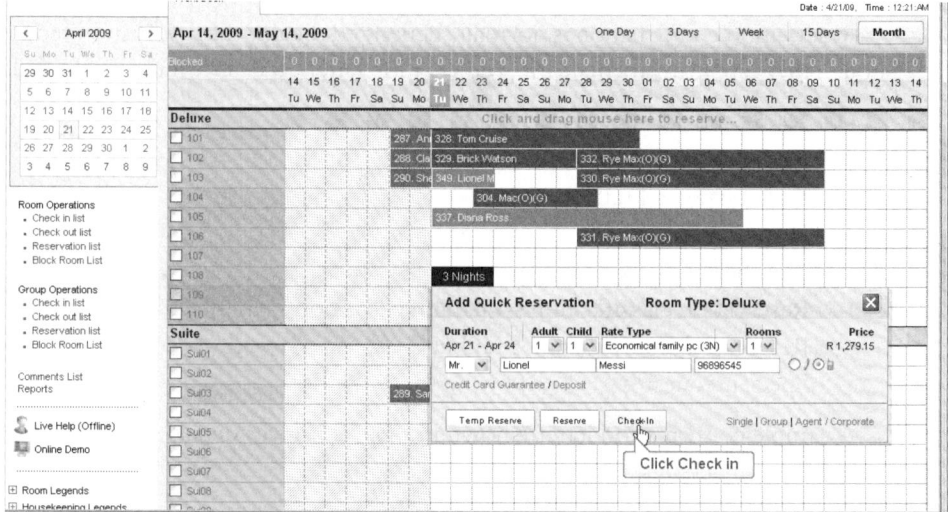

Fig. 7.11 Filling of reservation details and selecting check in

Source: http://www.hotelogix.com/popup_helpvideo/singlereservation/how_to_make_a_new_reservation.html

After Check in is selected, the screen displays the new reservation, checked-in guest, and room allotted (as shown in Fig. 7.12).

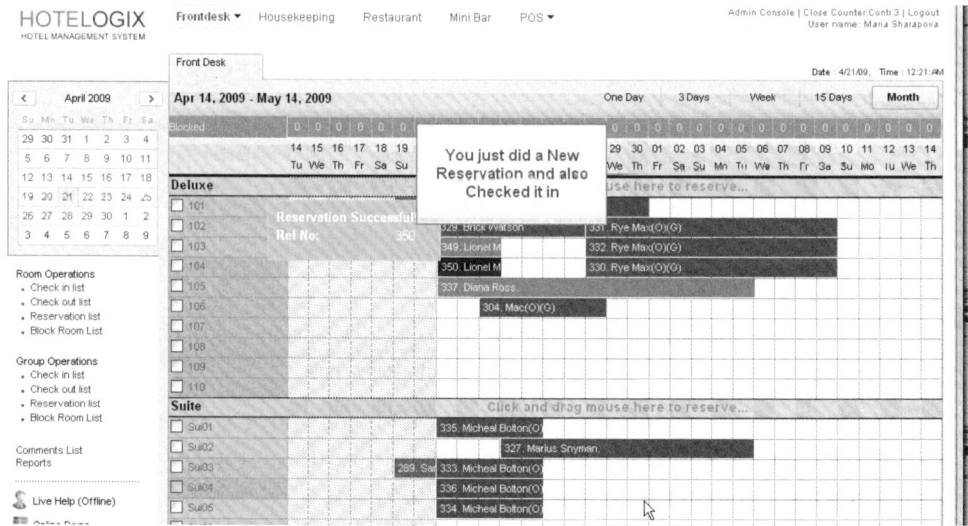

Fig. 7.12 New reservation and check in that has been completed

Source: http://www.hotelogix.com/popup_helpvideo/single-reservation/how_to_make_a_new_reservation.html

Reservation Confirmation

A reservation confirmation means that the hotel has acknowledged and verified the guest's room request and personal information by telephoning or mailing a letter of confirmation. A written confirmation from both the parties confirms the important points of agreement—names, dates, rates, type of accommodation, and number of guests. The guests are often asked to produce a copy of the letter of confirmation at check-in.

Reservation confirmation is normally generated the day the reservation request is received. Though there are many formats for confirmation letters, it usually includes the following points:

- Name and address of the guest
- Date and time of arrival
- Room type and rate
- Duration of stay
- Number of persons in a group
- Reservation classification—guaranteed or non-guaranteed
- Reservation confirmation number
- Special request

A sample confirmation letter (shown in Fig. 7.13) gives details about the reservation, rate, and number of persons.

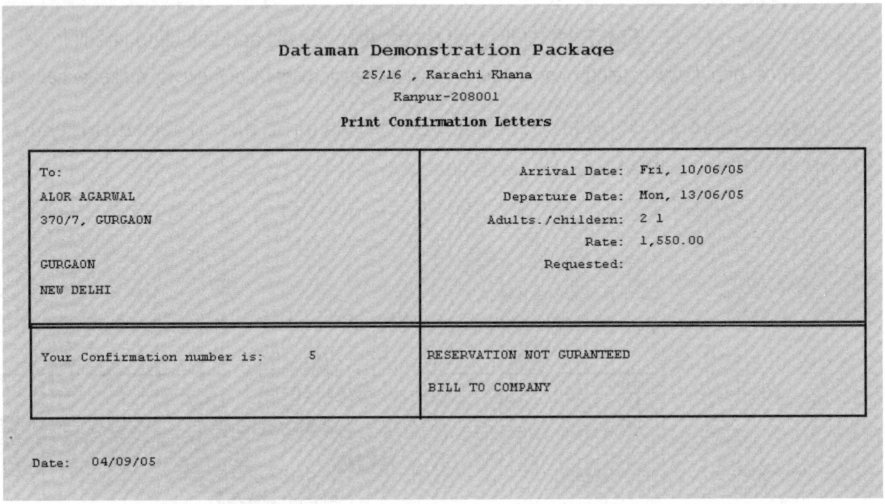

```
                Dataman Demonstration Package
                    25/16 , Karachi Khana
                       Ranpur-208001
                  Print Confirmation Letters

 To:                              Arrival Date:  Fri, 10/06/05
 ALOR AGARWAL                   Departure Date:  Mon, 13/06/05
 370/7, GURGAON                Adults./childern:  2 1
                                         Rate:  1,550.00
 GURGAON                              Requested:
 NEW DELHI


 Your Confirmation number is:    5     RESERVATION NOT GURANTEED

                                       BILL TO COMPANY


 Date:   04/09/05
```

Fig. 7.13   Sample confirmation letter (Datamannet copyright protected)
Source: www.datamannet.com/pps/presentation-aatithya.pps

Depending on the nature of the reservation, a letter of confirmation might also include a request for a deposit or prepayment, or an update of the original reservation detailing reconfirmation, modification, or cancellation.

Confirmation/Cancellation number   A hotel's reservation system typically uses unique methods of generating cancellation and confirmation numbers. These numbers include a portion of the guest's arrival date, the reservation agent's initials, a property code, and other relevant information. For example, in one system, the cancellation number 24012PPS120 represents the following facts:

240: A guest's scheduled date of arrival (from consecutively numbered days of the year)

12: Property code

PPS: Initials of the reservation agent issuing the cancellation number

120: Consecutive numbering of all cancellation numbers issued in the current year

Calendar dates are expressed in three digits when the days of the year are numbered consecutively from 001 to 365 (366 for a leap year). The number 240 in the aforementioned example corresponds to August 28 in a non leap year. Cross-checking reservation cancellation numbers with the scheduled date of arrival can also help agents perform other related front office functions. The cancellation of a reservation necessitates updating the reservation reports that assist management in staffing and planning.

Reservation Maintenance

Though an agent may be very attentive during the reservation process, there is simply no way of avoiding an occasional reservation change or cancellation. An agent's efficiency of organizing and retrieving reservation records and related files is vital to the reservation process. For this, an automated reservation system, which simplifies the tasks, especially in the areas of reservation recording, filing, retrieving, and modifying, is used.

Modifying non-guaranteed reservation   Some guests make non-guaranteed reservations assuming that they will arrive at the hotel before the cancellation hour. Sometimes, it may be almost impossible for the guest to arrive in time because of traffic problems, weather issues, flight delays, or personal emergencies. At the last minute, some guests may wish to change to guaranteed reservation thinking that the room will be available even after the cancellation hour. In this scenario, the reservation agent has to be very clear about the hotel's policies and take steps accordingly. If a reservation is changed, all the guest details are re-verified and a new reservation confirmation number is assigned.

Reservation cancellation   Cancellation of a reservation informs the hotel that a formerly reserved room is once again available and the front office can plan accordingly. The hotel should make the process of reservation cancellation easy and efficient. The reservation agents in charge of cancellation should be polite and courteous. The reservation that has to be cancelled may be guaranteed or non-guaranteed. To cancel a non-guaranteed reservation, the details of the guest are first obtained so that the correct cancellation is being carried out, after which a cancellation number is assigned. The guest would also be asked about alternative reservations and then the cancellation would be confirmed. A guaranteed reservation may be made by a credit card guarantee, by paying an advance deposit to the hotel, or by the travel agency. For a credit card-guaranteed reservation, the credit card companies usually want a no-show detail with proper cancellation numbers, and only then is the amount returned to the guest. In the case of advance deposits, the deposit would be returned to the guest only after proper cancellation of the reservation has been done. If a reservation has been made by a travel agency or through a corporate account, it is treated in a manner similar to that of a credit card transaction with an additional letter being sent to the corporate or travel agency.

Reservation Reports

An effective reservation system helps maximize room sales by accurately monitoring availability of rooms and forecasting revenue. Regardless of the degree of automation, the number and type of management reports available through a reservation system are functions of a hotel's needs and the system's capability and components. Common management reports include the following points:

Reservation transaction report   It summarizes the daily reservation activity in terms of record creation, modification, and cancellation. Reports on blocked rooms and no-shows are also prepared.

Commission agent report   Agents with contractual agreements may be given commissions for bookings they bring to the property. This report tracks the amount the hotel owes each agent.

Turn away report (or refusal report)   This report checks the number of requests refused by a hotel because rooms were not available on the requested dates. The report is usually helpful to those hotels that are running at full occupancy and are thinking of expansion.

Revenue forecast report   This report projects future revenue by multiplying predicted occupancy by current room rates.

ROOM MANAGEMENT MODULE

The room management module is an important information and communications division within a front office property management system (PMS). It is primarily designed to strengthen the communication links between the front office and housekeeping departments. Most room management modules perform the following functions:

- Identify the current room status
- Assist in assigning rooms to the guests at check-in
- Provide in-house guest information
- Organize housekeeping activities
- Generate useful reports for the management

A room management module alerts front-desk employees of the status of each room, similar to what room racks used to do in non-automated operations. A front-desk employee simply enters a room number using a keyboard, and the current status of the room immediately appears on the terminal's display screen. Once the room becomes clean and ready for occupancy, housekeeping staff change the room's status through a terminal in their work area, and the information is immediately communicated to all terminals at the front desk. Room status reports may also be printed (by printing the screen view) at any time for use by the management.

These modules are also capable of automatic room and rate assignments at the time of check-in. In addition, their ability to display guest data on terminals at the front desk, switchboard, and concierge stations eliminates the need for traditional front-office equipment such as room racks and information racks. They also enable management to efficiently schedule housekeeping staff and review detailed reports of housekeeping room attendants. In addition, automated wake-up systems and message-waiting systems can be interfaced with the room management module to provide greater control over these auxiliary guest services.

Room Status

Before assigning rooms to guests, front-desk employees must have access to current and accurate information about the status of rooms in the property. The current status of a room is determined by information about future availability (determined through reservations data) and current availability (determined through housekeeping data).

Information about future availability is important because it may affect the duration of stay of in-house guests. Access to room availability data, which extends several days into the future, gives front-desk employees reliable room status information and enhances their ability to satisfy the needs of guests while maximizing occupancy. Consider the following example:

Vikram checks in on Thursday for a one-night stay. However, during the course of his work on Friday, he finds that it is necessary to extend his stay through the weekend. The front-desk employee may be inclined to approve this extension based on the fact that Friday night's business is light. Later upon checking the reservations data, the employee learns that although the hotel has a low occupancy forecasted for Friday evening, all rooms are reserved on Saturday night. This obviously poses a problem that needs to be resolved according to the hotel's policy. However, it was better that the problem came up on a Friday than on a Saturday night, when the hotel is full and no alterations can be done.

The housekeeping's description of the current status of a room is crucial to the immediate selling position of that room. Common housekeeping descriptions of a room's status are as follows:

- Vacant but not clean
- Cleaned
- Out-of-order
- Ready for inspection

Information about current availability of rooms is absolutely essential for the front-desk employees to properly assign rooms to guests at the time of check-in. Computerized front office systems ensure timely communication by converting data input by the front-desk employees, housekeepers, or guest services personnel into messages that are available at several terminal locations across the lodging operation.

The hotel PMS channelizes data through the room management module and thereby helps coordinate the sale of rooms. Computer-based hotel technology is capable of instantly updating the housekeeping status of rooms, enabling front-desk employees to make quick and accurate room allocations to guests at the time of check-in. For example, when a housekeeping attendant informs the PMS that a room's status has been changed from 'vacant not clean' to 'clean', a notice is automatically printed at the room inspector's station. This notice informs the inspector that the room is ready for inspection. After inspecting the room, the inspector informs via the PMS that the room is ready for sale. A message conveying this change in room status is immediately relayed to the front desk terminals.

Room status discrepancy is a term that refers to situations in which the housekeeping department's description of a room status differs from the room status information that guides front-desk employees in assigning rooms to guests. A discrepancy could seriously affect a property's ability to satisfy guests and hence maximize room revenue. Non-automated properties experience room status discrepancies because of the time delay in communicating room status information from the housekeeping department to the front desk and also because of the cumbersome nature of comparing housekeeping and front desk room status information. Human error also increases discrepancies in non-automated systems.

The room management module generates a room discrepancy report, which signals to the management the specific rooms whose status must be investigated to avoid sleepers and skippers. The report notes differences between the front desk and housekeeping room status updates. This is an important dimension of the room management module.

Room and Rate Assignment

Room management modules can be programmed to assist front-desk employees in assigning rooms and rates to guests at check-in. The module can either make automatic assignments or require front desk personnel to input data to allocate rooms.

Automatic room and rate assignments are usually made according to fixed parameters specified by hotel officials. Rooms can be selected according to floors (similar to the way in which guests are usually seated at a dining room) or an index of the rooms that have been used. The computer system might track room histories (frequency of sales and ranking of rooms according to the usage data). The system uses the information to assign rooms so as to evenly distribute occupancy loads across the full inventory of rooms. A room can be sold at different rates (as shown in Fig. 7.14).

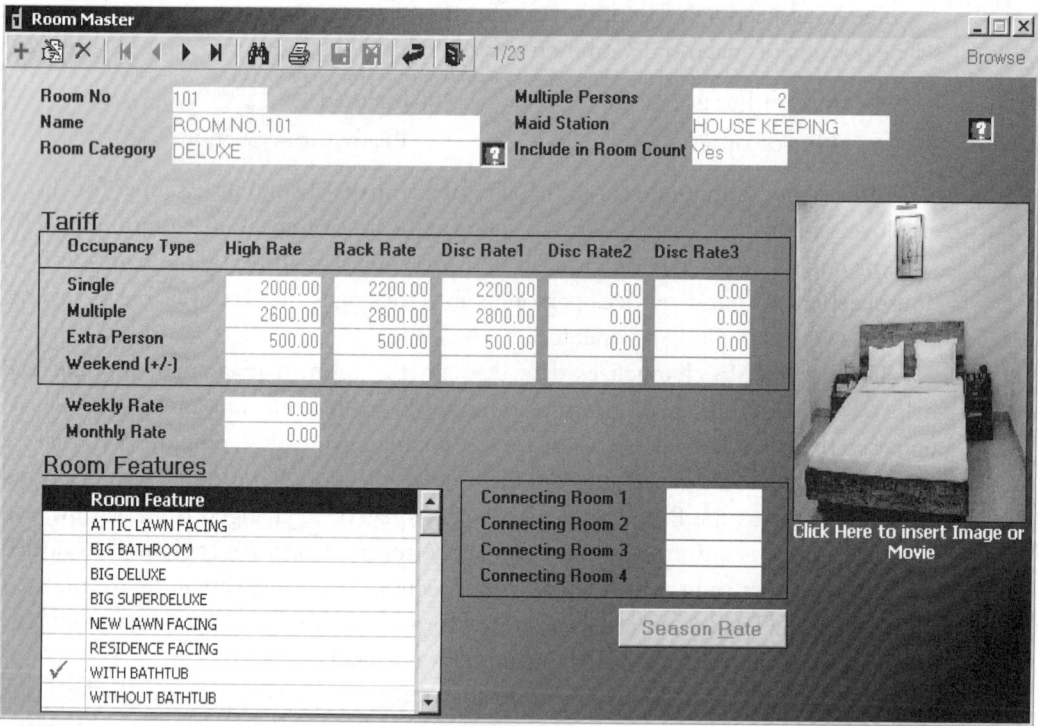

Fig. 7.14   Rooms with their rates (Datamannet copyright protected)

Source: www.datamannet.com/pps/presentation-aatithya.pps

Room and rate assignment programs are popular in the lodging industry. These programs give front-desk personnel direction in decision-making situations and help in increasing their control over actual room assignments. For example, in a property with 800 rooms, a front-desk employee can easily search for a guest's needs through a series of room and rate category queries. In addition, the front-desk employee may use the room management module to display a condensed list, which will enable the front-desk employee to quickly suggest a room to a guest waiting to check in. This enables a faster check-in process when compared to a random search through the entire room availability database.

To accommodate guest preferences and to ensure smooth check-in procedures, a room management module typically features an override function that front-desk employees can use to bypass the room or rate assignments automatically generated by the system. An override function is a useful feature. For example, most automatic room and rate assignment programs will assign guests only to those rooms whose status is clean and available for occupancy. However, when there is no room available, many times it might be necessary to assign a particular guest to a room whose status is 'vacant and not cleaned' or it is 'on a change'. For example, a guest may arrive for check-in and may have to leave immediately to attend an afternoon meeting. An override function of the module permits the front-desk employee to carry out the check-in procedures while informing the guest that the room will be available for occupancy sometime later in the day.

In-house Guest Information Functions

The room management module is also designed to provide review of the guest data. Guest data can be displayed on the terminal screen, enabling a guest services coordinator, switchboard operator, or front-desk employee to quickly identify the name, room number, and telephone extension of a particular guest. This function of the room management module also leads to the elimination of traditional information sources such as information racks, room racks, and telephone lists. The terminals might also be located at room service order stations, garage outlets, and other high guest-contact areas to enhance employee recognition of guests, thereby personalizing the services provided.

Guest data can also be transferred from a room management module to a point-of-sale area to speed up the verification and authorization process of purchases that a guest has made. When an electronic cash register system in a dining area is interfaced with the hotel's PMS, guest data can be reviewed before the charges are accepted. This capability allows cashiers to verify that a particular room is occupied and that the correct guest name is on the room record. Access to this data minimizes the likelihood of charges being accepted from the wrong guest folios, guests who have vacated their rooms, or those who are eligible for charge privileges.

Housekeeping Functions

Important housekeeping functions performed by the room management module include the following:

- Forecasting the number of rooms to be cleaned
- Scheduling room attendants
- Assigning workloads
- Measuring room attendants' productivity

A room management module forecasts the number of rooms that will have to be cleaned by processing the current house counts and the expected number of arrivals. After determining the number of rooms that will require cleaning, most modules can print out schedules for individual room attendants and then assign a specific number of rooms to each attendant on the basis of property-defined standards.

A housekeeping module displays room status, number of people checking in and out, and status of rooms (as shown in Fig. 7.15).

Fig. 7.15   Housekeeping module

Source: http://www.hotelogix.com/videotut/flash/Housekeeping/Housekeeping_Overview.html

The table row shows the room, its type, status, special instructions (if any), and name of the housekeeping staff allotted to the room to be cleaned (as shown in Fig. 7.16).

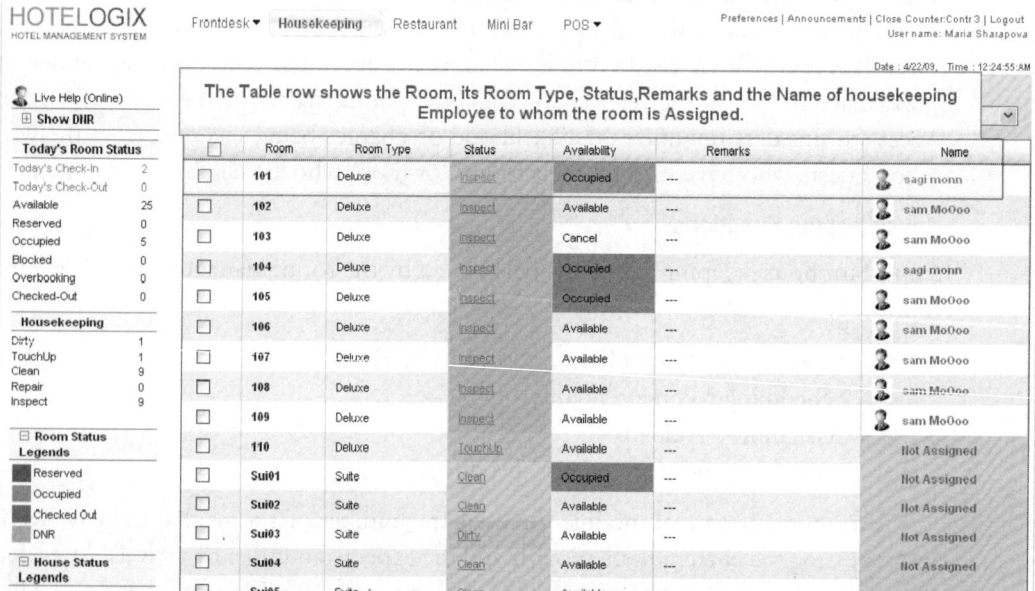

Fig. 7.16   Room number, room type, status, and housekeeping staff allotted for cleaning

Source: http://www.hotelogix.com/videotut/flash/Housekeeping/Housekeeping_Overview.html

The Housekeeping status can be changed by clicking the status button (as shown in Fig. 7.17).

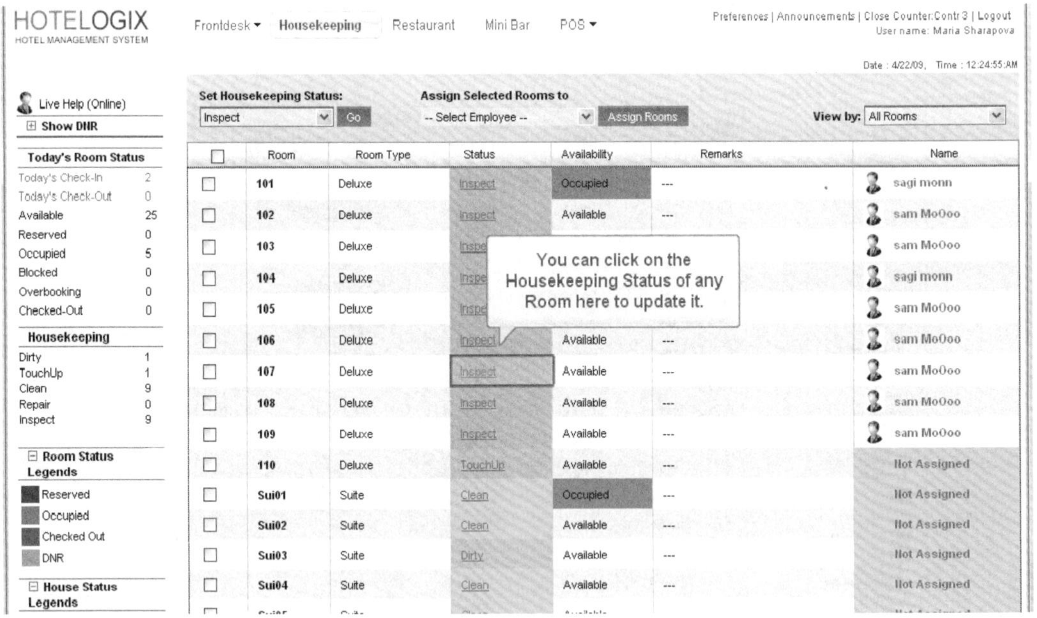

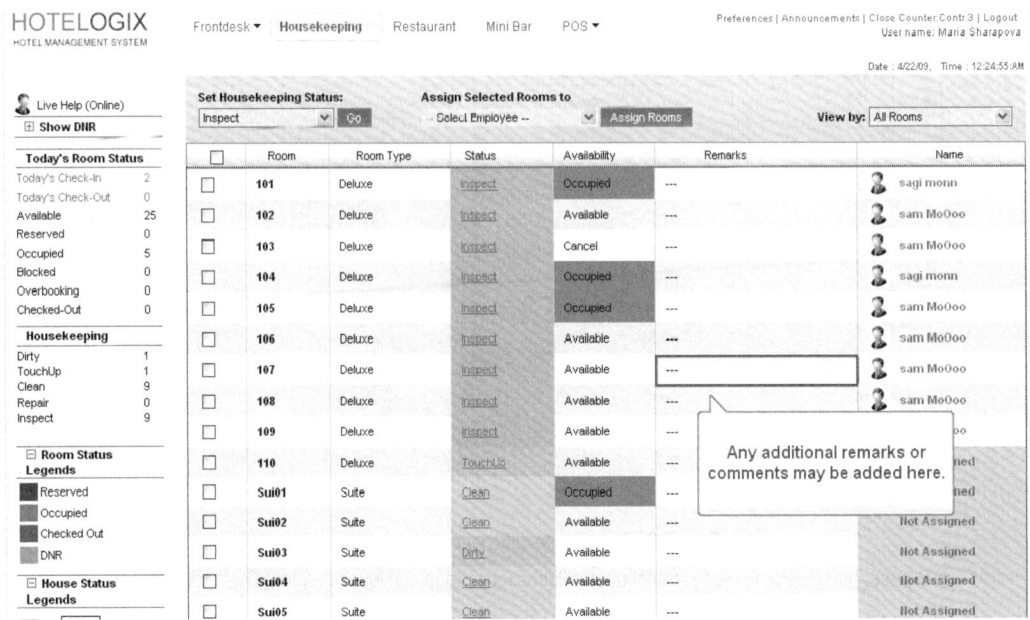

Fig. 7.17   Room status of a room being updated

Source: http://www.hotelogix.com/videotut/flash/Housekeeping/Housekeeping_Overview.html

Any special instruction to be given to the guest rooms are added in remarks (as shown in Fig.7.18)

Fig. 7.18   Special instructions for the housekeeping staff

Source: http://www.hotelogix.com/videotut/flash/Housekeeping/Housekeeping_Overview.html

Employees can be allotted (by name) to individual rooms or an employee can be allotted multiple rooms (as shown in Fig. 7.19 and Fig. 7.20).

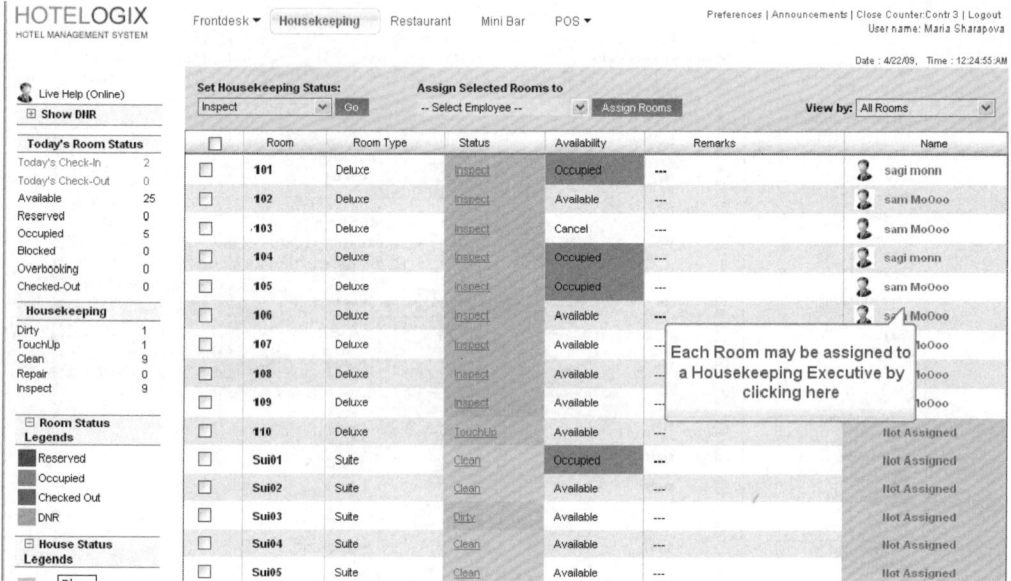

Fig. 7.19 Allocation of rooms to housekeeping executives

Source: http://www.hotelogix.com/videotut/flash/Housekeeping/Housekeeping_Overview.html

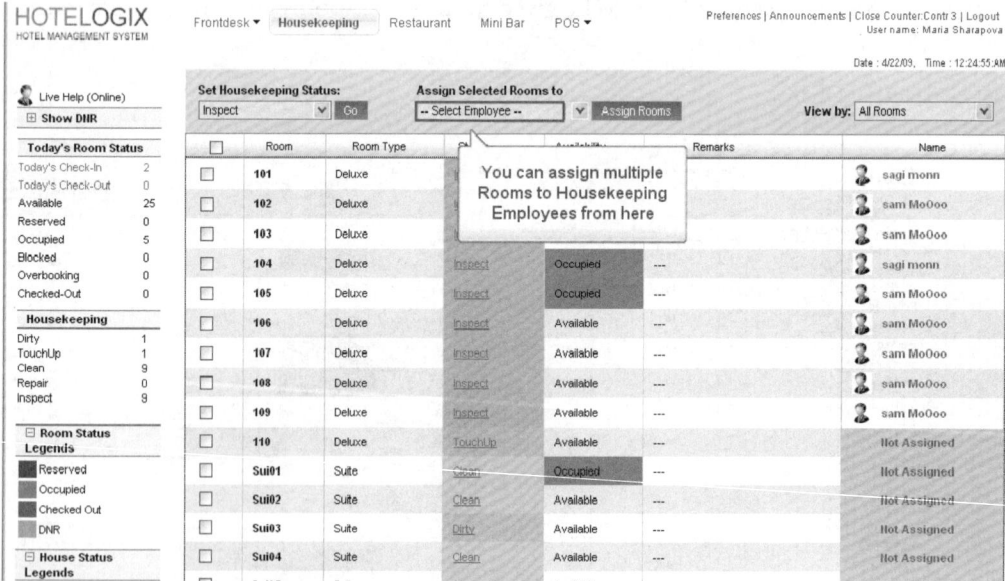

Fig. 7.20 Multiple allocation of employee for room cleaning

Source: http://www.hotelogix.com/videotut/flash/Housekeeping/Housekeeping_Overview.html

Upon entering a room that has to be cleaned, a room attendant may use the room's telephone interface to the PMS to enter his or her identification code number, room number (not always necessary), and the code identifying the room's current status. The computer system automatically logs the time of the call. When the room is clean and ready for inspection, the room attendant again uses the room's telephone interface to notify the inspector's station, and the computer system once again logs the time of the call. The log of time-in and time-out by room attendants enables the room management module to determine productivity rates. Productivity rates can be found out by calculating the average length of time an attendant spends in a room and the number of rooms attended to during a shift. The productivity report keeps the management apprised of potential inefficiencies and also helps in tracking the location of housekeeping personnel throughout a shift.

Generation of Reports

The number and types of reports that can be generated by a room management module are a function of the user's needs, the software capacity, and the contents of the room management file database. A large number of reports are possible because the room management module overlaps several key areas such as the rooms department, housekeeping department, and the auxiliary services. Most room management modules are designed to generate reports that focus primarily on room availability, room status, and room forecasting. These reports are designed to assist management in scheduling staff and distributing workloads.

A room allotment report summarizes rooms committed (booked or blocked) until a particular date. A registration progress report provides the rooms department with a summary of current house information. The report may list present check-ins, number of occupied rooms, names of guests with reservations who have not yet been registered, and the number of rooms available for sale. A registration progress report may also profile room status, room revenue, and average room rate. A room activity forecast provides information on anticipated arrivals, departures, and vacancies. This report assists the managers in staffing front desk and housekeeping areas. A departures report lists guests who have checked out, their room numbers, billing addresses, and folio numbers.

A housekeeper assignment report is used to assign floor and room numbers to room attendants and list room statuses. This report may also provide space for special messages from the housekeeping department. System-generated housekeeper productivity reports provide a relative productivity index of each housekeeper by listing the number of rooms cleaned and the amount of time taken to clean each room.

At the end of each month, quarter, and year room management modules are capable of generating room productivity reports, which rank room types according to percentage of occupancy and/or percentage of total room revenue. A room management module might also produce a room history report showing revenue history and use of each room based on room type. This report is especially useful for those properties that use an automatic room assignment function.

SUMMARY

The chapter talks about the computerized reservation system and room management module. Under computerized reservation system, the central reservation system, affiliate and non-affiliate systems, global distribution system, intersell agencies, and Internet distribution systems have been discussed. The chapter describes them in detail. It also explains why hotels use multiple reservation resources for better marketability of the property, thus leading to better revenue generation and increased percentage of occupancy.

The chapter throws light on a property level reservation system, how a reservation enquiry is made and processed, how room availability is determined, and how computerization helps in meeting guest needs by suggesting dates and rates of room availability in the hotel. Maintenance of reservation records, which help a hotel forecast its needs and give better services to guests, has also been discussed. The chapter also deals with reservation maintenance, focusing on cancellation, amendment or modification of reservation, and reservation maintenance.

The chapter analyses the room management module, which gives an idea about room status, thereby helping the front office in assigning rooms and updating information, which is instantly obtained by the front office, based on the room status updated by housekeeping. It also provides information regarding the number of guests and vacant rooms. The last part of the chapter deals with housekeeping functions that describe the number of rooms to be cleaned, manpower required, and also their productivity status.

KEY TERMS

Affiliate system   It refers to a hotel chain's reservation system in which all participating properties are contractually related.

Cancellation number   It is a number that is used for identification of a cancelled reservation.

Central reservation system (CRS)   It is a central (or computerized) reservation system that controls and maintains the reservations for several hotels from one location.

Central reservation office (CRO)   It is a call centre that deals with room reservation of a hotel chain or property.

Global distribution system (GDS)   It is a computerized system by which reservation-related information is stored and retrieved for multiple organizations. A GDS reserves hotel, airline, and car rental reservations.

Internet direct reservation (IDS)   It is a collection of Internet reservation systems, travel websites, online reservation systems, and travel portals, which specializes in Internet marketing of travel and related service for consumers

Intersell agencies   They are central reservation offices that are contracted to handle reservations for more than one product (hotels, airlines, cruises, and car rentals)

Non-affiliate system   A reservation network composed of a central reservation office, potential guests, and member-independent hotels

Over booking   Hotels book more rooms than the actual number of rooms present, taking into account that there will be cancellation.

Reservation availability   It refers to the enquiry process for finding out if there are free rooms in a hotel.

Reservation enquiry   It is a process by which the reservationist collects information connected to a guest's stay, proposed arrival and departure dates, and the type and number of rooms required.

Reservation maintenance   It refers to the procedure of modifying and cancelling reservations, filing documents, and retrieving them if a need arises.

Reservation record   It refers to a collection of data that identifies a guest based on his/her duration of stay in the hotel before the guest actually arrives at the property.

Room status   It refers to a code or description that indicates the room occupancy and housekeeping position of a room.

Skipper   It refers to a person who leaves a hotel without paying the bills.

Sleeper   It refers to a room that appears to be occupied but is actually vacant due to negligence of the staff.

REFERENCES

Connor. P., *Using Computers in Hospitality*, Cassell Publications, Sydney, 1996.

Kasavana, M. and R. Brooks, *Managing Front Office Operations*, 3rd edition, Educational Institute of American Hotel and Motel Association, Orlando, 1991.

Web References

http://www.hotelogix.com/single-reservation-video.php#123, last accessed on 29 September 2011.

http://www.hotelogix.com/videotut/flash/Housekeeping/Housekeeping_Overview.html, last accessed on 1 October 2011.

http://www.hotelogix.com/videotut/flash/Housekeeping/How_to_change_Housekeeping_Status.html, last accessed on 2 October 2011.

http://www.oberoihotels.com/, last accessed on 26 January 2012.

http://www.scribd.com/doc/29919138/Hotel-Property-Management-Systems, last accessed on 4 October 2011.

https://:www.ognxtra.com/ppt_online/Electronic, last accessed on 25 October 2011.

smileyplace.net/files/Chapter_4_-_Reservations.pdf, last accessed on 4 October 2011.

www.axses.com/encyc/archive/.../innadvance/innadvance-axses.ppt, last accessed on 20 September 2011.

www.datamannet.com/pps/presentation-aatithya.pps, last accessed on 20 October 2011.

EXERCISE

Concept Review Questions

1. Why is a computerized reservation system used in a hotel?
2. How does the operation of a central reservation system differ from a global distribution system?
3. Differentiate between affiliated and non-affiliated systems.
4. What is a reservation enquiry? What are the steps involved?
5. What is a reservation record?
6. How are reservation cancellations done?
7. What are the various reservation reports generated in a hotel?
8. What primary function does a room management module perform?
9. How does a room management module automatically allot rooms and rates to various rooms?
10. What are some of the reports that a room management module can generate?
11. How does a room management module help reduce room status discrepancies?
12. How can a room management module be used to effectively schedule housekeeping staff and measure their productivity?
13. How is a computerized reservation system used to increase the revenue of a hotel?

Multiple Choice Questions

1. Which of these is an Internet distribution system?
   (a) Travelocity
   (b) Expedia
   (c) Hotwire
   (d) All of these
2. Intersell agencies handle reservation of
   (a) hotels
   (b) restaurants
   (c) hospitals
   (d) none of these
3. Housekeeping functions performed by a room management module
   (a) forecast rooms to be done
   (b) schedule room attendants
   (c) both (a) and (b)
   (d) none of these
4. Which of the following are not present in a cancellation number?
   (a) Property code
   (b) Initials of reservation agent
   (c) Date of cancellation
   (d) All of these
5. Which department(s) update(s) the room status?
   (a) Housekeeping
   (b) Front office
   (c) Both (a) and (b)
   (d) None of these

Project Work

1. Using a property management system, do the following:
   (a) Make a reservation.
   (b) Cancel an existing reservation.
   (c) Allocate room attendants for cleaning.
   (d) Check room status by making a reservation and carry out a mock check-in of a guest.
   (e) Update room status after a guest checks in or checks out.
2. Open a central reservation system login of a hotel website having the facilities mentioned here:
   (a) Choose location/country/city
   (b) Choose property
   (c) Choose booking dates
   (d) Check availability
   (e) Check rates for the rooms available
   (f) Check facilities available in the property
   (g) Find the location on a map, and distance from the nearest airport or railway station

CASE STUDY 1

Punnet Sharma, general manager, and Kiran Bisht, front-office manager, of Hotel Hira International were discussing a conference they attended a week ago about non-affiliated central reservation systems. Hira International was a stand-alone property with no computerized reservation system. Both managers were impressed by the presentation and wished to implement the same in their hotel. They wanted to make a presentation for the managing director.

(a) Make a presentation for the managing director emphasizing the need, requirement, and benefits of a non-affiliate system.

CASE STUDY 2

Rakesh Pattanaik, the front-office Manager of Sea View Resort was annoyed with the housekeeping department. Two corporate guests on vacation had an unpleasant experience in the resort.

Sandeep Kapadia, a guest with a reservation, checked in to the resort. He was allotted Room no. 203. When he was taken to the room, he found that the room attendant was in the room. The room was not ready but the room attendant assured him that it would be done in ten minutes. However, it took twenty minutes to get the room ready while the guest waited in front of the room. Sandeep called the front desk from his room and asked for an explanation.

Madhusudan Jain had asked for a special flower arrangement and special amenities in his room at the time of reservation. He was allotted Room no. 312 but on reaching his room, he noticed that the arrangements were not in place. When he questioned the bell boy about it, he said that he would enquire and get back. The bell boy went to the front desk with the query and found that Jain was right. On further probing, it was found that the information was passed on to the housekeeping desk.

Rabi Narayan, the person manning the housekeeping desk, had forgotten to convey the details regarding special amenities to the hotel florist. Rabi had also wrongly updated the status of 'vacant ready' to room no. 203 instead 302.

(a) What went wrong in Sea View Resort?
(b) What steps need to be taken so that similar incident are not repeated?
(c) What functions in a housekeeping module could have avoided this situation?

CHAPTER 8

Account Management Module

LEARNING OBJECTIVES

After reading this chapter, you will be able to understand the following:

- Accounts receivable and account payable module
- Payroll module
- Inventory and purchase module
- Guest accounting module
- Types of accounts
- Settlement of accounts
- Night audit
- Reports related to accounting transactions

EXHIBIT 8.1 Streamlining the Ordering System

Sanjay Bakshi, the executive chef of The Green Palace hotel, was annoyed with Anurag Dixit, the purchase manager. A banquet party being hosted for 500 people had run out of salmon. The guest, a celebrity, had insisted that the salmon be served at the function. Though Sanjay had ordered for salmon, the total quantity he had asked for was not available in the stores.

Though Anurag had placed an order for the requisite amount, stocks had gone down as the coffee shop chef had taken some of the salmon from the stores. The storekeeper had issued the item but had not reordered as it was a slow-moving commodity at the hotel. The salmon would have normally arrived on the third day of ordering, but since the banquet was to be held in the evening, it was purchased from the market at a price much higher than the vendor price.

Since there was a communication gap between the executive chef and the coffee shop chef, the hotel ended up losing revenue. The hotel's ordering system was streamlined. The indenting was either done by the individual outlet's chef after checking with the executive chef or directly from each outlet to the purchase department. The purchase manager had to combine all the indents and then make a purchase order. Without an indent, no commodity could be issued from the stores. Requisitions were verified with the indent before the commodities were issued. A par stock had to be always maintained in the stores for items that were delivered after a long lead time. The inventory management system also had to alert the purchase manager if the stock level of the food item (e.g., salmon) went below the par stock, so that it could be reordered.

This chapter is divided into two parts: a guest accounting module (front-office accounting) and a general accounting module (back-office accounting) for the hotels. For the benefit of students, some accounting terminologies and basic accounting principles in hotels have been defined here:

Asset   The properties owned by a business firm are called assets. They include resources owned by a firm such as land, buildings, machineries, and bank deposits.

Account head   The description of an account (e.g., conveyance expenses, land and building) is called an account head.

Capital   The amount that the owner invests in the business is called capital.

Credit   It is derived from the Latin word 'credo' that means 'trust'.

Stock   Goods that lie unsold on a particular date are called closing stock. Stock can be categorized as opening and closing. Goods lying unsold at the beginning of an accounting period are called opening stock.

Creditor   A person to whom money is owed is called a creditor. 'Cr' is the abbreviation of creditor.

Current asset   These are assets that are meant to be converted into cash within a period of one year or during the normal operating cycle.

Current liabilities   These are debts that are normally paid within a period of one year.

Debit   The word debit is derived from the Latin word 'debeo' that means 'owed to me'.

Debtor   A person who owes an amount to somebody is called a debtor. 'Dr' is the abbreviation of debtor.

Equity   The residual interest when owners have more assets than liabilities is called equity.

Fixed asset   An asset that is required only for use and not for resale is called a fixed asset. They can be tangible assets such as land, building, furniture, and machinery.

Liability   Liability refers to the amount a business firm owes to others.

CLASSIFICATION OF ACCOUNTS

The different parts of a transaction (receiving or giving benefits) are grouped under three classes. They are personal, real or property, and nominal or fictitious accounts.

Personal account   Each individual and firm with whom a business has dealings with is recorded in this class of accounts. It shows how much the customers owe him/her and how much he/she owes his/her suppliers.

>   *Examples*: M/s Larsen & Toubro Ltd a/c; Mr Lawrence's a/c; M/s Ashok Traders a/c, Mr John's a/c, etc.

Real account   Each and every property or asset that a business owns is recorded. The acquisition, depreciation, and sale of assets are recorded in this class of accounts.

>   *Examples*: Building a/c, cash a/c, machinery a/c, bank a/c, land a/c, etc.

Nominal account   Incomes, gains, expenditures, or losses that are incurred during the conduct of businesses are recorded in these accounts.

>   *Examples*: Salaries a/c, rent a/c, sales a/c, interest received a/c, etc.

The procedure for posting in these three types of accounts, known as the golden rule of accounting, is easy to remember.

Personal account   Debit the receiver and credit the giver.

Real account   Debit what comes in and credit what goes out.

Nominal account   Debit all expenses or losses and credit all incomes or gains.

Account Books and Financial Statements

The major types of account books are described here:

Voucher   It is a document on which a transaction is recorded by debiting and crediting the two participating accounts.

Journal   It is an account book in which all the transactions are recorded in chronological order. It is maintained only manually by entering information from vouchers. It is not required in a computerized system.

Ledger   All accounts are recorded and maintained individually, account head-wise in a book called a general ledger or simply, a ledger. In a computerized system, the data from vouchers (input) is processed to prepare a ledger (output).

Cash book   It is a special type of ledger in which only cash transactions are recorded and maintained.

Bank book   It is a type of ledger in which only bank transactions are recorded and maintained.

Sales book   The credit sales of goods are recorded in a special ledger called a sales book.

Purchase book   The purchase of goods on a credit basis is recorded in a special ledger called purchase book.

Debtor's ledger   The transactions (credit sales) of all debtors are recorded and maintained in a ledger known as a debtor's ledger.

Creditor's ledger   The transactions (credit purchases) of all creditors are recorded and maintained in a ledger known as a creditor's ledger.

After preparation of all the aforementioned account books, the final statements of accounts are generated periodically (monthly or yearly). The major financial statements are described here.

Trial balance   It is a list of financial statements prepared monthly, quarterly, or annually to find out the balance in each account. In a trial balance, all debit balances are shown on one side and all credit balances on the other. The debit balance on one side should match the credit balance on the other.

Trading account   It is a financial statement that is prepared yearly to find out the gross profit or loss of the firm.

Gross profit/loss = total sales − cost of goods sold, where cost of goods sold = opening stock + purchases + direct (trading) expenses

Profit and loss account   After preparation of the trading account, a financial statement called a profit and loss account is generated to find out the net profit or net loss of the firm.

Net profit/loss = gross profit − indirect expenses

Balance sheet   A balance sheet is the most important financial statement of the company and shows its position on assets and liabilities on a particular date.

Accounts receivable statement   This statement lists the names of debtors and the amounts to be received by the company.

Account payable statement   This statement lists the names of creditors and the amounts to be paid by the company.

We will also discuss the accounts receivable and account payable modules in detail in the later part of this chapter.

Guest Accounting Module

The creation of electronic folios enables remote point-of-sale (POS) terminals to post charges directly to guest and non-guest accounts. The guest accounting module gives management considerable control over the financial aspects of a hotel. This front-office module is primarily responsible for online charge postings, automatic file updating (auditing) and maintenance, and folio display/printing on demand. In addition, guest accounting modules may provide electronic controls over areas such as folio handling, account balances, cashier reconciliation, food and beverage guest cheque control, account auditing, and accounts receivable. The following sections discuss guest accounting modules in relation to the points mentioned here:

- Types of guest accounts and non-guest accounts (also referred to as folios)
- Common procedures for posting charges to folios, updating accounts, and managing account settlements
- Typical reports generated for use by the management

TYPES OF ACCOUNTS

A computer-based property management system (PMS) ensures that preregistration folios are prepared for guests arriving with prior reservations. Preregistration folios are usually produced by the PMS reservations module when a reservation record is created. When guests arrive without reservations, front-desk employees enter the necessary data into the guest accounting module at check-in. The following information is entered in the folio—guest name, address, room number, and folio number.

If self check-in terminals are available at the property, guests may themselves enter all necessary data by responding to system-generated cues. Header information collected by these terminals is used for folio creation.

Folios

While not all hotel guest accounting modules offer the same folio formats, common types of folios include:

- Individual folios
- Master folios
- Non-guest folios

- Employee folios
- Control folios

Individual Folios

These folios are assigned to the in-house guests for the purpose of charting their financial transactions with the hotel. Figure 8.1 shows posting of charges to a folio.

Master Folios

These folios generally apply to more than one guest or room and contain a record of transactions that are not posted to individual folios. Master folios are commonly created to provide the kind of billing service that groups require. For example, consider the requirements at the 'International chefs conference'. While attendees at this conference are responsible for their personal expenses, the sponsoring organization pays all room charges. As participants dine at various food and

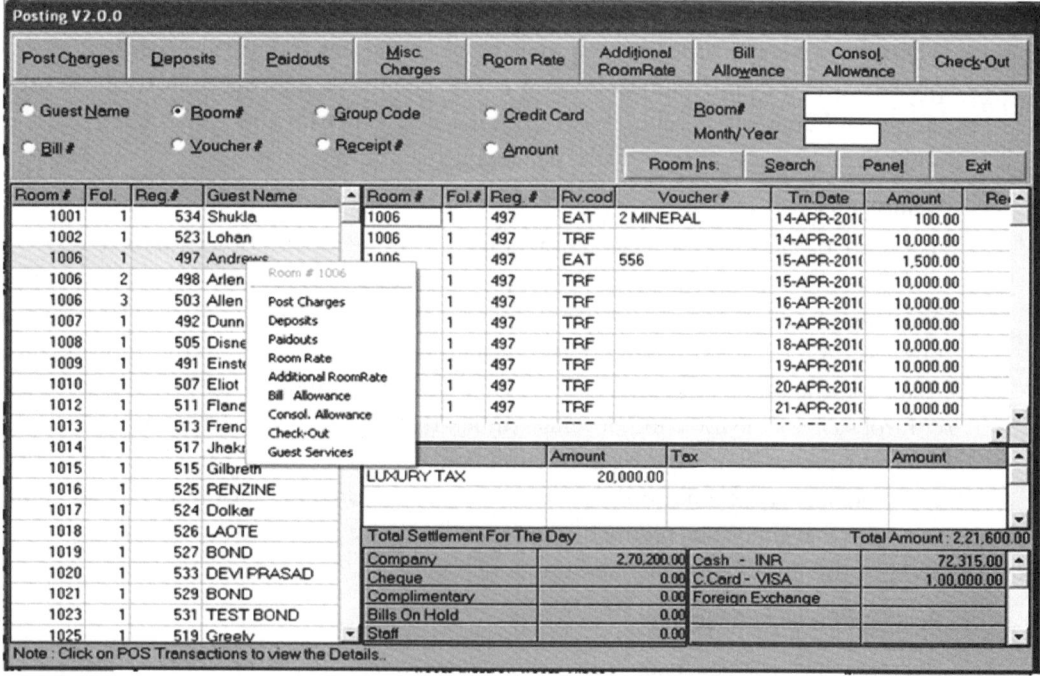

Fig. 8.1   Posting in a folio

Source: www.idsnext.com

beverage outlets in the hotel, their deferred payments are posted to their individual folios. Each night's room charges, however, are posted to the group's master folio. At checkout, each guest receives a folio documenting the charges that he/she is responsible for. The conference administrator is in charge of settling the master folio, which contains only room charges.

Non-guest Folios

These folios are created for individuals who have in-house charge privileges but are not registered as guests in the hotel. These individuals may include health club members, corporate clients, special club members, political leaders, or a local celebrity. Non-guest account numbers are assigned at the time of creation of the account and may be printed on specially prepared account cards. When purchases are charged to non-guest accounts, the cashier might request to see the account card to verify that a valid posting status exists.

Procedures for posting transactions to non-guest folios are similar to those required for online posting of transactions to guest folios. Instead of inputting a room number, the cashier or front-desk employee inputs the designated account number. The use of a unique billing number alerts the guest accounting module to the type of account being processed. For example, a six-digit account number may signal a non-guest account, while a four-digit number may signal an in-house guest account. The major difference between accounting for non-guest and in-house guest transactions lies in the area of account settlement. Guest folios are settled at the time of checkout. The terms for settlement of non-guest accounts are usually defined at the time of creation of these accounts. The term 'settlement' refers to bringing an active folio to zero balance by posting cash received, or by transferring the folio balance to the city ledger or to the company's credit card account.

Employee Folios

These folios usually offer charge privileges to employees and involve transactions that may be processed in a manner similar to non-guest accounts. Employee folios can be used to track employee purchases, compute discounts, monitor expenses and account activity, and separate authorized business charges form personal expenditures.

Control Folios

These folios are used to track transactions that are posted to other folios (individual, master, non-guest, or employee). Control folios provide the basis for double entry accounting and cross-checking of balances in all the electronic folios. For example, when an in-house guest makes a purchase at the hotel's restaurant, the amount is posted (debited) to the appropriate individual folio, and the same amount is simultaneously posted (credited) as a deferred payment to the control folio of the food and beverage outlet. It also tracks allowances, discounts, or correction vouchers prepared in a day. A correction voucher is a voucher that is used to rectify mistakes posted in a guest account on the same day, whereas, an allowance refers to an amount posted to a folio that cannot be rectified. Control folios serve as powerful internal control documents and greatly simplify ongoing auditing functions.

Posting Entries to Accounts

Account entries can be made from terminals at the front desk or from any remote POS terminal that interfaces with the PMS and guest accounting module. Account entries can also be made internally, that is, from within the guest accounting module itself. For example, during the system update routine, room charges and taxes might be automatically posted to all active guest folios. Although guest accounting modules vary in the specifications of their operation, most modules rely on specific data entry requirements in order to ensure that amounts are properly posted to the appropriate folios. Data entry requirements may contain the following sequence:

- Room number (or account number)
- Identification code
- Reference code
- Total charge

After a room number (or account number) is entered, a guest accounting module may require that an identification code be entered as well. This is generally done by inputting the first few letters of the guest's last name. An identification code enables the guest accounting module to process a charge to the correct folio when two or more separate accounts exist under the same room number, as shown in Fig. 8.1. In these situations, simply inputting a room number does not guarantee that the correct folio is retrieved and held ready to accept a charge. Therefore, a guest identification code is part of the required data entry sequence.

Before a charge can actually be posted to a folio, the guest accounting module may also require that a reference code be entered. This is typically done by inputting the serial number of a departmental source document. These documents are usually serially numbered for internal control purposes. This numbering system helps the guest accounting module to investigative searches and analyse account entries made by individual employees through POS terminals within the property. Guest accounting modules can also automatically transfer entries between two folios and perform multiple guest splits for any accounting transaction. Multiple guest splits involve charges that are to be divided among a group of guests, as shown in Fig. 8.2.

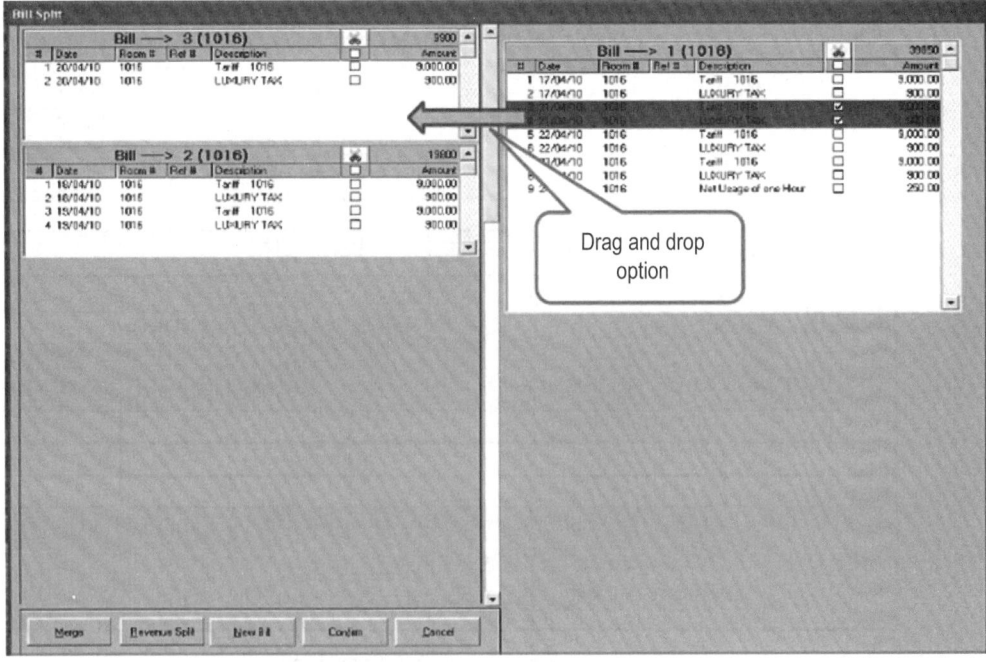

Fig. 8.2 Bill split

Source: www.idsnext.com

The final data entry requirement is the amount of the charge to be input. However, before accepting a charge and posting it to a folio, the guest accounting module initiates a credit monitoring routine. This routine compares the current folio balance with a predetermined credit limit (also called a house limit), which is determined by the management. Although most guest accounting modules allow managers to specify a single house limit, some provide for further options based on the guest history information, such as whether the guest is a repeat customer or a known business associate. Other options may include setting a house limit on the basis of the type of reservation or the credit authorization limits established by individual credit card companies (as shown in Fig. 8.3)

Regardless of how the credit limit is set, an attempt to post the charge to an account initiates a credit monitoring routine, thus ensuring that the outstanding balance during a guest's stay does not exceed the account's credit limit. When hotel policy commands that a line of credit is not to be extended to a guest, a folio can be set at a no-post status. The guest accounting module will not permit the charges to be posted to a folio with a no-post status.

Sometimes when in-house guests make purchases during their stay at the hotel, they are typically asked to present their room key to verify that a valid posting status exists for their individual folios. A few types of electronic locking systems interfaced with a PMS and guest accounting module, depend upon the insertion of plastic electronic keycards (containing strips of encoded magnetic data) to authorize the posting of charges from remote POS terminals. If a guest presents a keycard for an unoccupied room, an account with a no-post status, or a guest account that has already been closed (settled), the system will not permit the cashier to post the charge. Entering the guest's identification code (the first few letters of the purchaser's

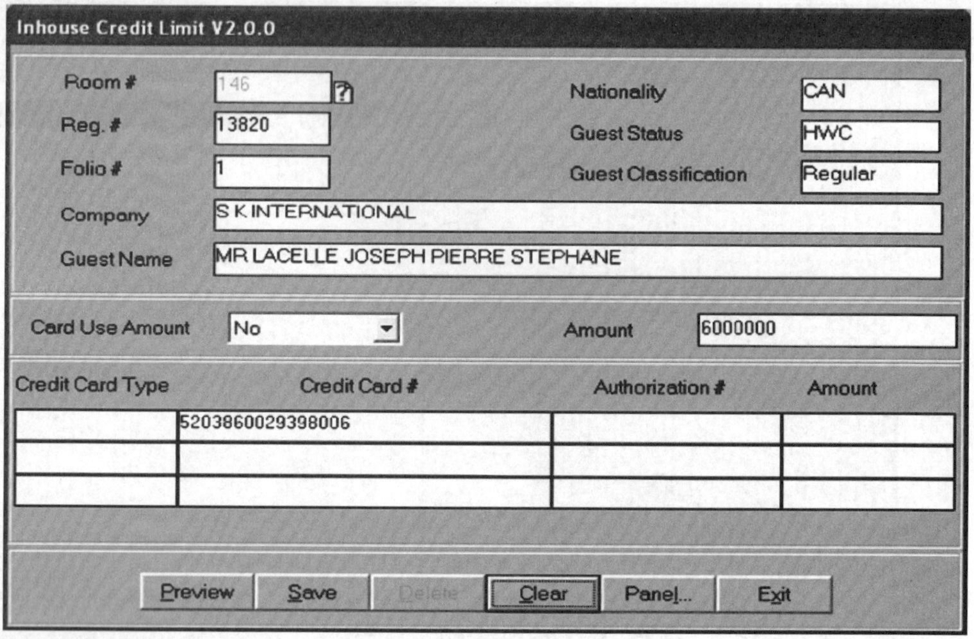

Fig. 8.3   In-house credit limit

Source: www.idsnext.com

last name) may provide further evidence that the person making the charge is not presently an authorized guest at the hotel.

NIGHT AUDIT

The night audit is a process by which all the financial transactions of a guest are recorded and tallied. By this, all guest payments and charges are cross-checked and verified to ensure that all postings have been done correctly. If any charges have not been posted or wrongly posted, they can be rectified in this audit process. Hotels usually operate 24×7 and the best time for any audit process to take place is during the night, considering that there are fewer check-ins and checkouts, hence the name night audit. The night audit usually happens after midnight when transactions in the hotel have reduced. Earlier when the audit process was manually done, it took many auditors and hours to finish the processes. Now with computerization, the process is usually carried out by the night auditor in less than an hour's time. The night audit process involves steps as mentioned here:

Posting room charges and taxes   Manually done, this process is very laborious and the chances of errors are high as a person has to carefully check all the folios and enter details regarding room rate, tax, and cross-check the name of the guest with the room number. However, in a computerized system where an electronic folio is maintained, the computer automatically updates room rates and taxes.

Assembling guest charges and payments   In a manual system, all the guest charges for the day have to be individually posted by guest name and room number and the bill amounts have to be verified. In a computerized system that has an interface with the POS system, as and when the guest is billed, the charges are automatically posted to the folio.

Reconciling departmental financial activities   All the revenue generating outlets submit their sale amounts either at the end of the day or at the end of the shift. The sale amount can be a payment by cash or credit card, or a signed bill. All these have to be submitted by the cashier. A cashier report lists all amounts in detail. Supporting documents are verified with the actual sales calculated by the computer and both amounts must be tallied to avoid any discrepancies between bills that have not been paid.

Running the trial balance   In accounts, whenever there is a debit, there also has to be a credit. Hence, the debit side should ideally always tally with the credit side. This is done as it is not feasible to make a balance sheet every day. By this process, errors in guest accounts or wrong postings (due to human error) can be rectified. The night auditor usually checks all this and the trial balance is run until all the postings are correct. A sample screenshot of a trial balance is shown in Fig. 8.4.

Reconciling accounts receivable   Usually a city ledger, which is part of accounts receivable, is maintained in a hotel (discussed earlier in this chapter). Similar to guest account charges, city ledger charges should also be verified. All outstanding amounts from companies and credit cards have to be obtained to maintain a proper flow of cash in the hotel, for which its accounting is necessary.

Preparing the night audit report   The night audit report is a concise report of the total transactions of a hotel in a day. The report is usually generated as per a hotel's requirement

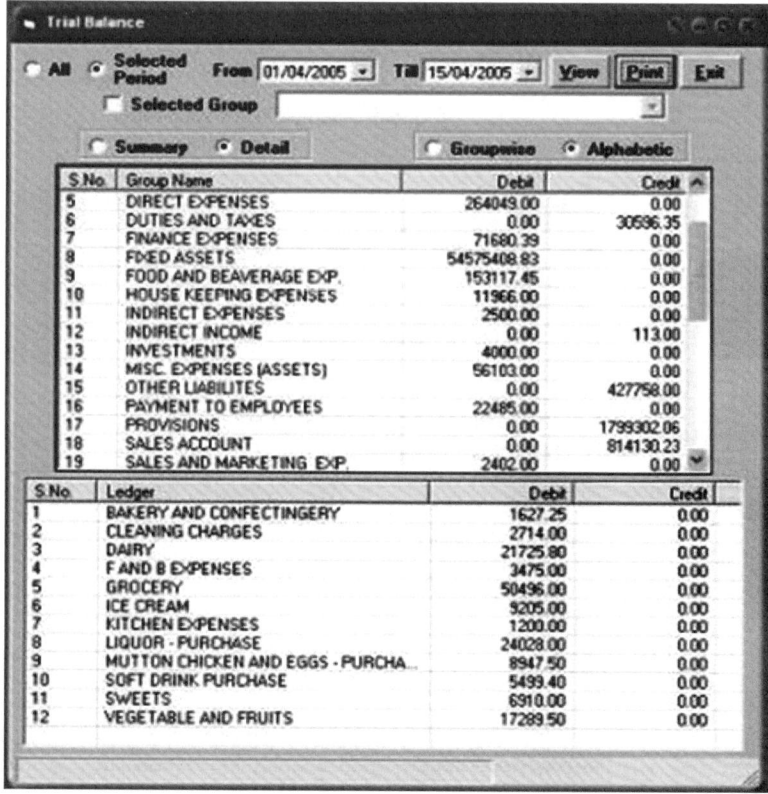

Fig. 8.4   Sample trial balance

Source: http://www.hotelmanagementsystem.co.in/images/screens/Trial%20Balance.jpg

and contain a flash report and department and outlet-wise sale reports. It also compares forecasted revenue with actual revenue. A sample night audit report is shown in Fig. 8.5 and a flash report in Fig. 8.6.

| | ←———— | For The Day ————→ | | ←———— | For The Month ————→ | |
Description	Amount	Allowance	Net	Amount	Allowance	Net
Tariff :-						
Tariff	206600.00	0.00	206600.00	619800.00	0.00	619800.00
Extra Bed	2000.00	0.00	2000.00	6000.00	0.00	6000.00
Retention Charge – FOM	0.00	0.00	0.00	0.00	0.00	0.00
Retention Charge – BQT	0.00	0.00	0.00	0.00	0.00	0.00
Total (A)	208600.00	0.00	208600.00	625800.00	0.00	625800.00
Food & Beverage :-						
MINGSING RESTAURANT	0.00	0.00	0.00	0.00	0.00	0.00
ROOM SERVICE	0.00	0.00	0.00	0.00	0.00	0.00
EAT	0.00	0.00	0.00	0.00	0.00	0.00
Banquets	0.00	00	0.00	0.00	0.00	0.00
PLAN SALES	810.00		810.00	2430.00	0.00	2430.00
Total (B)	810.00	0	810.00	2430.00	0.00	2430.00
Other Sales :-						
I.D.D	0.00			0.00	0.00	0.00
Internet Browse	0.00			0.00	0.00	0.00
ITV Movies	0.00			0.00	0.00	0.00
Local Call	0.00			0.00	0.00	0.00
S.T.D	0.00			0.00	0.00	0.00
Laundry	0.00			0.00	0.00	0.00
Tips	0.00			0.00	0.00	0.00
RoundOff	0.00	0.00	0.00	0.00	0.00	0.00
Exchg Gain/Loss	0.00	0.00	0.00	0.00	0.00	0.00
Total (C)	0.00	0.00	0.00	0.00	0.00	0.00
Taxes :-						
LUXURY TAX	23060.00	0.00	23060.00	69180.00	0.00	69180.00
VALUE ADDED TAX	0.00	0.00	0.00	0.00	0.00	0.00
Total (D)	23060.00	0.00	23060.00	69180.00	0.00	69180.00

> User-defined and system-generated night audit reports and MIS

Fig. 8.5　Sample night audit report

Source: www.idsnext.com

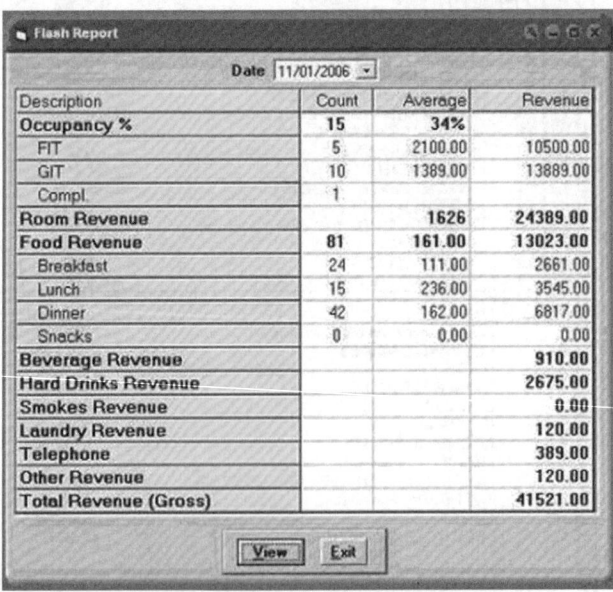

Fig. 8.6　Sample flash report

Source: http://www.hotelmanagementsystem.co.in/images/screens/FlashReport.jpg

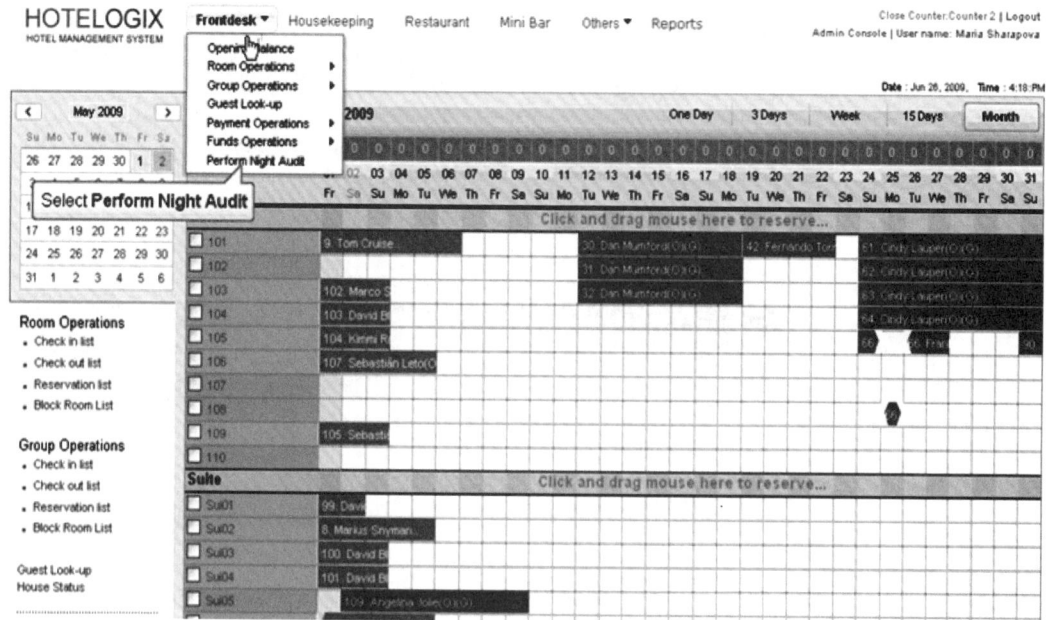

Fig. 8.7   Choosing night audit option

Source: http://www.hotelogix.com/videotut/flash/night_audit/night_audit_overview.html

In a computerized night audit process, the following steps are carried out:

- From the screen, choose the night audit option (as shown in Fig. 8.7).
- After the night audit option is chosen, the screen shows if any user is on the system. If it is so, they are forcibly logged out (as shown in Fig. 8.8).

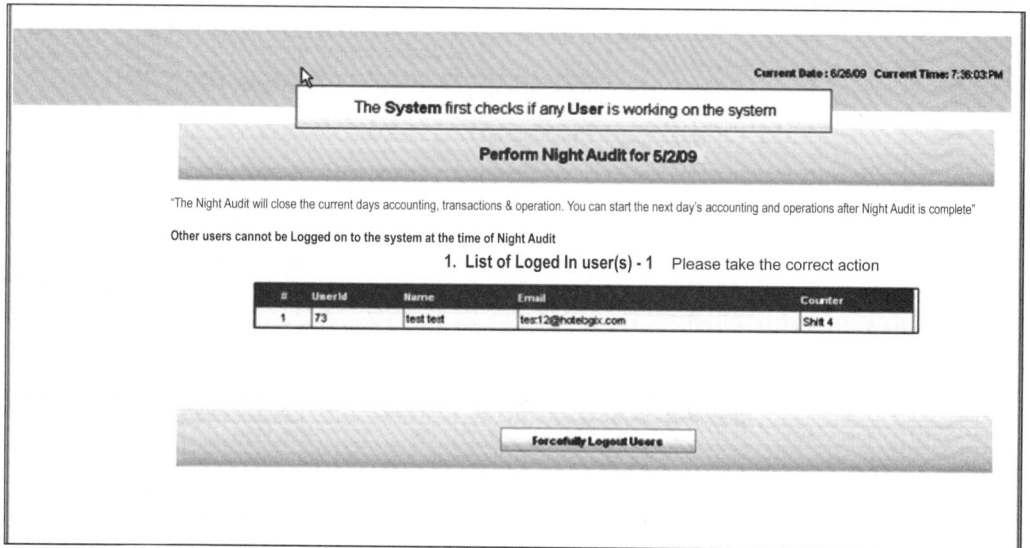

Fig. 8.8   Screen displaying users logged on to the system

Source: http://www.hotelogix.com/videotut/flash/night_audit/night_audit_overview.html

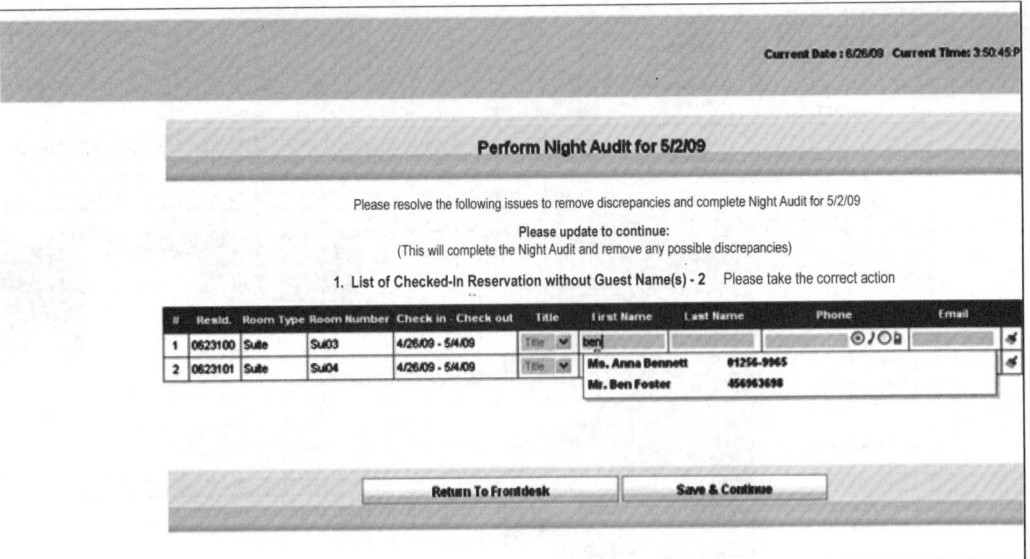

Fig. 8.9   Addition of names and contact details of guests

Source: http://www.hotelogix.com/videotut/flash/night_audit/night_audit_overview.html

- After logging out all users in the system, the night audit process begins with the check-in of a guest who has been allotted a room without inputting a name (as shown in Fig. 8.9).
- After check-in of the guest, the tentative checkout day of the guest is noted (as shown in Fig. 8.10).

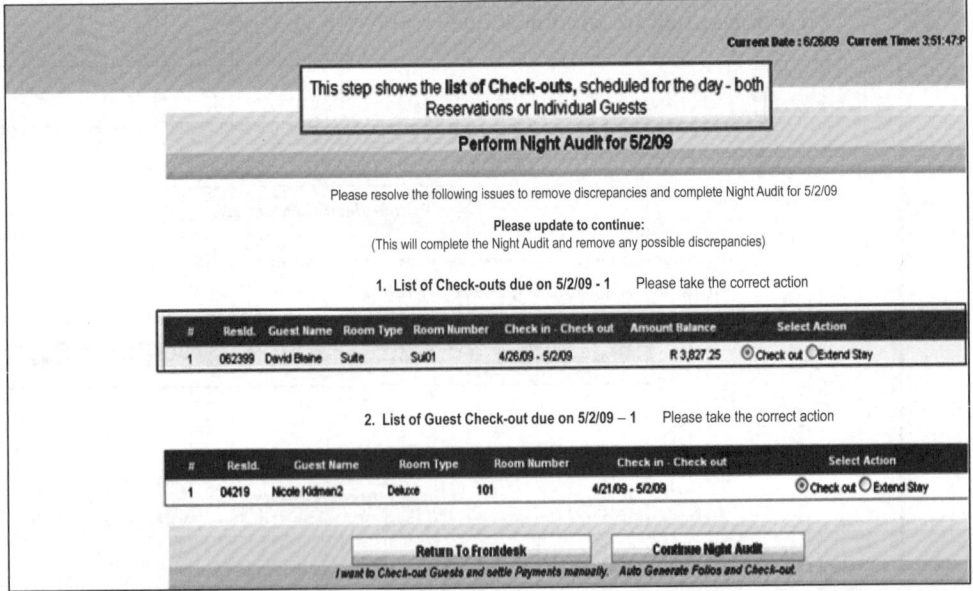

Fig. 8.10   Tentative checkout guest for the day

Source: http://www.hotelogix.com/videotut/flash/night_audit/night_audit_overview.html

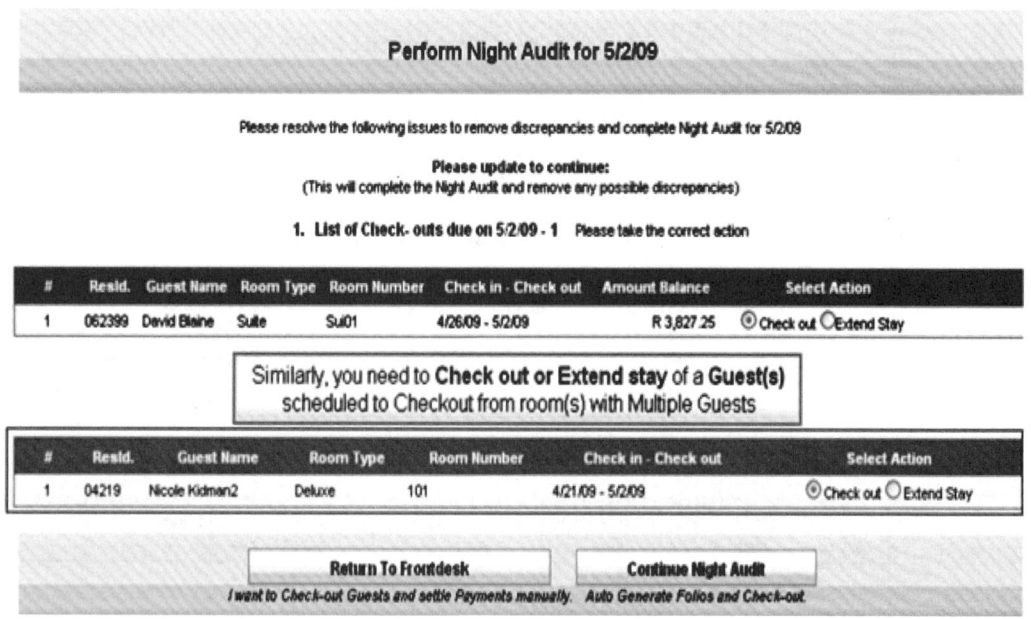

Fig. 8.11   Guest stay extended

Source: http://www.hotelogix.com/videotut/flash/night_audit/night_audit_overview.html

- If a guest is not checked out, the guest stay has to be extended (as shown in Fig. 8.11).
- If a guest is leaving the following day, the guest folio is generated automatically, as shown in Fig. 8.12.

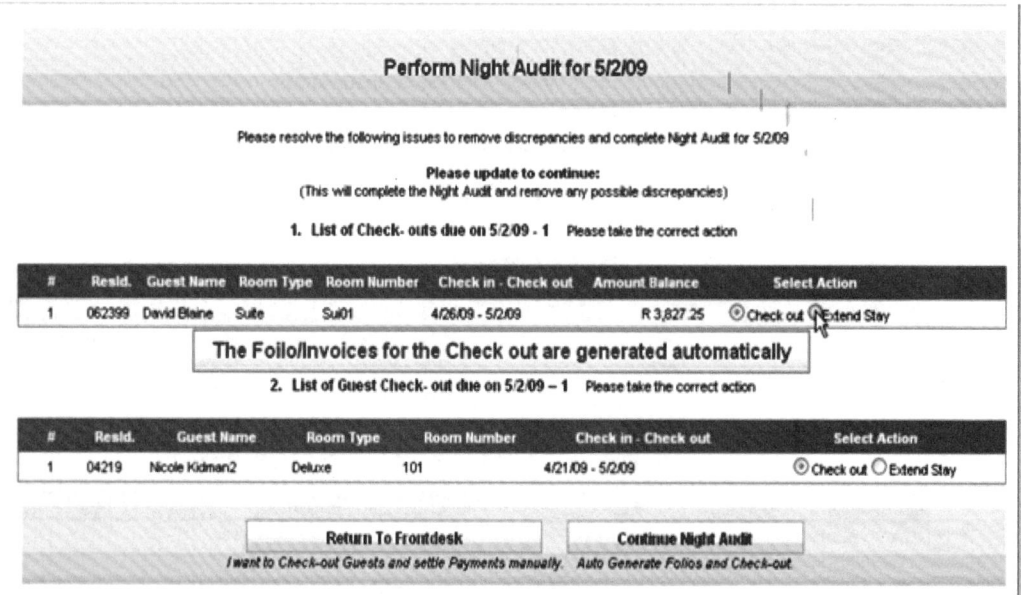

Fig. 8.12   Folio being generated for guest checkout

Source: http://www.hotelogix.com/videotut/flash/night_audit/night_audit_overview.html

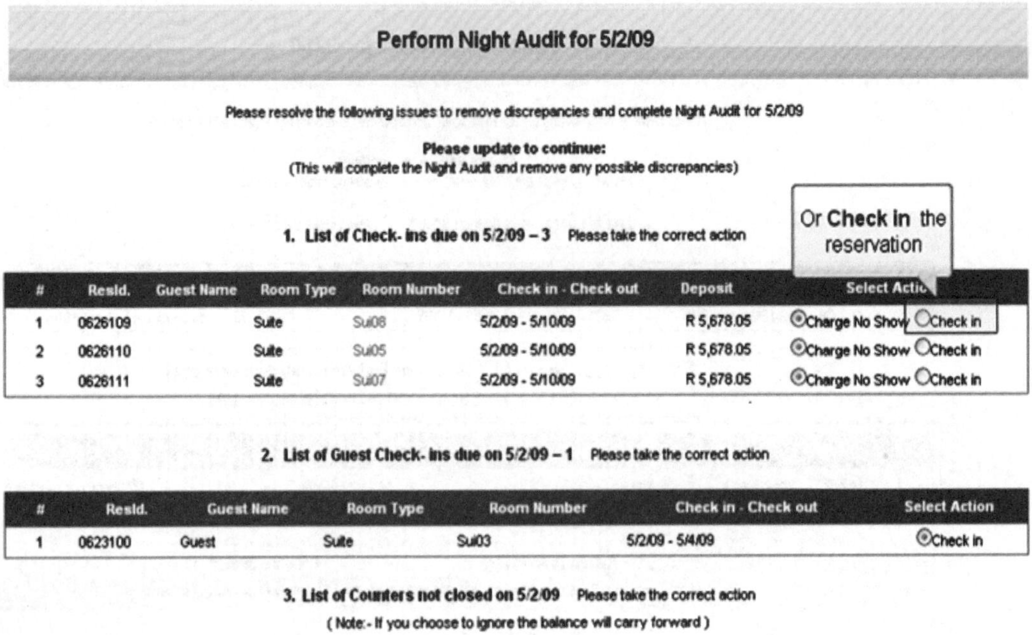

Perform Night Audit for 5/2/09

Please resolve the following issues to remove discrepancies and complete Night Audit for 5/2/09

Please update to continue:
(This will complete the Night Audit and remove any possible discrepancies)

1. List of Check- ins due on 5/2/09 – 3   Please take the correct action

> Or **Check in** the reservation

#	Resid.	Guest Name	Room Type	Room Number	Check in - Check out	Deposit	Select Action
1	0626109		Suite	Sui08	5/2/09 - 5/10/09	R 5,678.05	⊙Charge No Show ○Check in
2	0626110		Suite	Sui05	5/2/09 - 5/10/09	R 5,678.05	⊙Charge No Show ○Check in
3	0626111		Suite	Sui07	5/2/09 - 5/10/09	R 5,678.05	⊙Charge No Show ○Check in

2. List of Guest Check- ins due on 5/2/09 – 1   Please take the correct action

#	Resid.	Guest Name	Room Type	Room Number	Check in - Check out	Select Action
1	0623100	Guest	Suite	Sui03	5/2/09 - 5/4/09	⊙Check in

3. List of Counters not closed on 5/2/09   Please take the correct action
(Note:- If you choose to ignore the balance will carry forward)

Fig. 8.13   Check-in of guest

Source: http://www.hotelogix.com/videotut/flash/night_audit/night_audit_overview.html

- There may be some guests who may not have been checked in. When those guests arrive, they are either checked in or a no-show is charged (as shown in Fig. 8.13 and Fig. 8.14).
- After the necessary check-in, checkout, extended stay, and no-show charges have been posted, the revenue collection for each outlet is checked (as shown in Fig. 8.15).

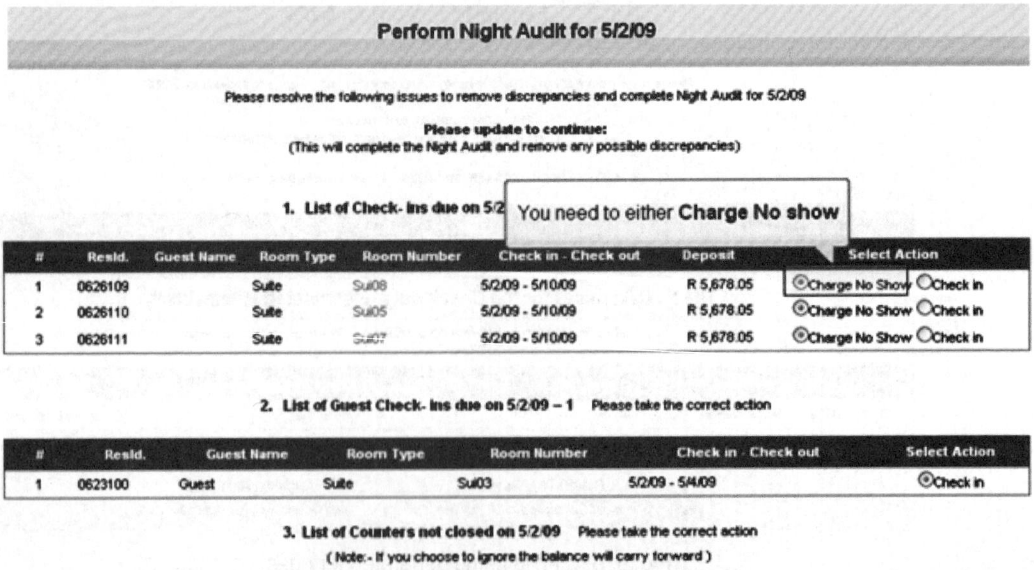

Perform Night Audit for 5/2/09

Please resolve the following issues to remove discrepancies and complete Night Audit for 5/2/09

Please update to continue:
(This will complete the Night Audit and remove any possible discrepancies)

1. List of Check- ins due on 5/2   You need to either **Charge No show**

#	Resid.	Guest Name	Room Type	Room Number	Check in - Check out	Deposit	Select Action
1	0626109		Suite	Sui08	5/2/09 - 5/10/09	R 5,678.05	⊙Charge No Show ○Check in
2	0626110		Suite	Sui05	5/2/09 - 5/10/09	R 5,678.05	⊙Charge No Show ○Check in
3	0626111		Suite	Sui07	5/2/09 - 5/10/09	R 5,678.05	⊙Charge No Show ○Check in

2. List of Guest Check- ins due on 5/2/09 – 1   Please take the correct action

#	Resid.	Guest Name	Room Type	Room Number	Check in - Check out	Select Action
1	0623100	Guest	Suite	Sui03	5/2/09 - 5/4/09	⊙Check in

3. List of Counters not closed on 5/2/09   Please take the correct action
(Note:- If you choose to ignore the balance will carry forward)

Fig. 8.14   No-show charge being carried out

Source: http://www.hotelogix.com/videotut/flash/night_audit/night_audit_overview.html

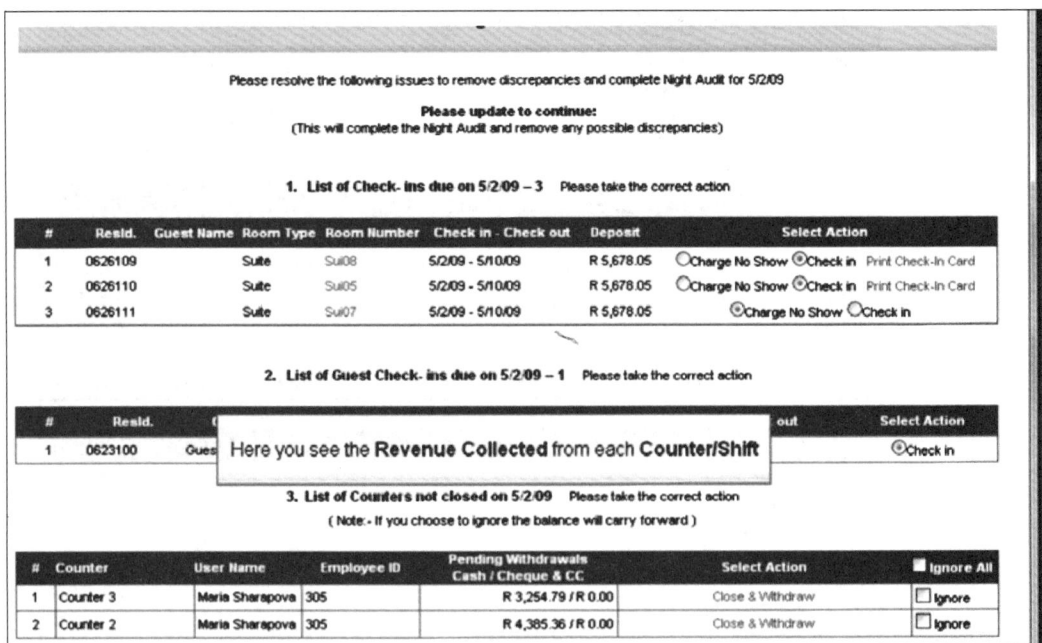

Fig. 8.15   Revenue collection from each outlet/counter

Source: http://www.hotelogix.com/videotut/flash/night_audit/night_audit_overview.html

- Finally, the summary of the night audit is obtained with number of rooms sold (as shown in Fig. 8.16).

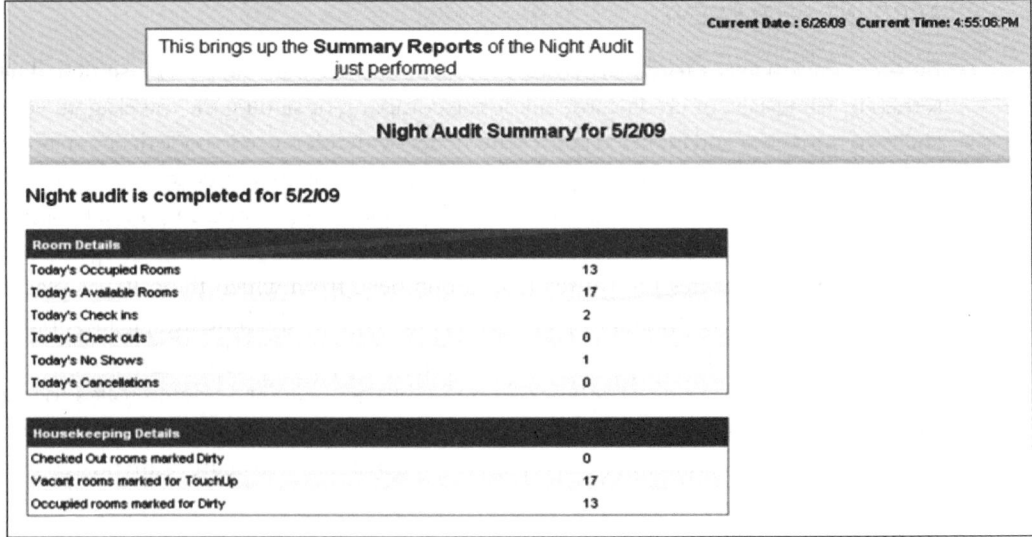

Fig. 8.16   Number of rooms sold summary report

Source: http://www.hotelogix.com/videotut/flash/night_audit/night_audit_overview.html

- The summary of the night audit is also obtained with the total accounts receivable (as shown in Fig. 8.17).

Housekeeping Details					
Checked Out rooms marked Dirty				0	
Vacant rooms marked for TouchUp					
Occupied rooms marked for Dirty			**Account Recievable** Summary		

Account Details

Counter	Revenue Received		Withdrawls		Balance
	Cash	Cheque & CC	Cash	Cheque & CC	
Counter 5	R 0.00	R 0.00	R 0.00	R 0.00	R 0.00
Counter 4	R 937.07	R 0.00	R 0.00	R 0.00	R 937.07
Counter 3	R 3,254.79	R 0.00	R 0.00	R 0.00	R 3,254.79
Counter 2	R 4,385.36	R 0.00	R 0.00	R 0.00	R 4,385.36
Counter 1	R 0.00	R 0.00	R 0.00	R 0.00	R 0.00
Total	**R 0.00**	**R 0.00**	**R 0.00**	**R 0.00**	**R 8,577.22**

Revenue List	Amount
Booking Revenue	R 9,710.88
Cancellation Revenue	R 0.00
No shows Revenue	R 3,548.78
Restaurant	R 0.00
Mini Bar	R 0.00
Confectionery	R 0.00
Book Store	R 0.00
Ice Cream Parlor	R 0.00
Total Revenue	**R 13,259.67**

Fig. 8.17   Total accounts receivable for a particular day

Source: http://www.hotelogix.com/videotut/flash/night_audit/night_audit_overview.html

ACCOUNT SETTLEMENT

Before a guest wishes to depart from the hotel, his/her bills have to be settled. The bills are settled by using cash or credit card, billing the company, or signing the travel agent voucher. The ability to print clear and itemized guest statements (with reference to code detail) may significantly reduce disputes about folio charges. For example, assume that at checkout Geeta has found a discrepancy with regard to a long-distance telephone charge appearing on her folio. The hotel cashier uses the reference code number on the folio to locate the proper telephone call record. The cashier then verifies the source (room number), from which the call was placed and the telephone number that was dialed. This procedure enables the hotel to quickly, objectively, and efficiently resolve disputes regarding amounts posted to guest folios.

System update routines can be programmed to generate pre-printed folios for guests who are expected to checkout on a particular day. Pre-printing folios significantly speed up the checkout process and minimize discrepancies. When additional charges are posted to folios, the pre-printed folios are simply discarded and updated folios, with the correct account balances, are printed at the time of checkout. Electronic folios are closed at the time of final settlement. Accounts that are accidentally closed can easily be reopened. Checkout triggers a communication to housekeeping and internally sets the account to a no-post status. Since the guest accounting module can be interfaced with other front-office modules, better communication among staff members is possible and more comprehensive reports are available for use by the management.

Accounting in Hotels

Usually in a hotel, the principle of double entry bookkeeping is followed. The accounts are maintained in an analytical manner and classified according to a uniform system of accounts. The uniform system of accounting is a manual, which uniformly classifies income, expenditure, assets, and liabilities for hotels and also standardizes a method of presenting financial results of an operation. In the uniform system of accounting, the activities of a hotel are divided into two main departments. The first is an operating or revenue earning department and the second is a non-operating and non-revenue earning department.

Basics of Bookkeeping

An account is a record of a financial transaction and all financial data is recorded and summarized. An account can experience an increase or decrease. For example, funds in a bank account (an asset), money owed as refunds to guests (a liability), and the profits made by a hotel owner can all either increase or decrease To understand bookkeeping better, it is written in the form of 'T' (as shown in Fig. 8.18). The increases are written on the left and right side of the 'T' format depending upon the type of account. The increases are written on the left-hand side of the transaction effect as an asset account and the decreases are written on the right-hand side. Standard terminology—debit and credit—are used to denote entries on the left-hand and right-hand sides of the 'T' format.

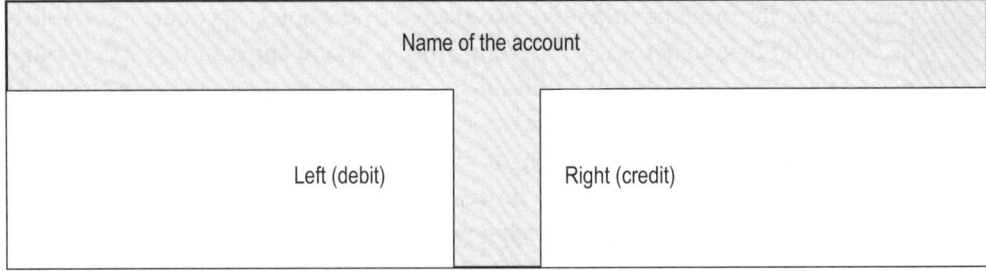

Fig. 8.18   T account

ACCOUNT RECEIVABLE MODULE

This module helps in tracking all the invoices that are awaiting payment from the guests. The following are the key functionalities of this module.

- Accounts classification for reconciliation and control
- Online credit management
- Reminder letters with varying degrees of severity
- Ageing analysis reports for review
- Interest for late payments
- Customer statements

An accounts receivable module monitors and maintains the account balances of guests. The accounts receiving module updates outstanding balances in guest accounts. Accounts receivable is a process of billing the guest and collecting the payments made. It takes into consideration amounts representing purchases made by guests in the hotel and those who have deferred

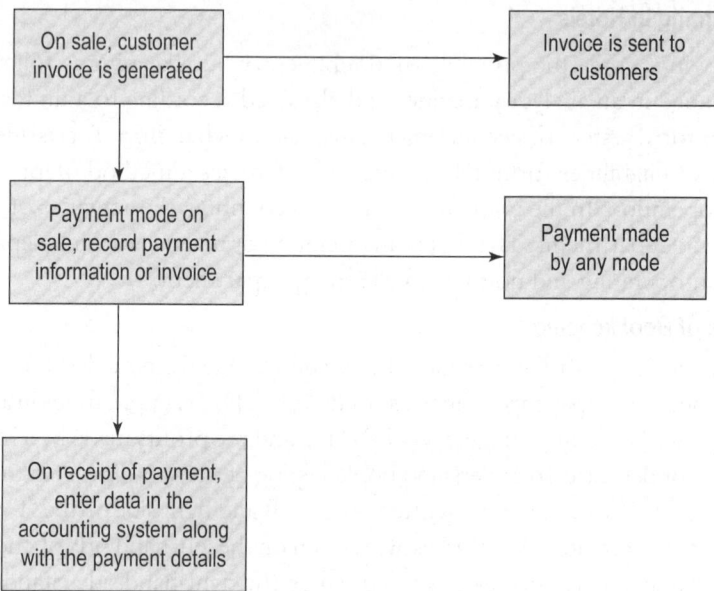

Fig. 8.19   Steps in accounts receivable

payment for the products and services rendered by the hotel. Accounts receivable balances can be automatically transferred to the front-office software applications or they can be manually posted directly to an accounts receivable program. Once entered into the back-office system, account collection begins. The accounts receivable module also generates ageing account receivable reports and audit reports of various account receivable transactions. In managing accounts receivables, the following procedures are recommended:

Establish a credit policy in which a potential guest's financial soundness can be reviewed prior to extending credit. For this, maintain a customer/guest master file. This file contains the guest's name, account number, address, telephone number, e-mail address, credit limit, last payment date, last payment amount, and the credit history.

Marketing factors must be noted since an excessively restricted credit policy will lead to loss of sale. Some hotels may offer more liberal payments than usual during slow periods in order to stimulate business by selling to company guests, groups, and travel agents who are unable to pay until later in the quarter. This policy is financially appropriate when the returns on the additional sales plus the lowering in inventory costs is greater than the incremental cost associated with the additional investment in accounts receivable. Figure 8.19 shows the steps involved in accounts receivable.

ACCOUNT PAYABLE MODULE

This module is provided with the functionality to enter, monitor, maintain, and process payment of invoices and credit notes. The following are the key features of this module:

- Immediate registration of incoming invoices
- Tracking and authorization of incoming invoices

- Entry of order-based and sundry invoices
- Automatic matching of invoices with receipts
- Self-billing of invoices that are suitable for environments where the receipt of goods automatically generates approved invoices in the system, which are paid through remittances. Suppliers need not send any invoices.
- Accounts classification for reconciliation

An account payable module tracks purchases, creditor positions, and the hotel's banking status. Accounts payable activities normally consist of posting supplier invoices, determining amounts due, and printing cheques for payment. The following are the three major files maintained by an accounts payable module:

- Vendor master file
- Invoice register file
- Cheque register file

Vendor master file   It contains an index of vendor names, addresses, telephone numbers, vendor code numbers, standard discount terms (time and percentage), and space for additional information.

Invoice register file   It is a complete list of outstanding invoices catalogued by vendor, invoice date, invoice number, or invoice due date. This file becomes especially important when the management wishes to take advantage of the vendor discount rates.

Cheque register file   The calculation and printing of bank cheques for payment to vendors is monitored through the cheque register file. Cheque production and distribution is summarized into a payables report and reconciled with bank statements.

The account payable module also generates reports regarding amount already paid, the amount to be paid to suppliers, and balance amount in a hotel's bank account.

PAYROLL MODULE

A payroll accounting module is an important part of a back-office package because of the complexities involved in properly processing time and attendance records, unique employee benefits, pay rates, deductions, and required payroll reports.

Whenever a new employee joins an organization, a master file is opened (as shown in Fig. 8. 20).

The employee master form is used to enter basic information of an employee when he/ she joins an organization. It has provisions for entering attendance, links to number of working days, personal details, contact details, and professional details.

The payroll module is dependent on parameters such as perks, deductions, and increments. Perks are entered in a perks master; it could be a percentage of gross salary, net salary, or the basic. The same applies to deductions in the deduction master. There are separate forms for deductions such as provident fund, professional tax, ESI, and income tax. The module can automatically calculate income tax deduction, taking into account the parameters set by the government.

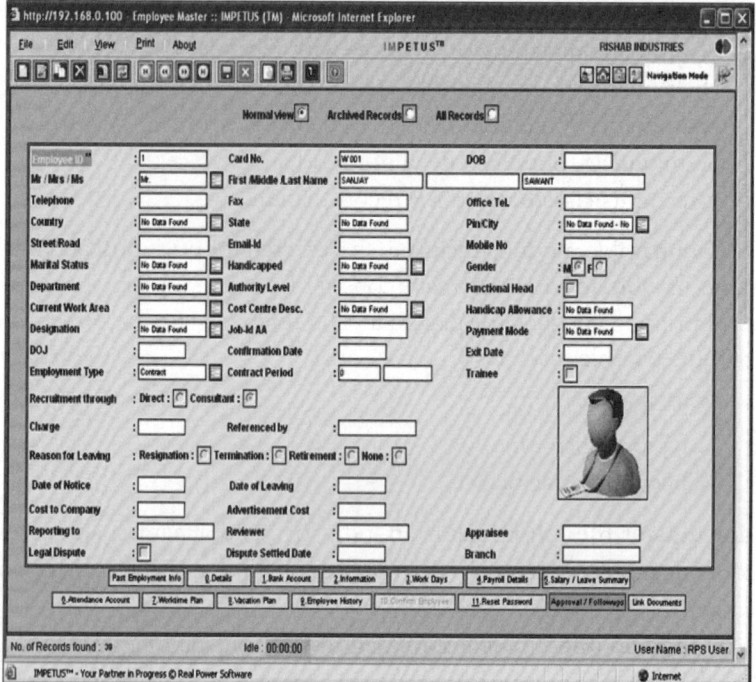

Fig. 8.20   Employee master

Source: http://www.realpowersoft.com/hr_module.html

The module also has provisions for employees who have either taken advances or loans from the organization. Figure 8.21 shows how advance or loan details are entered.

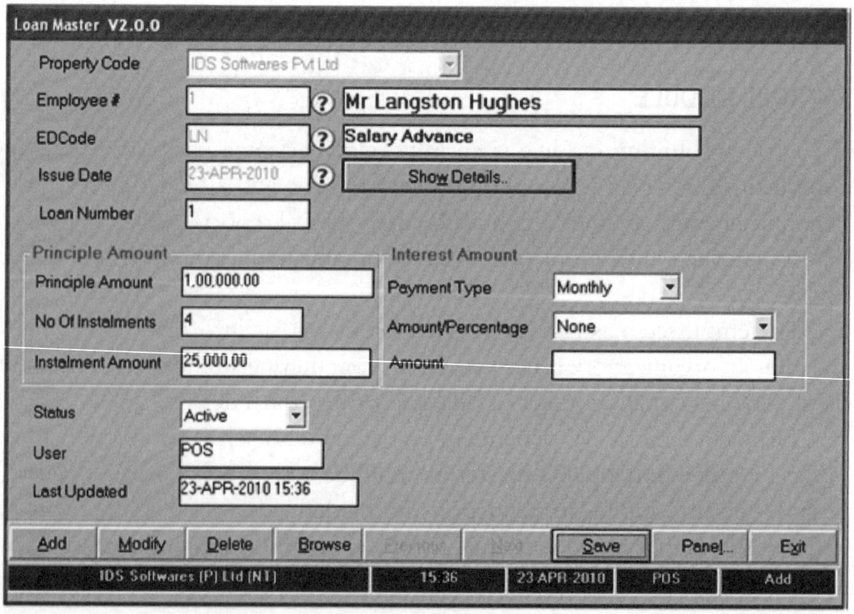

Fig. 8.21   Loan and advance taken by an employee

Courtesy: IDS

Attendance is recorded on a regular basis and each employee's leave, absenteeism, time in and out, breaks in duty, and if required overtime can also be accounted. A report can be generated on an individual's daily or monthly attendance. A monthly attendance report of all individuals working in an organization can also be obtained, as shown in Fig. 8.22.

Attendence Report

For Month: Aug/2005   Date   22/05/2006
For Category :All   Page No. 1

Name	Total Work Day	Sunday	Holiday	Absent	CL Taken	EL Taken
P C JOSHI	24.0	5.0	2.0	0.0	0.0	0.0
P.N.BHATT	25.0	4.0	2.0	0.0	0.0	0.0
BHUBAN C JOSHI	7.0	5.0	2.0	17.0	0.0	0.0
NAVEEN C JOSHI	24.0	5.0	2.0	0.0	0.0	0.0
RAM PRASAD	24.0	5.0	2.0	0.0	0.0	0.0
RUMAL SINGH	24.0	5.0	2.0	0.0	0.0	0.0
RUNWAR SINGH BISHT	-3.0	4.0	2.0	28.0	0.0	0.0
NATHU RAM	25.0	4.0	2.0	0.0	0.0	0.0
RHIM SINGH	25.0	4.0	2.0	0.0	0.0	0.0
SHAMSHER SINGH NEGI	25.0	4.0	2.0	0.0	0.0	0.0
MOHAN CHAND PANDEY	25.0	4.0	2.0	0.0	0.0	0.0
N C PANT	25.0	4.0	2.0	0.0	0.0	0.0
GOPAL RAM	25.0	4.0	2.0	0.0	0.0	0.0
KAILASH S NEGI	25.0	4.0	2.0	0.0	0.0	0.0
ANAND PRASAD	25.0	4.0	2.0	0.0	0.0	0.0
SURENDER	25.0	4.0	2.0	0.0	0.0	0.0
BABLU KUMAR	25.0	4.0	2.0	0.0	0.0	0.0
BHUPAL S PHARTIYAL	25.0	4.0	2.0	0.0	0.0	0.0
ROOP SINGH	24.0	5.0	2.0	0.0	0.0	0.0
SHIV BAHADUR	25.0	4.0	2.0	0.0	0.0	0.0
TIRAM S ADHIKARI	22.0	4.0	2.0	3.0	0.0	0.0

Fig. 8.22   Attendance report (Copyright protected)

Source: http://www.datamannet.com/products/hospitality/hotel.php

Salary cheques can be generated and printed with the help of a payroll module. Salary details can also be automatically transferred to banks from where it can be credited to the employee account. Salary slips of individuals can be generated and also e-mailed. Figure 8.23 shows the format of an individual's payslip.

Pay Slip

For Month: Mar/2006
For Category :All
For Employee : N C PANT

PARTICULARS		EARNINGS		DEDUCTIONS	
Name	N C PANT				
Father's/Husband Name		Basic	2500.00	P.F.	
Designation	BILLING CLERK	D.A.		E.S.I.	200.00
Category	BILLING CLERK	H.R.A.		Loan Instt.	
P.F. No.		Conveyance		Advance	
Bank A/C No.		L.T.A.		Others Ded.	
Working Days	27.00	Medical		Income Tax	
Earn Leave Availe		Other Allow			
Loan Balance		OT Amt	0.00		
		Total	2500.00	Total	200.00
Net Salary	2300.00				
ACCOUNTANT				MANAGER/AUTHORISED SIGNATORY	

Fig. 8.23   Format of a payslip (Copyright protected)

Source: http://www.datamannet.com/products/hospitality/hotel.php

INVENTORY AND PURCHASE MODULES

Since the inventory module, purchase module, and stock control system are interlinked, we will discuss this together. Inventory refers to the stock of raw materials and finished goods available in an organization for production and sale. An inventory module ensures that each item is maintained at a proper level. An improper stock level can lead to the following problems:

- Low inventory of raw materials or unavailability of a material may lead to idle time in a production process and result in wastage of manpower resources. It may also lead to a guest being dissatisfied because of non-availability of items and hence loss of goodwill.
- High inventory of raw materials or prepared food leads to unnecessary investment, thereby increasing the financial burden of a hotel.

Therefore, maintenance of optimum level of inventories (neither high nor low) becomes critical for an organization. The following are the major objectives of implementing a computerized inventory management in an organization:

- Maintaining an optimum level of raw materials and finished products (food and beverage items)
- Preparation of purchase orders and inventory status reports accurately and on time
- Preparation of analysis reports
- Generation of management information system (MIS) reports that help management in effective decision-making.

Each item in a stock has to be accounted for. For this, an inventory master file is maintained. Details pertaining to product name, a brief description, product code, product purchase unit, unit price, order lead time, minimum–maximum stock level, and the date of last purchase are mentioned.

Stock Control Systems

A stock control system manages and controls the flow of stocks through an organization by recording the quantity and value of each stock at different locations of the hotel.

Stock control is based on the accounting principle that an item's opening stock and purchase must be equal to the item's closing stock and consumption. If both the results are not equal, it shows there is a variance, which means that some stock is missing.

It is very laborious and time-consuming to conduct stock control manually since every product received and issued has to be checked; also the previous closing figures have to be found and the theoretical closing stock calculated. In addition, the actual quantity has to tally with the theoretical tally. If both figures match, there is no variance. A computerized system removes repetitive, and error-prone calculations as the tasks are done automatically.

A stock control system is involved with receiving, issuing, and taking stock of items. This system can be used when each item in the stock is registered on the system. Unit prices, maximum and minimum stocks, and the reorder levels of items can be entered (as shown in Fig. 8.24).

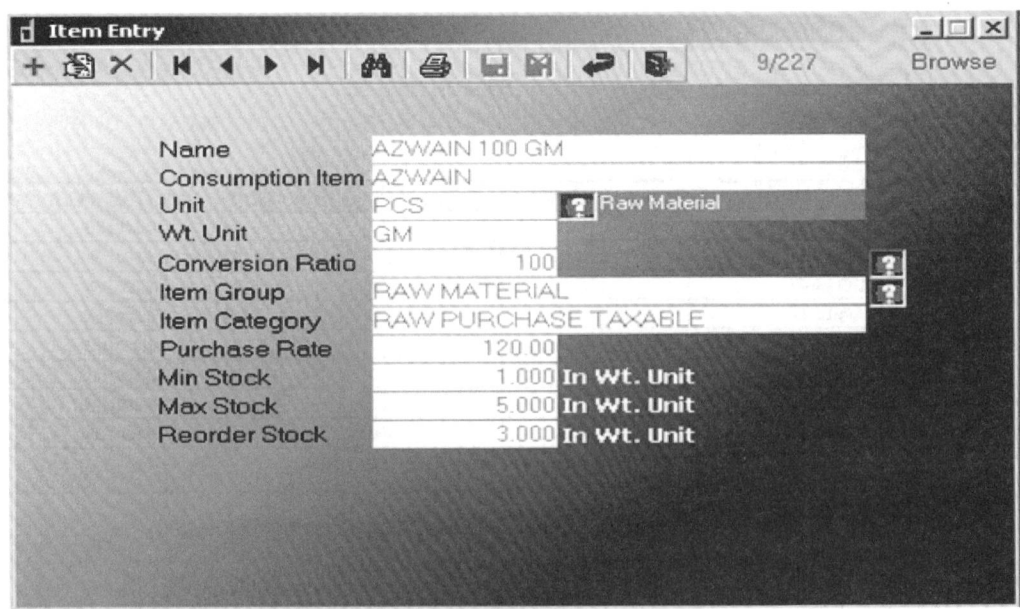

Fig. 8.24 Addition of a new item in the master (Copyright protected)

Source: http://www.datamannet.com/products/hospitality/hotel.php

When a store receives goods, a goods received note (GRN) is generated. As and when a stock item code is entered, the system displays its description and purchase unit. The quantity delivered and price are entered into the system. A material receipt report is generated, as shown in Fig. 8.25.

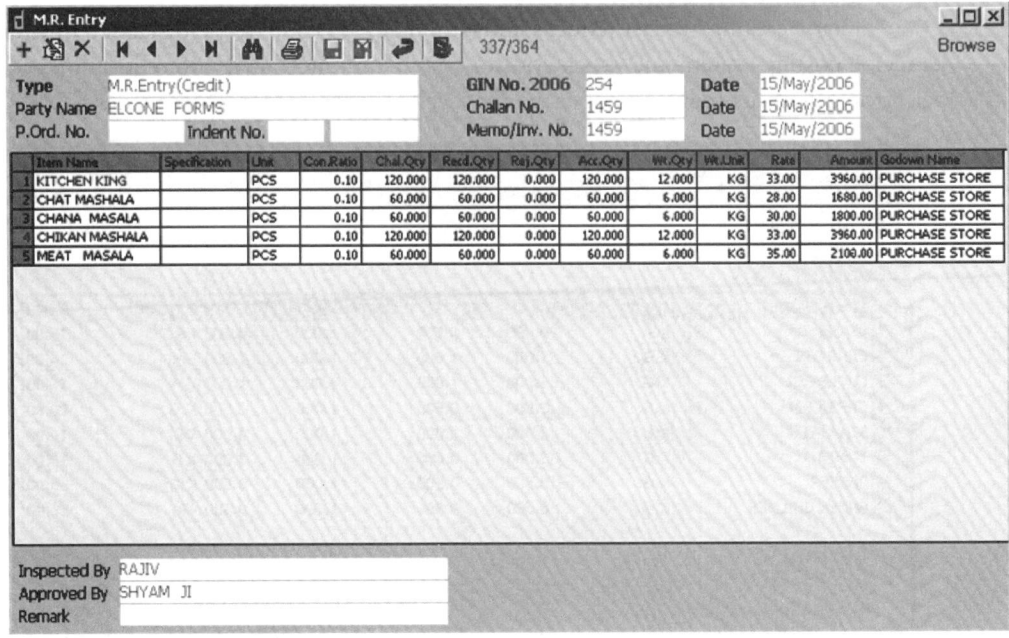

Fig. 8.25 Entry of material received (Copyright protected)

Source: http://www.datamannet.com/products/hospitality/hotel.php

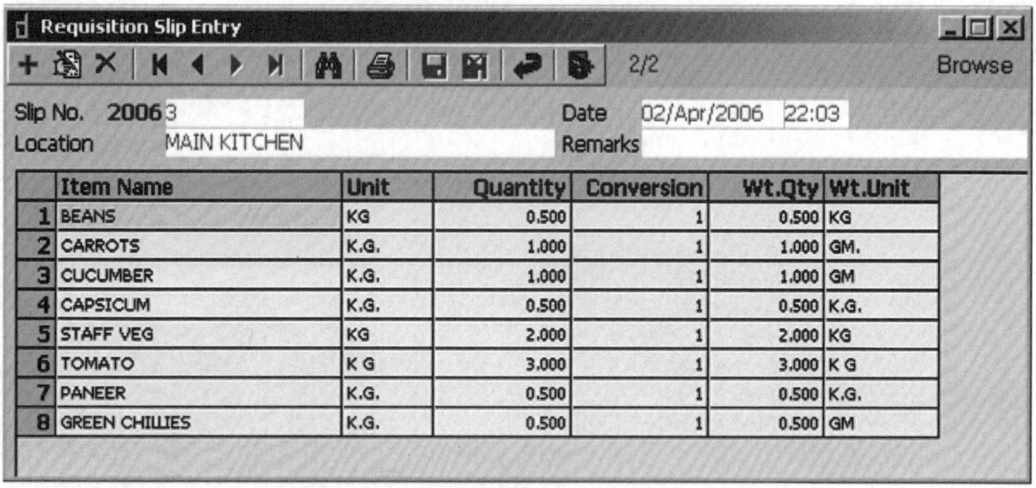

Requisition Slip Entry

2/2 Browse

Slip No. 2006 3 Date 02/Apr/2006 22:03

Location MAIN KITCHEN Remarks

	Item Name	Unit	Quantity	Conversion	Wt.Qty	Wt.Unit
1	BEANS	KG	0.500	1	0.500	KG
2	CARROTS	K.G.	1.000	1	1.000	GM.
3	CUCUMBER	K.G.	1.000	1	1.000	GM
4	CAPSICUM	K.G.	0.500	1	0.500	K.G.
5	STAFF VEG	KG	2.000	1	2.000	KG
6	TOMATO	K G	3.000	1	3.000	K G
7	PANEER	K.G.	0.500	1	0.500	K.G.
8	GREEN CHILLIES	K.G.	0.500	1	0.500	GM

Fig. 8.26 Sample requisition slip (Copyright protected)

Source: http://www.datamannet.com/products/hospitality/hotel.php

Stock is usually issued from a central store as and when a requisition is received from the kitchen or food production area. These requisitions were earlier drawn manually by the head chef who estimated the amount of each ingredient required to prepare the menu for a particular day. Sometimes these requisitions tended to be overestimated, resulting in wastage and increased costs. Figure 8.26 shows a computerized requisition slip.

As and when a requisition is received, the stores department issues the item. This is shown in Fig. 8.27.

A recipe costing system reduces and eliminates wastage (to be discussed in Chapter 9). If a recipe costing system and stock control system are interfaced, the requisition could be sent electronically, thus reducing paper work as well as helping in updating stock.

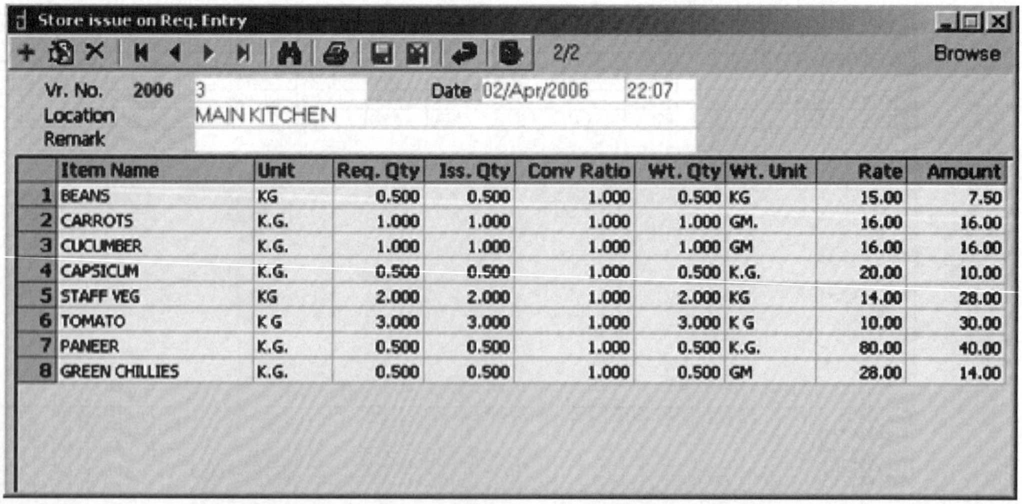

Store issue on Req. Entry

2/2 Browse

Vr. No. 2006 3 Date 02/Apr/2006 22:07

Location MAIN KITCHEN

Remark

	Item Name	Unit	Req. Qty	Iss. Qty	Conv Ratio	Wt. Qty	Wt. Unit	Rate	Amount
1	BEANS	KG	0.500	0.500	1.000	0.500	KG	15.00	7.50
2	CARROTS	K.G.	1.000	1.000	1.000	1.000	GM.	16.00	16.00
3	CUCUMBER	K.G.	1.000	1.000	1.000	1.000	GM	16.00	16.00
4	CAPSICUM	K.G.	0.500	0.500	1.000	0.500	K.G.	20.00	10.00
5	STAFF VEG	KG	2.000	2.000	1.000	2.000	KG	14.00	28.00
6	TOMATO	K G	3.000	3.000	1.000	3.000	K G	10.00	30.00
7	PANEER	K.G.	0.500	0.500	1.000	0.500	K.G.	80.00	40.00
8	GREEN CHILLIES	K.G.	0.500	0.500	1.000	0.500	GM	28.00	14.00

Fig. 8.27 Store issue on requisition (Copyright protected)

Source: http://www.datamannet.com/products/hospitality/hotel.php

The stock control system also provides other facilities to help manage the stocks in a business. The amount of each item issued during a period is recorded and calculated to obtain its usage rate. This helps in calculating which items are maximally and sparingly used. Items that are used in larger quantities can be procured in bulk at competitive prices, while the inventory of stocks that are used in lesser amounts can be reduced, helping in lowering holding inventory costs. The other function of a stock control system is automatic reordering. If a minimum stock level is set for each stock item, the stock control system can automatically identify the item/items to be reordered. This helps to ensure that all items are never out of stock by providing details about stock items and their tentative reordering quantities.

Purchase Module

Purchases in hotels are usually made by cash or on credit. The purchases are made by the purchase department on the basis of a requisition, which is duly sanctioned by the appropriate authority. The purchase of raw material or direct consumable goods is based on standard specifications such as description of quality, size, and weight from the purchasing agents or suppliers. Figure 8.28 depicts a purchase order. Usually, hotels employ purchasing agents for a year. The purchase policy depends upon size, type, and style of the hotel. Generally, the following methods are taken into consideration:

- Contract purchasing
- Quotations and tender
- Periodical purchasing
- Market purchasing

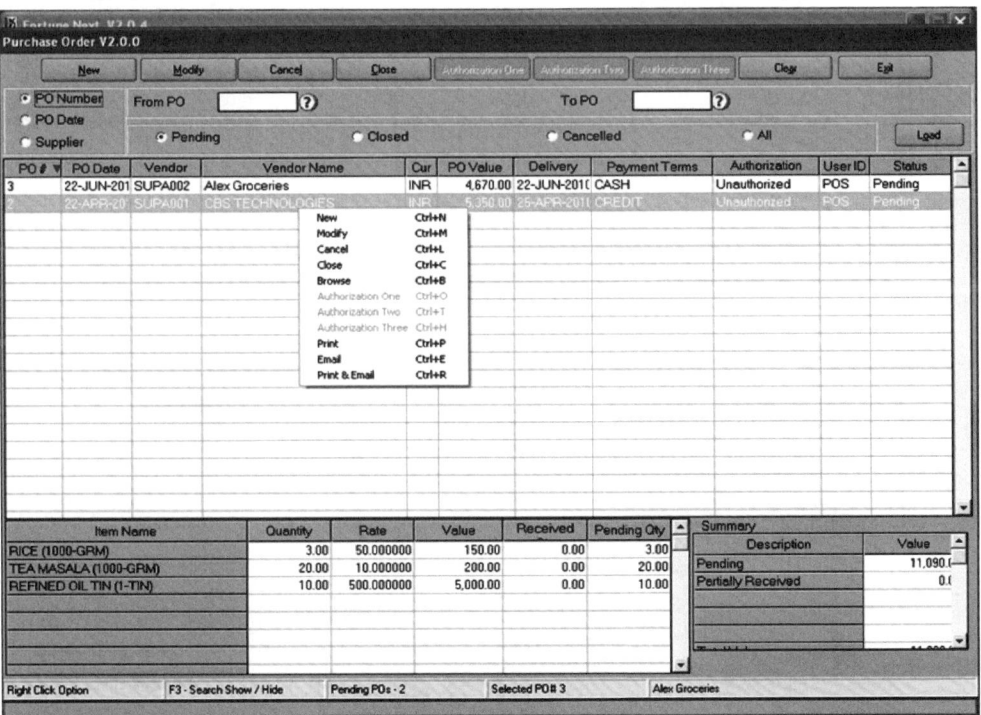

Fig. 8.28 Purchase order

Courtesy: IDS

In a purchasing module, a purchase order file is maintained. If the rates have been fixed by the tender system, then they are also fed into the system. This helps the management have control over the purchasing, ordering, and receiving practices since proper documentation is carried out. Using the minimum–maximum inventory level data transferred from the stock control module, the purchasing module generates purchase orders based on an order point established through usage rates and delivery time periods.

Generation of Reports

Guest accounting modules are capable of producing formatted statements and reports summarizing financial transactions that occur between guests and the hotel, monitor activities within revenue centres, and audit findings. The analytical capacity of this module has simplified traditional hotel auditing procedures while providing increased control over guest accounting procedures. The ability to print guest and non-guest folios at any time during the guest cycle provides management with important accounting information on a timely basis. Room statistics reports (as shown in Fig. 8.29) and revenue reports detailing occupancy loads and revenue generated from room rent give managers essential information related to occupancy percentages, average rate (per room and per guest), and departmental revenue summaries.

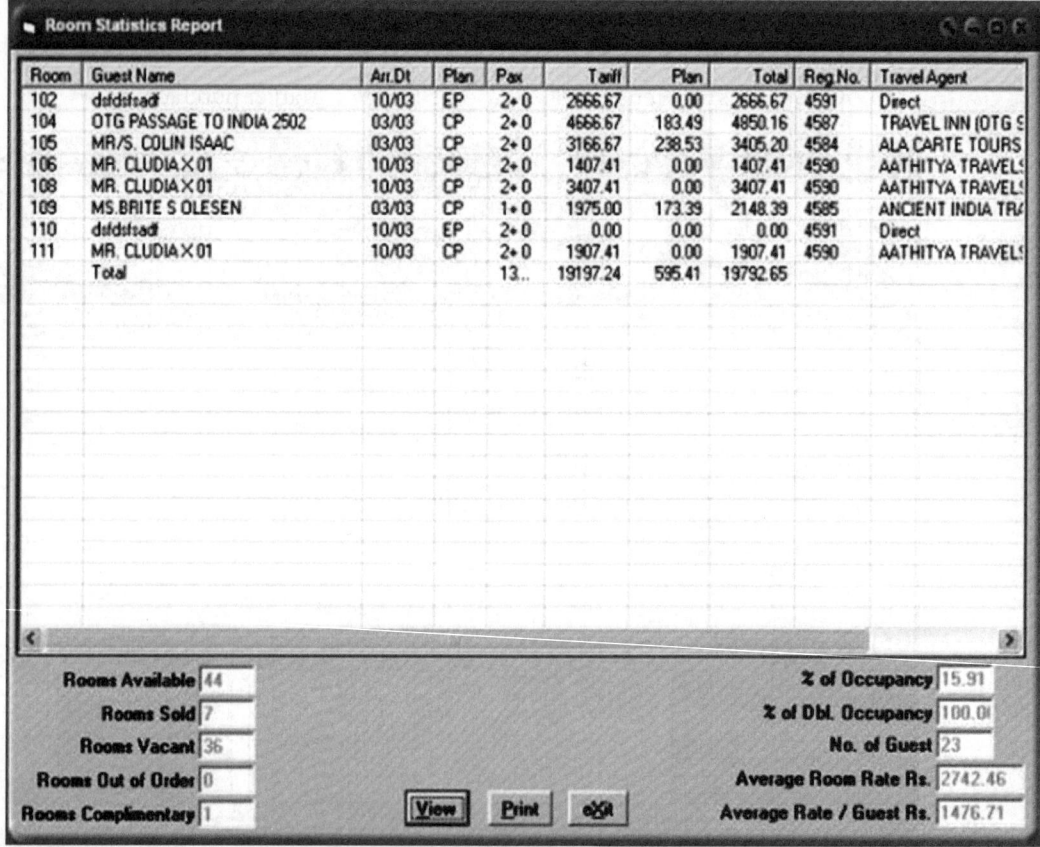

Fig. 8.29   Room statistics report

Source: http://www.hotelmanagementsystem.co.in/images/screens/Room%20Statistics.jpg

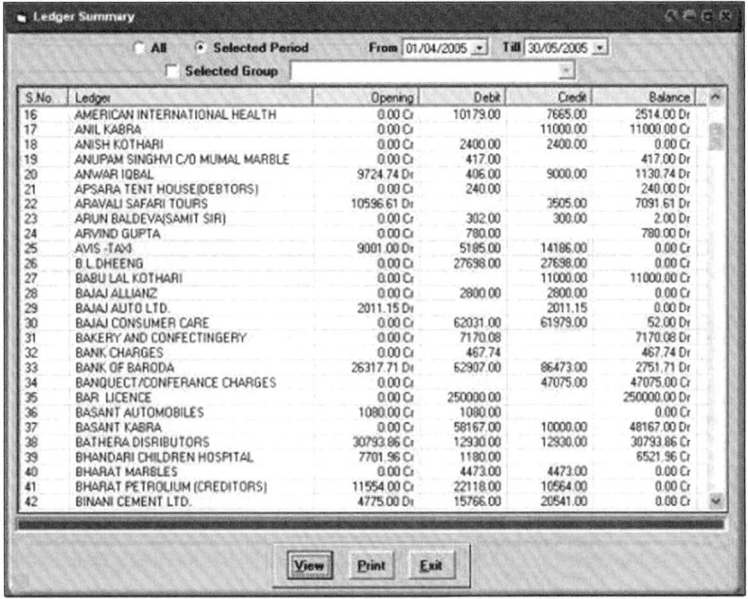

Fig. 8.30   Ledger summary report

Source: http://www.hotelmanagementsystem.co.in/images/screens/LedgerSummary.jpg

A ledger summary report is shown in Fig. 8.30. It shows presence of guests and non-guests, and credit card activity on the basis of beginning balance, cumulative charges, and credits. A credit card report is given in Fig. 8.31.

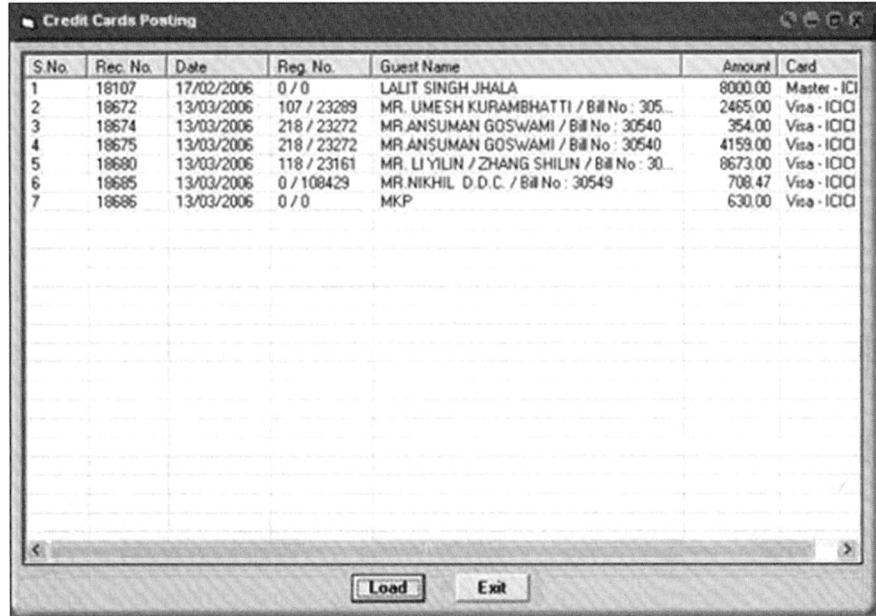

Fig. 8.31   Credit card report

Source: www.hotelmanagementsystem.co.in/images/screens/Credit%20Card%20Posting.jpg

Revenue forecast reports analyse expected revenue based on reservations in a hotel with the actual revenue. A daily revenue report gives a brief idea about this.

Guest cheque control reports compare guest cheques used in revenue outlets (food and beverage outlets) with source documents to identify discrepancies. Besides this, guest origin reports analyse foreign guests by the country of origin and also compare guest turnover in the present year with the previous year. A sample cashier report is shown in Fig. 8.32.

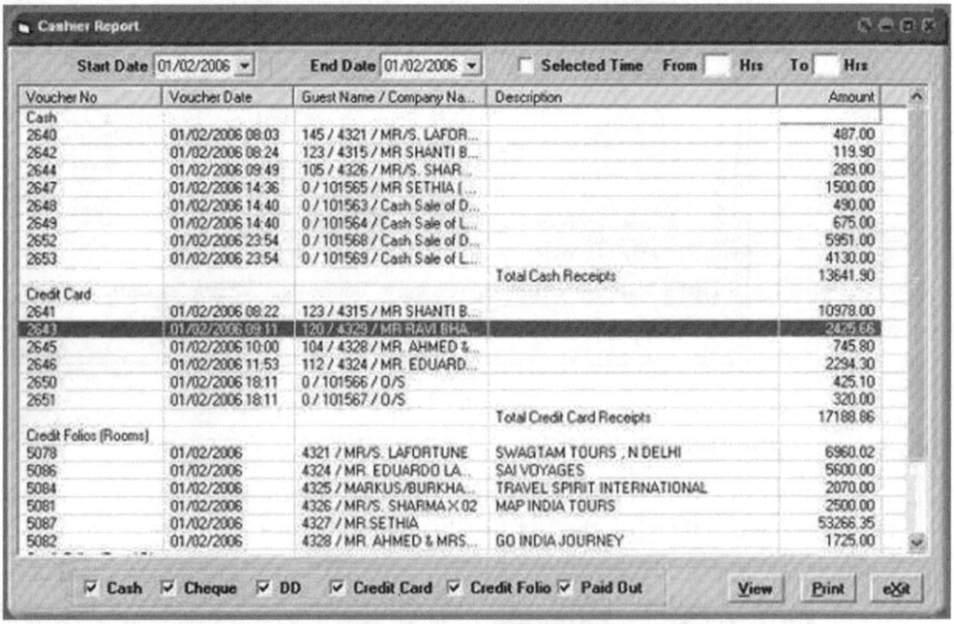

Fig. 8.32   Cashier's report

Source: http://www.hotelmanagementsystem.co.in/images/screens/CashierReport.jpg

SUMMARY

This chapter has been divided into two parts—guest accounting and general accounting modules. A guest accounting module increases a management's control over guest accounts and significantly alters the night audit routine. Guest accounts are maintained electronically, thereby eliminating the need for folio cards, trays, and mechanical posting equipment. The guest accounting module monitors predetermined guest credit limits and is capable of creating several types of folio formats. When revenue centers are interfaced with the property management system (PMS), remote electronic cash registers communicate with the front desk and guest charges are automatically posted to the appropriate folios. At checkout, outstanding account balances are automatically transferred to the city ledger (accounts receivable) for collection.

The general accounting module discusses accounts receivable and payable. The two modules talk about money owed and money to be obtained. The general accounting module also explains the payroll module using which, the attendance of each employee can be documented, and his/her leaves, absenteeism, and weekly offs can also be recorded. An employee's salary break-up, as stored in the master automatic salary is generated for each individual. The inventory module discusses how goods are kept in the stores as per requirements and the area of storage required (dry store, walk-in chiller, or deep freezer). The module also touches upon a par stock level and its interface with other systems for automatic changes in inventory after an issue (from any department of a hotel) for an item is registered. The purchase module discusses the various methods by which purchases are done in a hotel and how purchases are monitored so that an item is neither overstocked nor understocked.

KEY TERMS

Account payable module   This module is provided with the functionality to enter, monitor, maintain, and process payment of invoices and credit notes.

Accounts receivable   This module helps in tracking all invoices that are awaiting payment from customers.

Balance sheet   This is a financial statement of a company that shows the position of assets and liabilities on a particular date.

City ledger   It refers to a set of accounts that is used to record charges and payments from non-registered guests in a hotel.

Control folio   It is a folio that is used to track all transactions posted to other folios (individual, master, non-guest, or employee).

Employee folio   It is a folio that offers charge privileges to employees of the hotel.

Guest folio   It is a bill sheet that is kept updated by the front-office cashier and contains details of services availed by each guest.

Identification code   It is a code number that is generated by a computer for each resident guest and is used whenever a transaction is made.

Master folio   It refers to a folio, usually for more than one guest or room, and contains a record of transactions, which are not posted to individual folios. It is usually made for groups.

Night audit   It refers to a process of reviewing the accuracy and completeness of accounting transactions for a day.

Payroll module   This module generates the salary of each individual working in an organization.

Purchase module   This module helps in identifying the needs and thereby orders and documents the requirements after receiving requisites from respective departments.

Stock control   It is a process by which stocks are maintained in a store.

Split folio   If a group or two individuals sharing a room want the bills to be separated, then the guest folio is divided.

Trial balance   It is a system to check that all transactions are recorded correctly and the debit balance matches with the credit balance.

Uniform system of accounts for lodging industry (USALI)   It is a set of standard accounting procedures that are used for financial transactions in a hotel.

REFERENCES

Baker, S., J. Huyton, and P. Bradley, *Principles of Hotel Front Office Operations*, Continuum, London, 2001.

Connor, P., *Using Computers in Hospitality*, Cassell Publications, Sydney, 1996.

Gupta, S. and S. Gupta, *Computer Aided Management*, Excel Books, New Delhi, 2004.

Rawat, G.S., J.M.S. Negi, and N.K. Gupta, *Elements of Hotel Accountancy*, Aman Publications, New Delhi, 2005.

Woods, R., J. Ninemeir, D. Hayes, and M. Austin, *Professional Front Office Management*, Pearson Education, New Delhi, 2008.

Web References

http://datamannet.com/presentation, last accessed on 18 October 2011.

http://support.resortdata.com/RDPWin/Help/Content/Res/Options/Folio.htm&docid, last accessed on 6 November 2011.

http://www.accounting4manager.com/wp-content/uploads/2008/04/system-of-controls-for-computerized-accounts-payable.jpg, last accessed on 1 November 2011.

http://www.anandsystems.com/hotel_software/reports.html, last accessed on 1 February 2012.

http://www.datamannet.com/products/hospitality/hotel.php, last accessed on 13 February 2013.

http://www.docstoc.com/docs/12578160/Tax-Calculator-2007-08-

09-Version-2-and-3&usg, last accessed on 12 February 2013.

http://www.hotelmanagementsystem.co.in/images/screens/CashierReport.jpg, last accessed on 6 November 2011.

http://www.hotelmanagementsystem.co.in/images/screens/FlashReport.jpg, last accessed on 6 November 2011.

http://www.hotelmanagementsystem.co.in/images/screens/LedgerSummary.jpg, last accessed on 6 November 2011.

http://www.hotelmanagementsystem.co.in/images/screens/Room%20Statistics.jpg, last accessed on 6 November 2011.

http://www.hotelmanagementsystem.co.in/images/screens/Trial%20Balance.jpg, last accessed on 6 November 2011.

http://www.hotelogix.com/videotut/flash/night_audit/night_audit_overview.html, last accessed on 1 November 2011.

http://www.managementstudyguide.com/erp-accounts-payable-receivable.html, last accessed on 29 October 2011.

http://www.realpowersoft.com/hr_module.html, last accessed on 16 October 2011.

http://www.scribd.com/doc/29919138/Hotel-Property-Management-Systems, last accessed on 4 October 2011.

www.hotelmanagementsystem.co.in/images/screens/Credit%20Card%20Posting.jpg, last accessed on 13 February 2013.

www.idsnext.com, last accessed on 14 April 2012.

www.softnology.com/PPT/Download_ERMS_PRESEN... - United States, last accessed on 10 October 2011.

EXERCISES

Concept Review Questions

1. What is the difference between the accounts receivable and account payable modules?
2. How does the payroll module generate cheques for salary?
3. What are the functions of the inventory management module?
4. How is the stock control module interfaced with an account management module?
5. How does computerization help in the night audit process?
6. What is a purchase module? How is it linked with the inventory module?
7. How does a guest accounting module operate?
8. What is the difference between a guest ledger and city ledger?
9. What are the kinds of reports generated by an accounting module?
10. What are the points to be kept in mind while designing an inventory module for a hotel?
11. In having a computerized night audit, what points should be considered?
12. What fields must be considered while setting up a payroll module?

Multiple Choice Questions

1. A guest who pays a bill through a credit card is posted to
   (a) a guest ledger
   (b) a city ledger
   (c) either (a) or (b)
   (d) none of these
2. An inventory management module helps in
   (a) reducing inventory cost
   (b) maintaining par stock
   (c) better accountability
   (d) all of these
3. A payroll module generates payslips on the basis of
   (a) basic salary and perks
   (b) attendance of the employee
   (c) both (a) and (b)
   (d) service period
4. In the night audit process
   (a) departmental financial reconciliation is done
   (b) posting of room and tax is done
   (c) trial balance is run
   (d) all of these
5. An account payable module
   (a) registers incoming invoices
   (b) prepares and prints cheques
   (c) does credit management
   (d) none of these

Project Work

1. In a guest accounting module, carry out the following actions:
   (a) Check-in and checkout a guest
   (b) Check-in a guest and create a folio
   (c) For group check-in, create separate folios for each guest
   (d) Checkout and settle the accounts of a guest
   (e) Settle the accounts and generate bills of the guest
2. Perform the following actions in an inventory management module:
   (a) Add a new item in the item master
   (b) Mention par stock of the item
   (c) Issue and receive an item
   (d) Generate a purchase order
   (e) Make a goods received note
   (f) Make a vendor list

CASE STUDY 1

Neeraj, a waiter at Krishna International, has been working for the last three years. Being in the food and beverage service, he is occasionally allotted a break-shift, wherein he can rejoin in the evening at 7 p.m.

It was found by Dhiraj, his duty manager that he usually reports late for the evening shift. On further enquiry, it was also found that he usually forgets to mark time-out when he goes for a break or somebody else signs for him even when he is actually late. The hotel has a swipe card system (magnetic card), which is maintained by the time office. When a person joins duty, he/she has to sign in (swipe in console) and when he/she goes for a break or completes duty, he/she has to sign out. For a break duty, an individual has to sign in and sign out twice a day. You have been appointed as the HR manager of the hotel.

(a) What steps will you take to ensure proper time in and time out is done by all employees of the hotel?
(b) What type of system or rectification in the current system would you suggest, so that errors in noting time in and time out are minimized?
(c) What is the role of security in maintaining proper documentation of attendance?

CASE STUDY 2

Sujit, a front-office assistant at Hotel Vikram, a three-star property in the heart of the city, was on duty on the night of 29 September. A family of five without prior reservation came to the hotel. As the guests were about to carry out the check-in formalities, one of the ladies suddenly fell sick. Since it was around 2' o clock in the morning, Sujit hurriedly allotted rooms 301 and 306 to the family, without asking for any advance deposit or credit card guarantee. Sujit was alone on the shift with a trainee. The night auditor had completed his audit and had gone for a break. Sujit neither passed on the information nor mentioned it in the system. The personnel who came in the following day were satisfied to see the registration card and the generated folio. During the night audit process on 30 September, when the room rents and taxes were posted and verified, no amount was found against rooms 301 and 306. On checking with housekeeping, it was found that both the rooms were vacant. The contact details mentioned in the registration card were also found to be fake. Besides not paying the room rent, the guest had also ordered breakfast and lunch, which had been signed too. The total bill amount was around ₹15,000.

(a) What went wrong in Hotel Vikram?
(b) If you were the front-office manager, what would you have done?
(c) What steps should be taken in the account management module so that a similar incident is not repeated?

CHAPTER 9

Food and Beverage Management

LEARNING OBJECTIVES

After reading this chapter, you will be able to understand the following:

- Electronic cash register
- Point-of-sale (POS) hardware and software components
- Recipe management system
- Menu management system
- Sales analysis
- Automated beverage management system

Computer-based hotel property management systems (PMS) consist of modules like computer-based food and beverage management systems, which function with specific hardware components and a few application software packages. This chapter will discuss the service-oriented

---

EXHIBIT 9.1   Efficient Use of Software

Tirthankar Bose, the coffee shop manager, had a brilliant career. He was selected as a management trainee from a catering college and within five years became an outlet manager.

Tirthankar had a pleasing personality, the reason for his out-of-turn promotion. Though the sales of the coffee shop were good, the cost of the food sold was considerably high. Executive chef Hemant Kalra had left the organization two months ago and the hotel was in the process of hiring a replacement. In the meanwhile, measures had to be taken to arrest the rising food costs. Vivek Kulbhaskar, the general manager of the hotel, summoned Tirthankar and asked him to look into the matter and come up with solutions.

The hotel had a computerized recipe management software, which stored the master recipes for all the dishes. Therefore, Tirthankar was under the impression that everything was fine, atleast theoretically. However, one day when food was actually being served to the guest, to Tirthankar's surprise he noticed that neither was the food as per the recipe nor was it portioned accurately. On further investigation it was also found that the wastages were also very high. Since individual chefs were preparing and plating dishes as they wished, there was no proper control mechanism.

Tirthankar came to the conclusion that without proper supervision and training, software alone could not solve all issues until human beings used it efficiently.

applications of a computer-based food and beverage management system. These applications rely upon electronic cash register (ECR) and point-of-sale (POS) technology to monitor service area transactions through pre-check terminals, remote work station printers, displays, printer controllers, and registers.

Similar to other computer hardware components, ECRs and POS terminals require various software programs that instruct them on what to do, how to do, and when to carry out tasks. ECR/POS software not only direct internal system operations, but also maintain various files and produce reports for the management's use. In this chapter, we will also discuss the types of data stored in major ECR/POS files and the information contained in some of the reports generated by them.

ECR/POS HARDWARE COMPONENTS

Although some food service operators and computer system vendors use the terms 'register' and 'terminal' interchangeably, the terms actually refer to different equipment functions. In this chapter, the term 'register' refers to an ECR/POS device that is connected to a cash drawer whereas all other ECR/POS devices are called terminals.

Since ECR/POS devices are sold as modular units, everything other than the basic terminal is considered optional equipment. The cash drawer is no exception. An organization can connect up to four cash drawers to a single register. Multiple cash drawers enhance the management's cash control system as several cashiers work on the same register during a shift. Each cashier can be assigned a separate cash drawer so that at the end of the shift, cash drawer receipts are individually consolidated and tallied. This helps the management calculate the sale during each drawer (shift) in detail.

A terminal not having its own cash drawer is called a pre-check terminal. Such terminals are used to only enter orders but not settle guest accounts. A waiter can use a pre-check terminal located in a dining area service station to relay the orders to the kitchen and bar production areas; however, the same terminal cannot be used to settle guest bills.

An ECR/POS device with a cash drawer can support both pre-checking and cashiering functions. For example, an employee at a cashier desk in a hotel or restaurant might serve as a cashier for the food service outlet as well as an order-taker for the room service. When a room service employee is answering calls, he/she uses the register as a pre-check terminal. The register there relays the room service orders to the appropriate kitchen and bar production areas. When the order has to be delivered by room service, the room service employee picks up the printed guest bill from the cashier's desk. After the order is delivered, the room service employee presents the settled guest bills to the cashier, who then uses the register to close the guest cheque within the system. POS and ECR are explained in Chapter 10.

The components of an ECR/POS system hardware include keyboards, display screens, various printers, and a printer controller. The first part of this chapter talks about the hardware components. Keyboards are examined by keyboard design, types of keys, and keyboard overlays. Sections on display screen addresses, size and function of the operator, and customer displays have been included. The characteristics of guest cheque printers, receipt printers, remote workstation printers, and journal printers have been discussed next. The final section focuses on the functions of a printer controller.

Keyboards

There are two primary types of keyboard surfaces: micro-motion and reed style. The micro-motion keyboard designs have a flat, water-proof surface, whereas the reed style keyboard design contains water-proof keys, which are raised above the surface of the keyboard. The keyboard is provided with a number of hard and soft keys. Hard keys are dedicated to a specific function as they have been programmed by the manufacturer. On the other hand, soft keys can be programmed by users to meet the specific operational needs of a restaurant/hotel.

Both keyboard designs can support interchangeable menu boards. A menu board overlays the keyboard surface and then identifies functions performed by a key during a specific meal period. Menu boards can be developed to meet the precise requirements of each outlet. Menu boards for both micro-motion and reed style keyboard designs have a number of keys. The key types may include the following:

- Preset keys
- Price look-up (PLU) keys
- Function keys

- Settlement keys
- Modifier keys
- Numeric keypad

A waiter usually enters orders by using preset keys and price look-up (PLU) keys. Modifier keys may be used in combination with preset and PLU keys to list the preparation instructions for various steaks (rare, medium, or well-done) for the food production departments. Modifier keys may also be used to alter the rates according to the portions available (small, medium, and large). A numeric keypad, on the other hand, is used for data-entry operations, thus helping cashiers find items by price if the cost of an item is not identified by the preset or PLU keys. The function keys and settlement keys are used to correct and complete the transaction. Touch screen ECR/POS devices and magnetic strip readers will soon replace traditional keyboard entry procedures. These automation advances have been described in detail later in the chapter.

Restaurant managers or outlet managers usually only determine and fix the positioning of most keys on the keyboard overlay. By posting keys for similar items and functions together, and arranging the groups logically, managers can improve system performance and increase employee output. The following sections discuss the types of keys found on ECR/POS system keyboards.

Preset Keys

These keys are programmed to maintain the price, descriptor, department, tax, and inventory status for a limited number of menu items. Automatic menu pricing speeds up guest service, eliminates pricing errors, and permits greater menu flexibility. The term descriptor refers to the abbreviated description of the menu item, for example, 'DLMAKH' for *Dal Makhani* or 'TNDCHK' for *Tandoori Chicken*. However, some systems vary the length of descriptors to make it 8–10 characters long.

The preset key is linked to a department code and the printer routing code. The department code refers to the menu category to which the preset item belongs, such as starter, main course, dessert, and so on. A printer routing code is used to direct preparation instructions to the proper food production area or section. For example, if a tandoori item is ordered, it is printed in the tandoor section printer, whereas a dessert is printed in the dessert section. This helps each section to sequentially prepare and serve the orders in time without misplacing any of them.

Once a preset key is pressed, the description of the item and its price are retrieved from the memory, and appears on the operator's display screen. This data can be relayed (along with preparation instructions) to the appropriate production station and can also be printed on the guest bill. In addition, the sales represented by this transaction can be retained for revenue reporting and for tracking inventory levels. Sales data of individual items are important for guest bill totaling as well as for management reports. Figure 9.1 shows a POS screen with various keys.

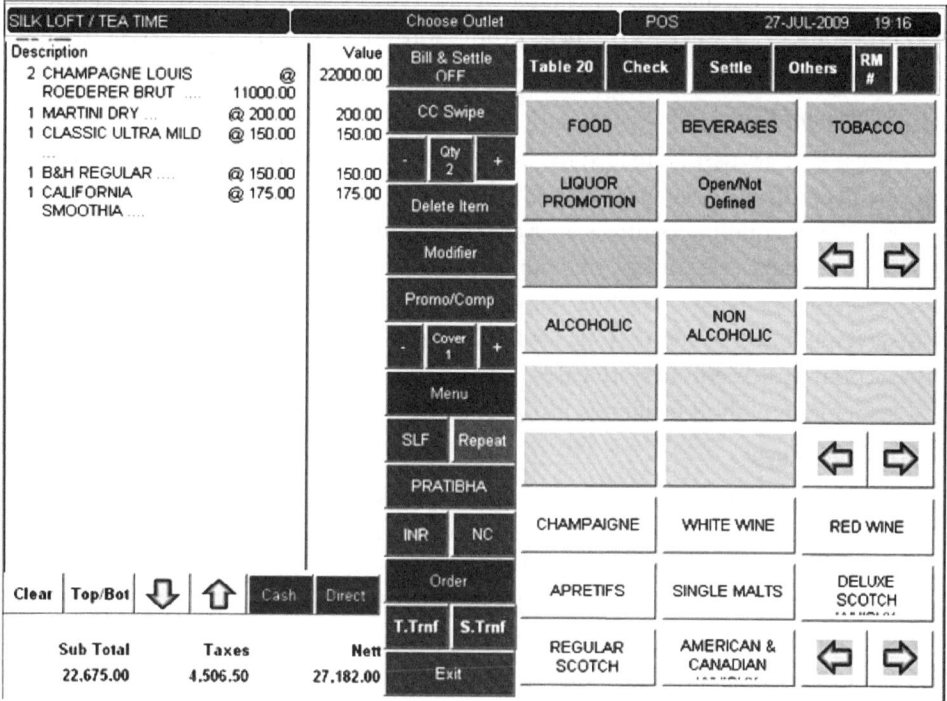

Fig. 9.1   POS screen with options such as choose outlet and settle

Courtesy: IDS

Price Look-up Keys

As terminals have only a limited number of preset keys, PLU keys are also used to supplement various transaction entries. These keys operate just like the preset keys; the only difference is that they require the user to identify the menu item by its reference code number (up to five digits) rather than name or descriptor. A waiter entering an order for coffee would merely press the item's designated key. In the absence of a coffee preset key, the order-taker would enter the item's code number (e.g., 025) and then press the PLU key. PLU keys perform the same functions as preset keys. Preset keys and PLU keys enable the system to maintain a file of most menu items in terms of price, descriptor, tax, department, and even inventory status.

Function Keys

While the preset and PLU keys are used for the purpose of entering orders, function keys assist the user in processing the transactions. The sample function keys are as follows: clear, discount, void, and no-sale. Function keys are important for error correction (clear and void), legitimate price alteration (discount), and proper cash handling (no-sale). A restaurant may sometimes attempt to increase its weekly lunch sales by issuing coupons to nearby local businesses, give discounts to regular guests, or on special occasions (anniversaries or birthdays). When such a coupon is used at the time of settlement, the cashier enters the value of the coupon/discount amount or percentage to be redeemed at the time of settlement; the cashier typically enters the value of the coupon and then presses the 'discount' key. The value of the coupon is credited to the guest bill and the remainder of the bill is settled through normal standard settlement procedures. The success of these promotions in a day can be tracked if the system also prints itemized discounts and daily discount totals.

Settlement Keys

These keys are used to record the methods by which accounts are settled; they could be via cash, credit card, house account, charge transfer, or other payment methods. Settlement keys increase revenue accounting controls as it classifies transactions at the time of settlement. Although a restaurant may use any one of a number of revenue accounting methods, most operations either use server banking or cashier banking. Server banking assigns responsibility for guest cheque settlement on the server. Cashier banking, on the other hand, involves a non-server handling account settlement. In both cases, tracking the identification of the banker and the transaction settlement method facilitates fast and accurate sales reconciliation.

Modifier Keys

These keys allow waiters to relay special preparation instructions (e.g., spicy or less spicy) to remote workstation printers or video display screens located in the food production department. A waiter enters the item ordered and then presses the appropriate preparation modifier if required. Modifier keys may also be used to alter the menu items' prices. For example, these keys may be useful in a restaurant that sells a dish as full portion and half-portion; a single preset key can be designated for a dish per portion and a modifier key can be programmed as a half-portion modifier. When a half-portion is sold, the waiter simply presses both the dish preset key and the half-portion modifier key to register a half-portion sale. The system maintains the total dish revenue by adding the total rupee amount for the half-portion sale only. The system after the sale also adjusts the inventory records accordingly.

Numeric Keypad

It is a set of keys that is used to assign menu items by price, access PLU data by menu item code number, access open guest cheque accounts by serial number, and perform other data entry operations. For example, if a register can also be used to record and store payroll data, an employee identification number can also be entered as and when an employee begins and completes his/her work shift. In addition, menu item code numbers can also be entered through the numeric keypad to access various files in order to make the management approve the adjustments. A numeric keypad is also used to enter report codes, which also initiate the production of management reports.

A step-by-step process shows how an order is added up and a bill generated from a POS system.

• A POS system's opening screenshot of a sample restaurant is shown in Fig. 9.2.

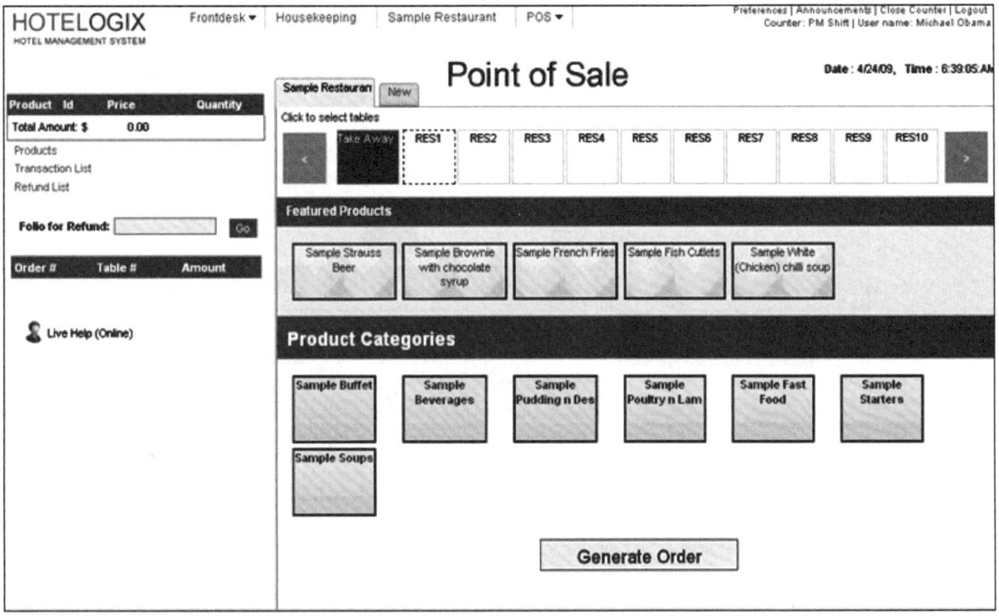

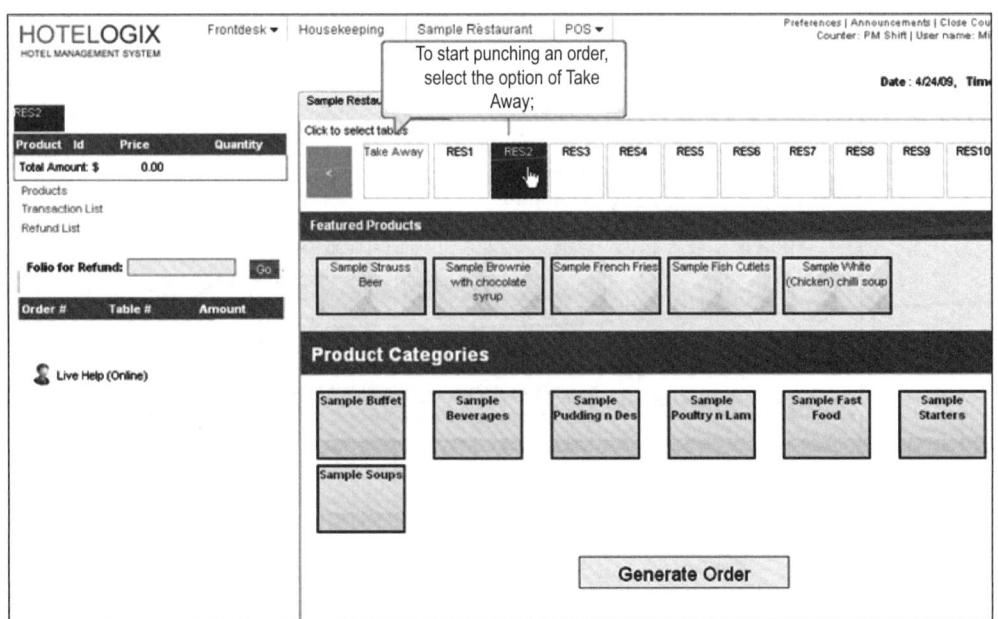

Fig. 9.2 POS of a restaurant

Source: http://www.hotelogix.com/videotut/flash/POS/How_to_make_a_Sale.htm

• A restaurant may have both dine-in or takeaway options, as shown in Fig. 9.3.

Fig. 9.3 Choosing from takeaway or dine-in options

Source: http://www.hotelogix.com/videotut/flash/POS/How_to_make_a_Sale.htm

If a guest opts for the dine-in option, he/she is allotted a specific table in the restaurant whose number is mentioned, as shown in Fig. 9.4. The same is mentioned in the right-hand column also.

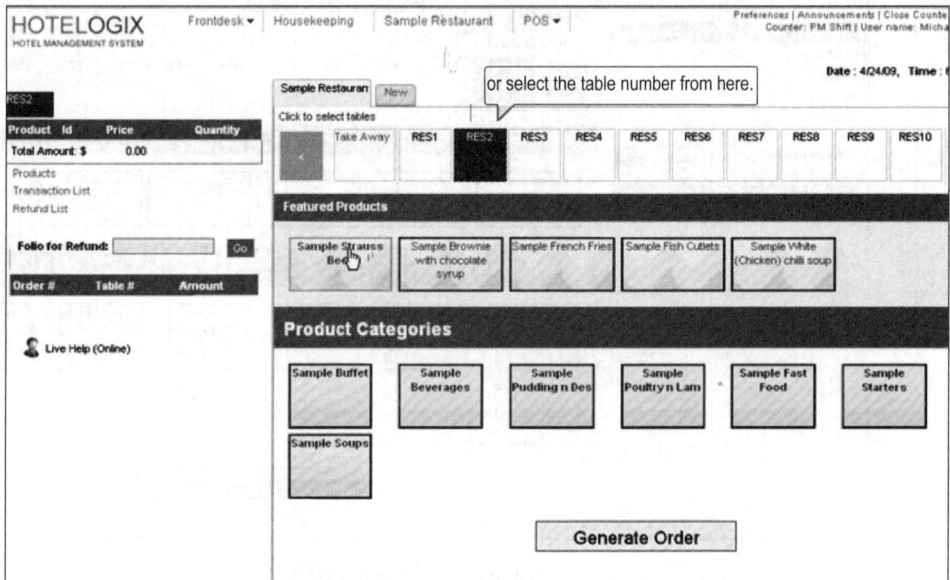

Fig. 9.4   Choosing table number in a particular restaurant
Source: http://www.hotelogix.com/videotut/flash/POS/How_to_make_a_Sale.htm

- After the table number is chosen, the products or dishes are chosen and entered by the server (as shown in Fig. 9.5).

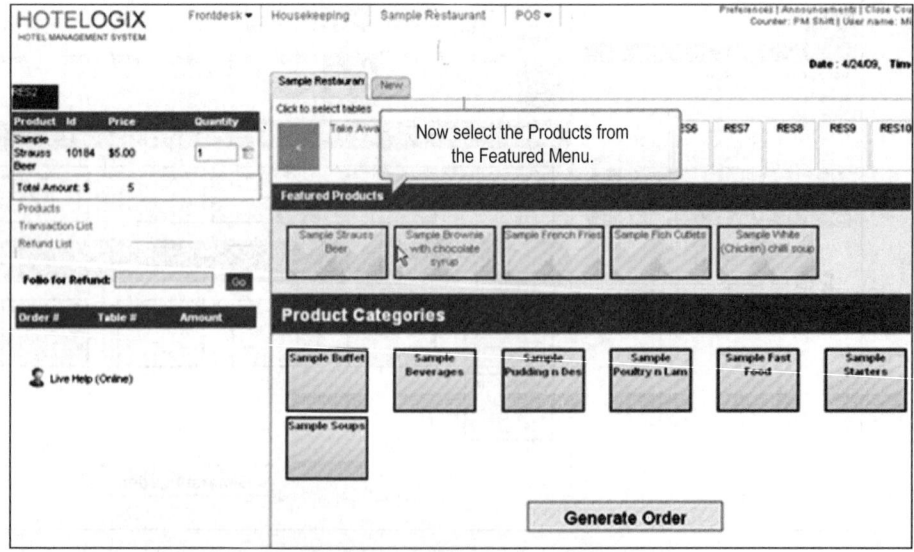

Fig. 9.5   Selection of products
Source: http://www.hotelogix.com/videotut/flash/POS/How_to_make_a_Sale.htm

- As the items are added with their respective quantities, the same gets displayed on the left-hand side column (as shown in Fig. 9.6).

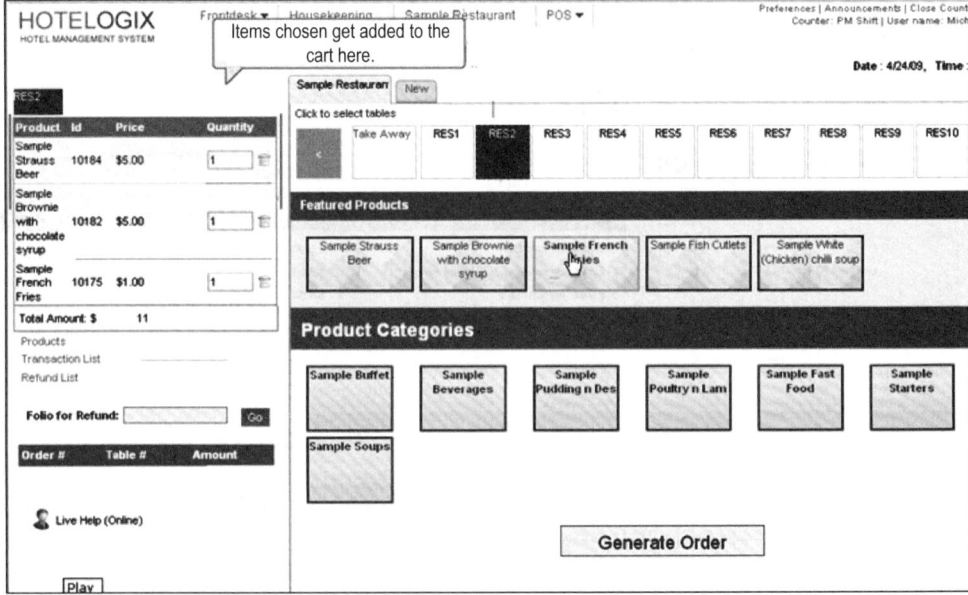

Fig. 9.6   Addition of items

Source: http://www.hotelogix.com/videotut/flash/POS/How_to_make_a_Sale.htm

- After all the items have been added, the order is generated (as shown in Fig. 9.7).

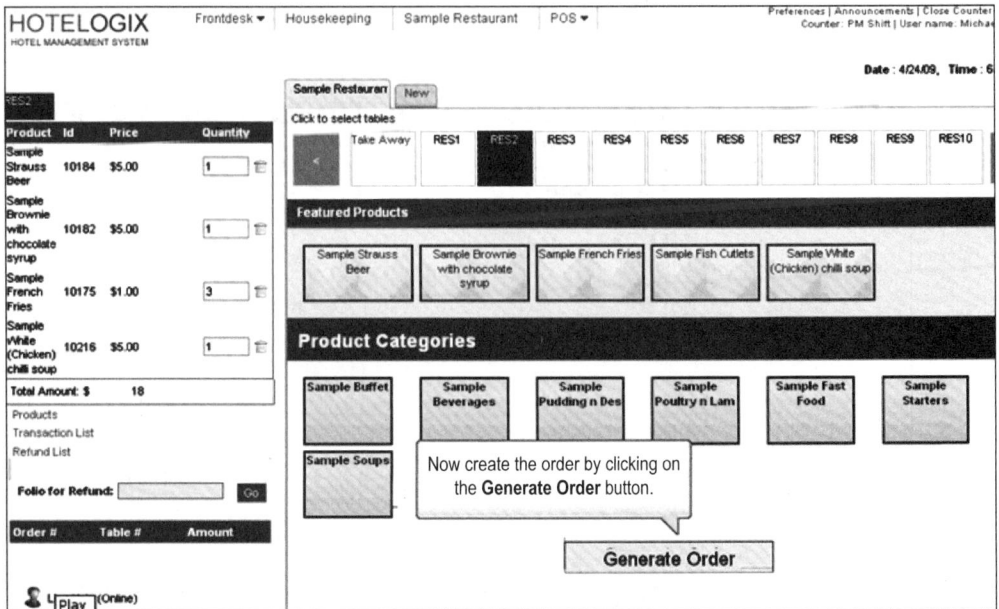

Fig. 9.7   Creation of an order

Source: http://www.hotelogix.com/videotut/flash/POS/How_to_make_a_Sale.htm

- After the order is generated, each item's unit price, number of units (qty), price, discount (if any) and the total amount, are shown (as shown in Fig. 9.8).

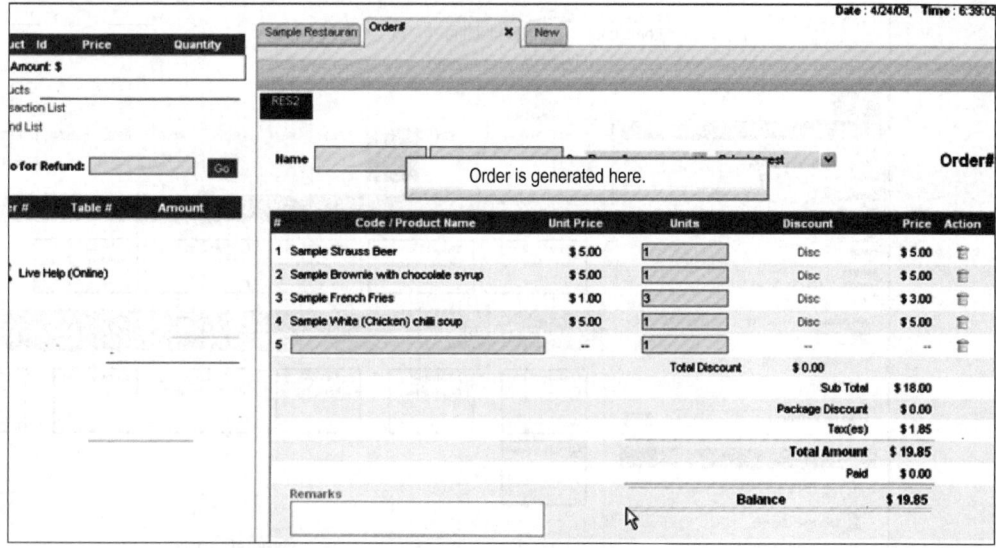

Fig. 9.8   Generation of an order

Source: http://www.hotelogix.com/videotut/flash/POS/How_to_make_a_Sale.htm

- After the order has been made, it is possible that there could be some alterations, as shown in Fig. 9.9.

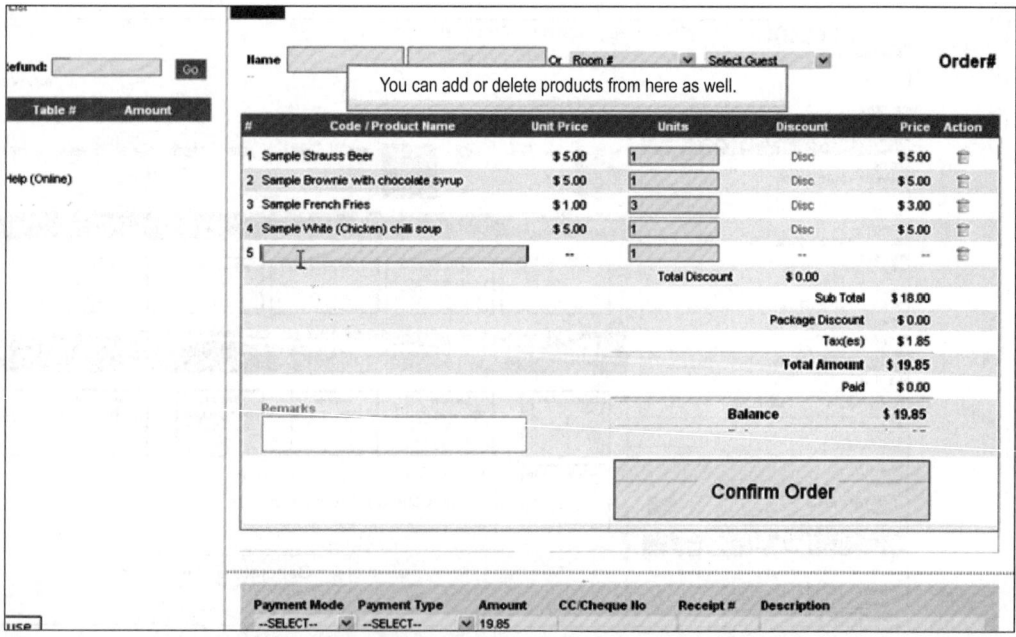

Fig. 9.9   Option to add or delete items in the order

Source: http://www.hotelogix.com/videotut/flash/POS/How_to_make_a_Sale.htm

- After alterations are done, the order is confirmed (as shown in Fig. 9.10).

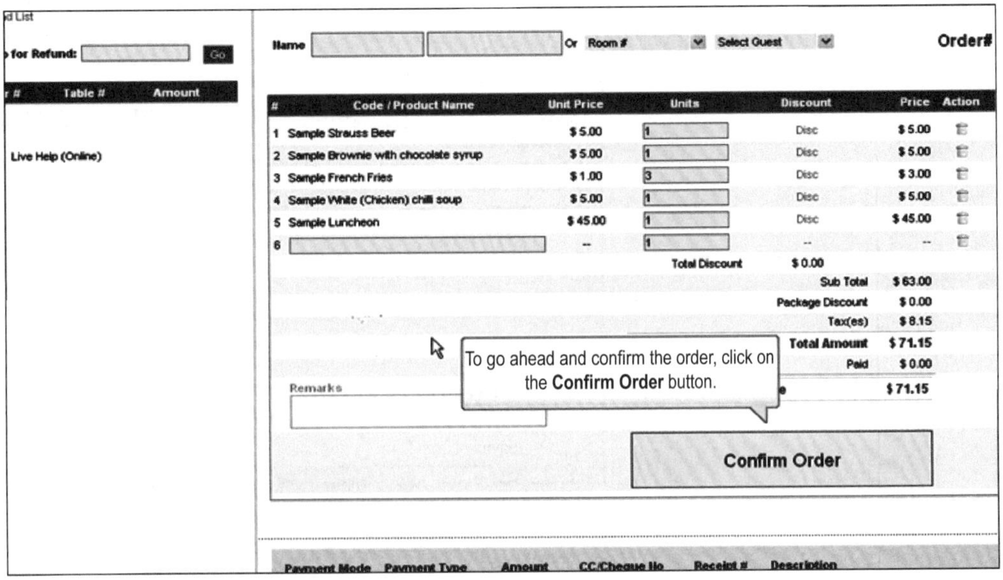

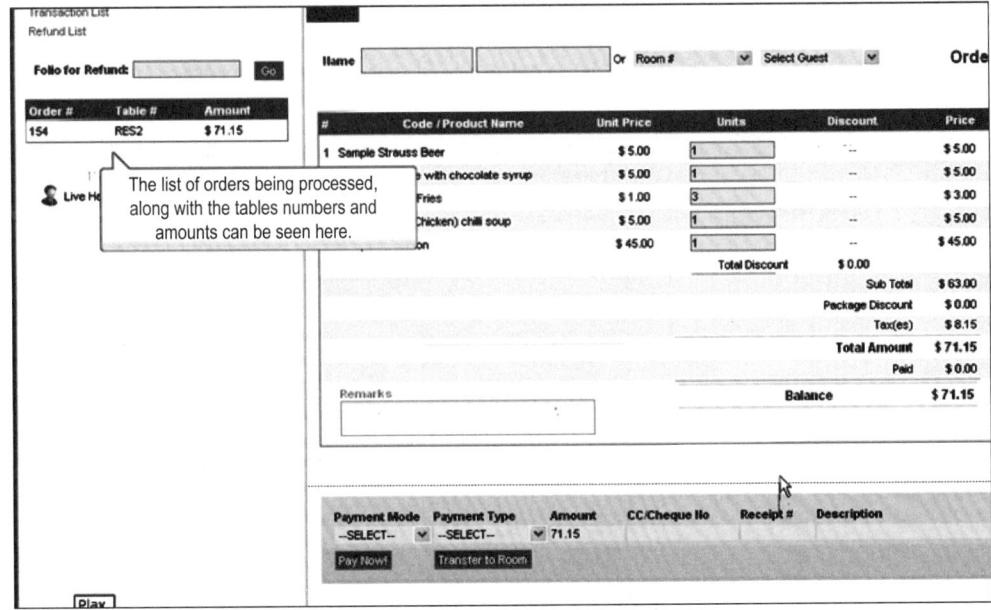

Fig. 9.10   Confirmation of the order

Source: http://www.hotelogix.com/videotut/flash/POS/How_to_make_a_Sale.htm

- After the order has been confirmed, the order number, table number, and the amount is shown in the left-hand column (as shown in Fig. 9.11). Only one order has been shown here but if there are multiple orders, those could be viewed on the screen as well.

Fig. 9.11   Viewing of the order, table number, and amount

Source: http://www.hotelogix.com/videotut/flash/POS/How_to_make_a_Sale.htm

- After the order has been prepared and served to the guest, it is time for the bill/cheque to be prepared. To pay the bill, guests are asked what mode of payment they would choose, as shown in Fig. 9.12.

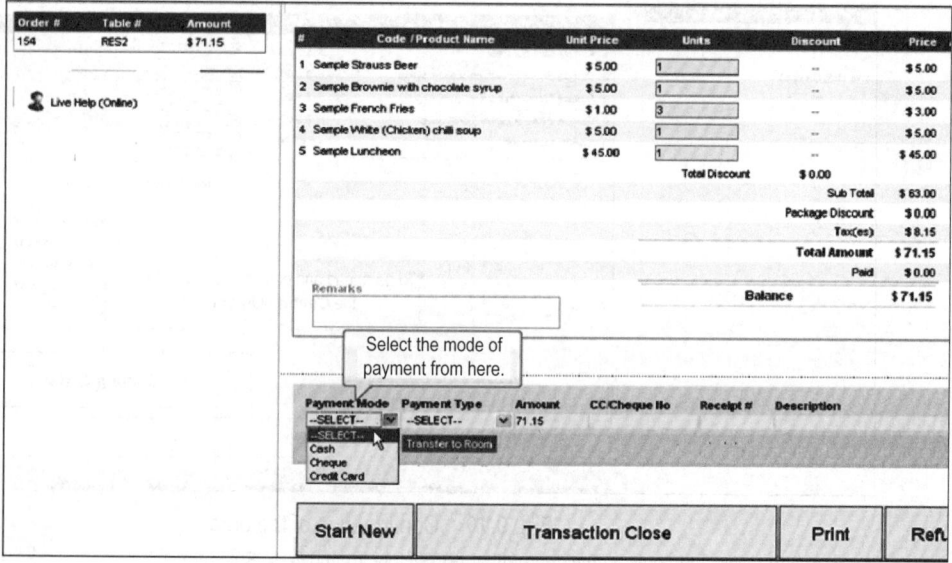

Fig. 9.12   Modes of payment

Source: http://www.hotelogix.com/videotut/flash/POS/How_to_make_a_Sale.htm

- A guest staying at the hotel may opt to pay the bill later. In that case, it is transferred to the room account. Sometimes, the guest may wish to pay the bill by cash (as shown in Fig. 9.13) via the 'pay now' option.
- After the payment has been made, the file for a particular transaction closes and is not visible.

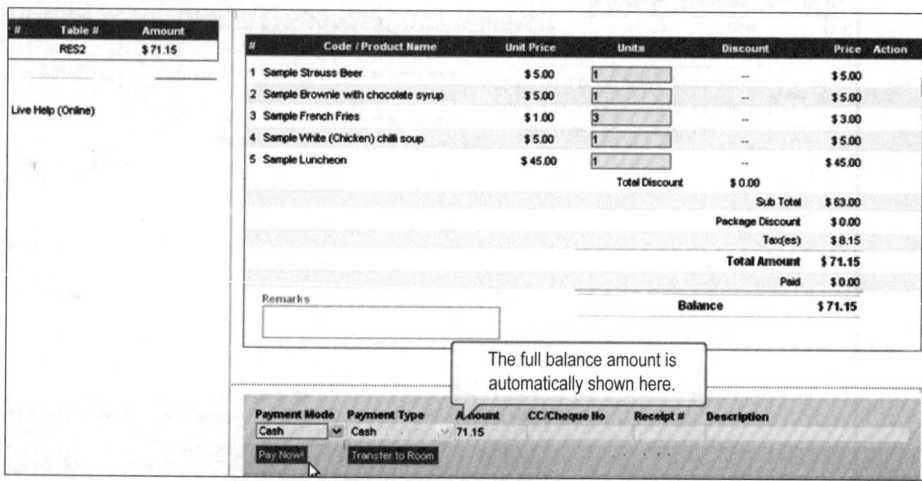

Fig. 9.13   Mode of payment selected with instructions to pay

Source: http://www.hotelogix.com/videotut/flash/POS/How_to_make_a_Sale.htm

Display Screens

Other than the keyboard, a register terminal typically also contains an operator display screen that supports a customer display unit as well. An operator display screen is generally a standard system component, which enables the operator to view and edit, if necessary, the transaction entries.

The unit allows the user to monitor transactions in progress and may also serve as a prompt for various system procedures. The length and number of lines that are displayed are often important considerations when selecting ECR/POS devices. Line lengths generally range from 7 to 80 characters whereas, the number of lines usually varies from 1 to 24. An operator display screen is encased in the primary housing of the ECR/POS device. This is not always true for customer display units.

The designs of customer display units consist of those which rest atop, inside, or besides the ECR/POS device. A customer display unit is more restricted in size and range when compared to operator display screens; it permits the guest to observe the operator's entries. In many table service restaurants, the settlement activities often take place in front of the guests; therefore, a customer display unit may not be required.

Customer display units also permit management to spot-cheque cashier activities. For example, an employee operating a cash register without a customer display unit might ring up a ₹500 transaction as ₹50. Later, to balance the register's cash, the employee might take the ₹450 for personal use. This kind of theft is riskier for the cashier if the terminal contains a customer display unit because the manager might observe the bogus ₹450 and take appropriate corrective action. Customer display screens are often more important for this purpose rather than for the assurance they offer to the guests.

Printers

Register printers are sometimes also classified as on-board or remote printing devices. On-board printing devices are normally located within six feet of the terminals they are connected to. These devices are usually guest cheque printers and receipt printers. Remote printing devices are workstation printers and journal printers, which are usually located more than six feet from the terminals they support. These printing devices usually require separate cabling from the terminal to the printer.

Guest Cheque Printers

These on-board printing devices are sometimes called slip printers. The guest cheque printers of most ECR/POS systems are capable of the following:

- Immediate cheque printing
- Delayed cheque printing
- Retained cheque printing

Immediate cheque printing refers to the ability of the system to print items as soon as it is fed into the terminal; delayed cheque printing prints items after the whole order has been entered in the system whereas, retained bill printing prints the guest bills at any time following order entry and before the settlement of guest bills. Sophisticated guest cheque printers can be equipped with an automatic format number reader (AFNR) and possess automatic slip feed (ASF) capabilities also.

An automatic slip feed (ASF) capability prevents overprinting of items and amounts on the guest bills. ECR/POS systems without an ASF capability require a waiter or cashier to manually insert a guest bill into the printer's slot and then align the printer's ribbon with the next blank printing line on the guest bill. This could be an unpleasant procedure for waiters to follow during a busy lunch or dinner period. If the alignment is improper, the guest bill would appear disorganized and messy with items or amount printed over one another or with gaps between the lines. A system with ASF capability retains the number of the last line printed for each open guest cheque with the top edge of the printer's slot, and the terminal automatically moves the cheque/bill to the next available printing line and prints the order-entry data. Since the guest bills are placed within the printer's slot the same way every time, the waiters may spend lesser time manipulating machinery and more time meeting the guests' requirements.

Receipt Printers

They are on-board printing devices that produce hard copy on a narrow register tape. Although the usefulness of a receipt printer is somewhat limited, these devices may help control the production of menu items that are not prepared at the departments receiving orders through the remote display or printing devices. For example, when the kitchen prepares desserts and the pantry area is not equipped with a remote communication device, desserts could be served without ever being entered into the system. If this happens, it is also possible that desserts could be served without amounts ever being posted to guest bills. This situation can be avoided with a receipt printer. Waiters preparing desserts are required to deliver a receipt tape to the dessert pantry area as proof that the items are properly posted to guest bills for final settlement. This procedure ensures that every menu item served is printed somewhere in the system, enhancing the management's internal control of the inventory.

Work Station Printers

They are remote printers usually placed at kitchen preparation areas and on service bars. As orders are entered at a pre-check terminal, they are sent to a designated remote workstation printer to imitate production.

The printouts correspond to the items printed on the guest bill. This communications system enables waiters to spend more time in meeting guest needs, significantly reducing traffic in the kitchen and bar areas, as waiters do not have to personally deliver the orders to the respective sections.

If there is no necessity for a hard copy output in the kitchen areas, video display units (also called kitchen monitors) can be viable alternatives to workstation printers. Since these units display several orders on a single screen, kitchen employees do not have to handle numerous pieces of paper, reducing the probability of error or misplacement. An accompanying cursor control keypad enables the kitchen employees to easily review previously submitted orders by scrolling down, full screen at a time. Figure 9.14 shows a screenshot of a kitchen display unit mentioning the name of the restaurant, number of covers, table numbers, time of the order, and also the status of the order (pending or in the process).

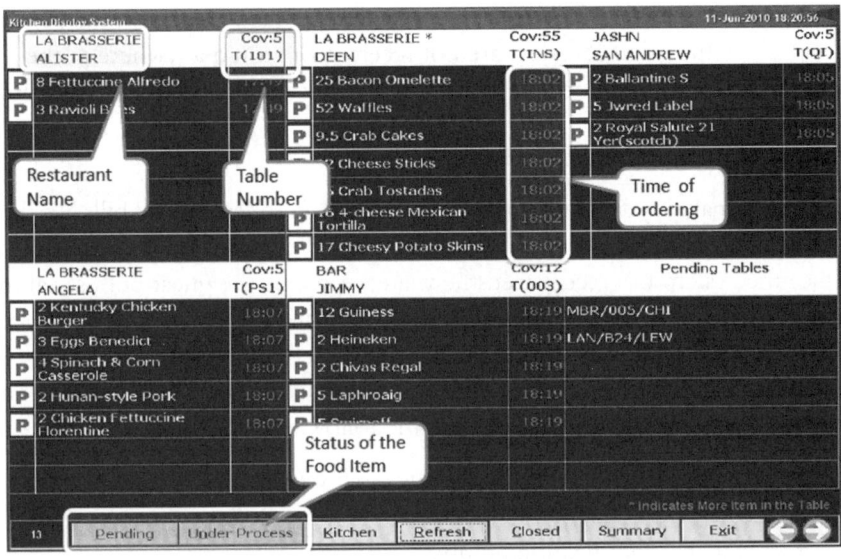

Fig. 9.14   Kitchen display screen

Source: www.idsnext.com

Journal Printers

Remote printers produce a continuous detailed record of all transactions entered anywhere in the system. Journal printers are usually located in secured areas, away from service and production areas. Hard copy is produced on a narrow register tape (usually 20 columns wide) and provides the management with a thorough system audit. In addition to providing an audit trail, journal printers also print a variety of management reports. The management routinely reviews the journal printouts to verify that the system is being properly used and there is no pilferage or fraud.

Printer Controllers

One of the most important peripheral devices in an ECR/POS system with remote workstation devices is the printer controller, also called a network controller. The printer controller coordinates communications between the cashier or pre-check terminals and workstation printers or kitchen monitors, while ensuring that waiters need to enter their orders only once.

When several pre-check terminals send data to the same workstation printer or kitchen monitor *simultaneously*, the printer controller immediately processes data from one of the terminals and temporarily stores (buffers) other communications until the printer becomes available. As the remote printer or kitchen monitor outputs data sent from one terminal, the printer controller sends the next set of data, and continues until all orders are printed or displayed. The printer does not get jammed and prints out orders in the same sequence as sent by the waiters. Since remote workstation units are typically very fast, the time delay between order entry and printout is minimal, even for those orders that are temporarily held by the printer controller.

Without the printer controller, a remote workstation unit would be able to receive and print only a single set of data at a time. When the remote printer is receiving data from one terminal, waiters entering orders at other pre-check terminals would receive a response similar to a telephone busy signal since the system would be busy. Orders would then have to be entered again, since the original orders are not received or stored anywhere in the ECR/POS system. The printer controller stops the repetition of work and the delay in placing orders to the kitchen.

COMPUTER-BASED GUEST BILL

Many automated systems use pre-printed, serially numbered guest bills like those used in manual guest billing systems. Before entering an order, the waiter 'opens' the guest bill serial number. Once the system has recognized the waiter, it opens the guest bill's serial number. After this, orders can be entered and relayed to remote printers in the production areas. The same items (with their selling prices) are printed on the waiter's guest bill.

Once a guest bill has been opened, it becomes part of the system's open bill file. For each opened guest bill, the file may contain the following data:

- Terminal number where the guest bill was opened
- Guest bill serial number
- Waiter's identification number
- Time guest bill was opened
- Menu items ordered
- Price of each menu item ordered
- Applicable tax
- Total amount due

A waiter adds orders to the guest bill at the pre-check terminal by first inputting the guest bill's serial number and then entering additional items.

Newer systems have eliminated the traditional guest bill altogether. These systems maintain only an electronic file of each open guest bill. A narrow, receipt-like guest bill can be printed at any time during the service, but is usually not printed until after the meal is served and when the waiter presents it to the guest for settlement. Since no paper forms are used, the table number is often used in tracking multiple bills per table. While presenting the bills to guests for settlement, the receipt-like guest bills can be inserted in high-quality paper, vinyl, or leather presentation jackets.

Electronic cash registers and POS technology simplify the guest bill control functions and eliminate the need for many time-consuming manual audit procedures. Automated pre-checking functions eliminate mistakes the waiter might make in pricing items on the guest bills or in calculating the total. When an item has to be voided, a supervisor (with a special identification number) accesses the system and deletes the item. Generally, automated systems produce a report, which lists the entire guest bills with voided or returned items, the waiters responsible, and the supervisors who voided the items. It is important for an automated system to distinguish voided from returned items because a returned item should be included in inventory usage reports while voided items should not be. If an item is voided after it has been prepared, the item would be classified as 'returned'. When all daily reports are generated, void bill reports are also generated, as shown in Fig. 9.15.

At any point, the managers and supervisors can access the system and monitor the status of any guest bill. This bill-tracking capability can help identify potential walkouts, reduce frauds in order taking, and tighten guest bills and sales income control.

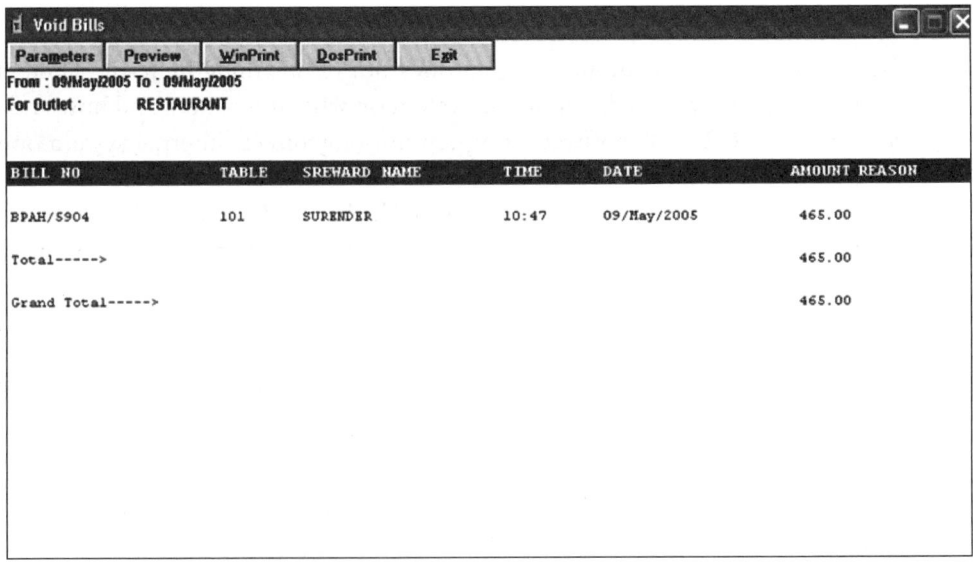

Fig. 9.15   Void bills report (Copyright protected)

Source: www.datamannet.com/pps/presentation-aatithya.pps

The status of a guest bill changes from open to close as and when the payment is received from the guest and is recorded in the system. Most automated systems produce an outstanding cheque report that lists all guest bills (by waiter) that have not been settled. These reports may have a list containing the guest bill number, waiter identification number, time at which the guest bill was opened, number of guests, table number, and guest bill amount. This makes it easier for managers to determine responsibility for unsettled guest bills. Figure 9.16 shows an unsettled bill report.

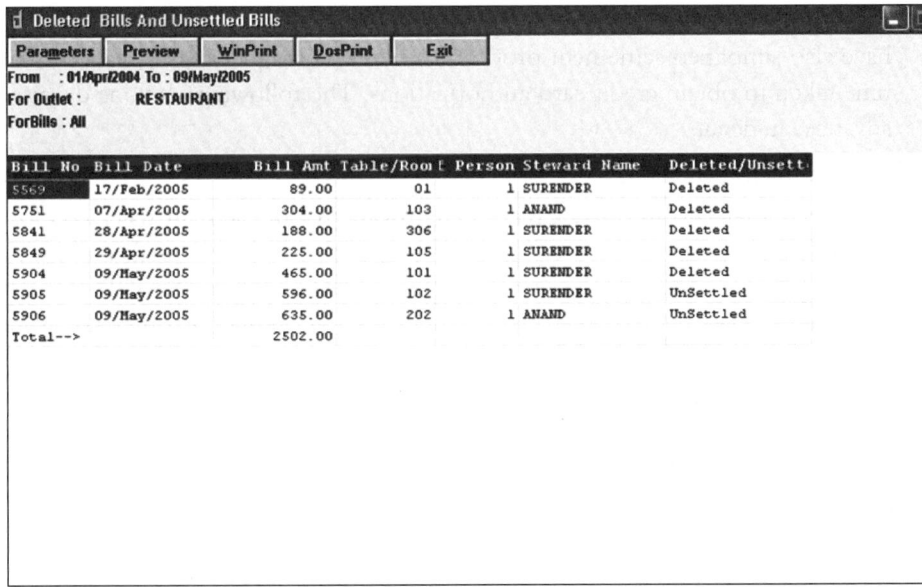

Fig. 9.16   Unsettled bill report and deleted report (Copyright protected)

Source: www.datamannet.com/pps/presentation-aatithya.pps

ECR/POS SOFTWARE

The hardware of any computer system does not do anything by itself. There should be a set of software programs directing the system on what it must do, and how and when to go about it. The ECR/POS software program not only directs internal system operations but also maintains files and produces reports for the management if necessary.

Data maintained in the files can be accessed by ECR/POS terminals and formatted reports can also be printed out on a narrow register tape. The menu item file briefly examines the types of data stored by the major ECR/POS files.

Menu Item File

A menu item file usually contains data of all the menu items sold in the restaurant. Records within this file may contain the following data:

- Identification number
- Descriptor
- Price
- Tax
- Applicable modifier keys
- Printer routing code

This file is generally used to monitor menu keyboard operations. The management can control information about the current menu items for the various meal periods. Reports can be produced for each meal period identifying menu item descriptor, price, and applicable tax. When menu items, prices, or taxes are to be changed, the menu item file is accessed and appropriate changes are entered according to procedures indicated in the user's manual provided by the system's vendor.

Automation Advances

The importance of practical, easy-to-use, fast, and reliable input devices have prompted the development of touch screen, barcode, and wireless server terminals. Advances in automation have also simplified settlement procedures by using magnetic strip readers, which reduce the time taken to obtain credit card authorizations. The following sections discuss each of these advances in detail.

Touch Screen Technology

There is perhaps no area of ECR/POS hardware that has actually received more research and development than touch screen technology. Touch screen terminals are nowadays replacing traditional keyboard as order-entry devices in many POS systems. Touch screen terminals have been developed for fast food operations that allow customers to place their orders without interacting with counter employees.

A touch screen terminal contains a unique variation of a cathode ray tube (CRT) screen and a special microprocessor to control it. The self-contained microprocessor displays data on areas of the screen that are sensitive to touch. Touching one of the sensitized areas produces an electronic charge, which is translated into digital signals that specify what area was touched, for transmission to the microprocessor. This signal also instructs the microprocessor to display the screen if the instructions of the present screen are complete.

The terminal designs vary from vendor to vendor. Flat touch screen terminals that require significantly lesser space than the traditional POS terminals are now available. Flat screens measure only three-and-a-half inches in thickness and can be mounted on walls, ceilings, counters, or shelving units. The design offers restaurants more flexibility in determining where to locate the terminals.

Touch screens and POS systems   Touch-sensitive screens simplify data entry and can be used in place of traditional CRT screens and POS keyboards. In POS system keyboards, PLUs must often be used to enter orders because many systems maintain a limited number of preset keys. PLUs generally require additional order-entry procedures related to the product code and the number assigned to each menu item. In some instances, the order takers memorize these codes, or the management tapes a list of the code numbers at keyboard terminals for the waiter to be quick in his/her work. Most touch screen terminals eliminate the need for PLUs altogether, and hence decrease the time necessary to enter orders into the system.

Touch screen terminals are also very interactive, that is, the system provides on-screen prompts guiding waiters or customers through the order-entry or settlement procedures. When a waiter enters an order for a menu item that needs preparation instructions, the touch screen technology eliminates the possibility of a waiter sending incomplete orders to the production areas. The interactive nature of these systems also decreases the time taken to train new employees in these operations.

Touch screen terminals can also be equipped with magnetic strip reader devices that allow waiters and managers to use barcoded company identification cards to sign in and out of the system. This type of system ensures that as soon as an employee signs into the system, a message screen is displayed. The message screen also enables the management to deliver different messages to different categories of employees or to individual employees as well. For example, employees with job codes corresponding to waiters might receive a message from a supervisor about their work schedule changes. Since all employees must touch the message screen to complete sign-in procedures, the management is assured that all employees have received their messages.

Touch screens and customer order-entry systems   Some fast food operations have also installed counter-top recessed touch screen terminals by which the customers can place their orders without interacting with counter employees. This new self-service option helps in reducing labour costs and hence speeds up the services. Some systems have coloured graphic components, which encourage customers to use these terminals. For example, icons (graphic images) can be used—caricature drawings representing chicken, fish, French fries, burgers, etc., and company logos representing specific soft drink choices can also be added.

Another system enables customers to place orders by following six simple steps. The customer activates the terminal by pressing the Start feature on the screen. The screen then shifts to a display asking the customer to indicate whether the order will be take-out (packed and parceled) or whether the customer will dine at the premises only. Next, the screen shifts to display the menu options. To order, the customer simply touches the desired item on the screen. As items are touched, a 'video receipt' appears on the right side of the screen that keeps a running total of the items and the amount during the ordering process. When the order is complete, the customer touches a 'finished' box on the screen. At this point, a suggestive selling display appears asking the customer if he or she would like to add on some soft drinks, cheese for burgers, or desserts (if not ordered). The final screen displays the total amount due. The order is then placed and saved at the terminal and the guest pays for the same.

Barcode Terminals

Barcode terminals also simplify data entry and can be used in place of traditional keyboards or touch screen terminals. With this system, waiters can use hand-held, pen-like barcode readers to enter orders at service station terminals from a laminated bar-coded menu.

Orders can be entered quickly through the barcode terminals since no keystrokes are involved. In addition, waiters do not have to keep switching from one screen to another as is the case with touch screen terminals. Waiter training time can also be reduced since all the orders are entered from one barcoded menu only.

Wireless Terminals

Wireless order-entry terminals have changed ECR/POS technology. The terminals are small enough to be held in one's hand; hence they are called hand-held terminals (HHTs). When they are as large as the size of a normal terminal keyboard, they are called portable server terminals. These devices perform most of the functions of a pre-check terminal. Wireless technology can be a major advantage for large establishments with drive-through facilities, when the distance between two services stations is long, in outdoor dining areas, or at very busy lounges where it is difficult to reach a pre-check terminal. In any establishment, service can be quicker because waiters do not have to wait to use a pre-check terminal during peak business periods and orders can be entered from the tableside itself in front of the guest.

Wireless terminals also have facilities for two-way communications. Such communications not only allow a waiter to include special instructions such as 'no salt' or 'more spicy' as part of an order, but also immediately alerts a waiter if an item/dish is out of stock. Typically, when an order is ready for pick-up, the waiter receives a signal on the hand-held unit sent by the production section. In some cases, appetizers and drinks can be ready within a few seconds of the waiter entering the orders. For this reason, two-way communications are needed so that as soon as the order is ready, it is served to the guest.

Since all items have to be entered through the waiter's hand-held unit, the frequent problem of coffee or desserts being left out of guest cheque can be eliminated. Some units enable managers to monitor services with their own hand-held devices.

Hand-held units have low-frequency FM radio transmitters and receivers. As orders are entered at the guest's table, analog signals are sent to antenna units located within the dining area. These antenna units relay the analog signals to a radio base station where the modem converts the analog signals into digital signals, which are cabled to the restaurant computer system processing unit. From the restaurant's computer processing unit, signals are then relayed to remote workstation printers or kitchen monitors.

Up to four antenna units can be connected to one radio base station. Before installation, a site survey should be conducted to determine the optimum location for each antenna unit. The amount and location of metal structures in a restaurant are also important for installation since they could obstruct some of the signals. Generally, each antenna unit requires separate cabling to the radio base station.

A charged battery pack is used for powering each hand-held server terminal. A fully charged battery pack can last for about eight hours. It is recommended that two fully charged battery packs be available for each hand-held unit.

Magnetic Strip Readers

A magnetic strip reader is an optional input device that connects to a register. Magnetic strip readers do not replace keyboards, touch screen devices, or barcode terminals. Instead, they extend their abilities. Magnetic strip readers are capable of collecting data stored on a magnetized film strip typically located on the back of a credit card or the house account card. As explained earlier, terminals equipped with magnetic strip readers can be used by employees with plastic, barcoded identification card to sign into the system. In addition, managers might use specially encoded cards to access ongoing transactions and other operational data.

With magnetic strip reads, credit card and house account transactions can be handled directly within an ECR/POS system. The connection or an inbuilt magnetic strip reader with the cashier terminal allows rapid data entry and efficient settlement processing.

Power platforms   Processing of credit card transactions is simplified when a power platform is used to consolidate the electronic communications between a hospitality establishment and a credit card authorization centre. An ECR/POS power platform connects all ECR/POS terminals to a single processor for the transaction settlement. This eliminates the need for individual telephone lines at each ECR/POS cashier terminal. Power platforms can capture credit card authorizations in three seconds or less. This swift retrieval of data helps reduce the time, cost, and risks associated with the credit card transactions.

Smart cards   Smart cards are usually made of plastic and are about the same size as that of credit cards. Microchips embedded in the smart cards store information that can be accessed by a specially designed card reader. Smart cards can store information in several files, which can be accessed for different functions. The security of information stored in smart cards is controlled by a personal identification number (PIN), which must be used to access the files.

Since smart cards contain necessary information for authorizing credit purchases, a specially designed card reader helps to secure credit authorizations quickly. There is no waiting for telephone lines to clear, as is the case with the use of magnetic strip readers and power platforms.

Debit cards   Debit cards differ from credit cards in that the cardholder must deposit the money in order to give the card a value. The cardholder deposits money in advance of purchases through a debit card centre or an electronic debit posting machine. As the cardholder makes purchases, the balance on the debit card falls. For example, a cardholder who has deposited ₹30,000 to a debit card account has a value of ₹30,000 encoded on the magnetic strip section of the plastic debit card. As the cardholder makes purchases, the value in the debit card decreases. To settle the transaction, the establishment bills the debit service centre, which is identified through the information recorded in the magnetic strip section of the user's card.

Nowadays we have a debit card system that can be used along with the automatic teller machine (ATM) cards for settling bills. A restaurant outlet that accepts ATM card payment has specially designed equipment at cashier stations. After the amount of payment is entered into an ECR, a display on the back of the register asks the guest to swipe the ATM card through a card reader. In some systems, the guest may then have to enter his or her personal identification number (PIN) on a numeric key pad, which is out of the cashier's sight. Usually within eight to ten seconds, cash is transferred from the guest's account to the restaurant's bank account.

RECIPE MANAGEMENT SYSTEM

The recipe management software is an excellent tool that has been designed keeping in view the key requirements of individual chefs. Recipe management software comes to our aid in the following cases: preparing a salad, preparing a menu for guests, or changing menus at restaurants as per the changing palates of customers. This software is efficient.

The recipe management software has been designed to be user-friendly so that all its features are very simple to use; it hardly requires any technical knowledge and it is simple to follow instructions even for the most intricate dishes. Chefs at restaurants/hotels get the much required backup as the software elegantly handles every single aspect in great detail beginning with a starter, going on to the main course, till the handling of desserts.

This kind of software is in use since a hotel may be using hundreds or thousands of ingredients, sauces, spices, etc., for preparing dishes. Remembering all the recipes and ingredients is a difficult and uphill task for the chefs, purchase department, and the food and beverage (F&B) controller. As new and innovative recipes are added every day, a database of recipes is a necessity. Since a hotel uses standardized recipes for preparing dishes (to maintain uniformity), a recipe software is essential. This is described here:

- The opening screenshot of a sample recipe software is shown in Fig. 9.17. The screenshot shows how to go about creating a new database, opening an existing database, and viewing standardized recipes with sample pictures and standard ingredients.

Fig. 9.17 Opening screenshot of Resort Recipe
Source: www.resortsoftware.com/downloads/recipe/default.aspx

- On clicking the first option, d:\My Documents\RecipeData\StandardRecipe.rsd, the screenshot shown in Fig. 9.18 appears. The standard recipes are listed on the left-hand side and are arranged alphabetically. On clicking the desired recipe (e.g., apple sauce), the full description of the dish appears on the screen on the right, as shown in Fig. 9.18. It shows the ingredients, the quantity of ingredients, method of preparation, category it belongs to, such as accompaniment or sauces, and the yield of the item.

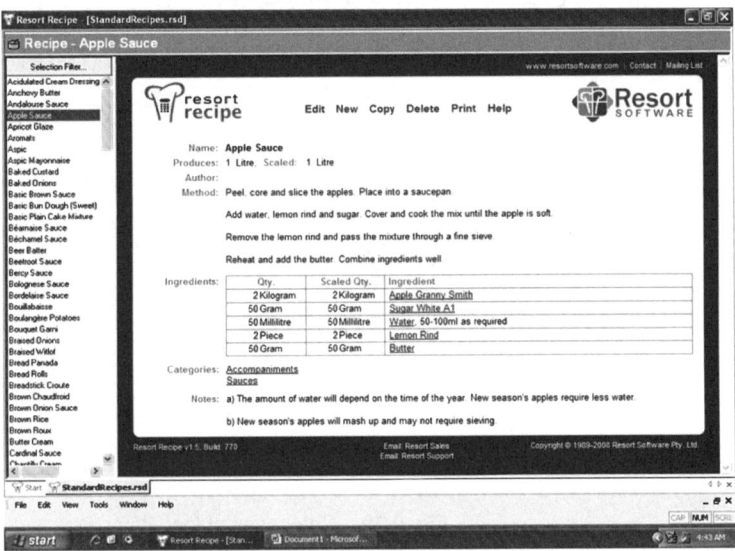

Fig. 9.18   Standard recipe and details about apple sauce

Source: www.resortsoftware.com/downloads/recipe/default.aspx

- On clicking the second option, d:\My Documents\RecipeData\ RR_samples_with pics. rsd and further selecting the dish baked loin of lamb with crispy pancetta, the recipe appears with its photograph and category (as shown in Fig. 9.19). This helps the chef to plate the dish in a more professional manner and maintain uniformity.

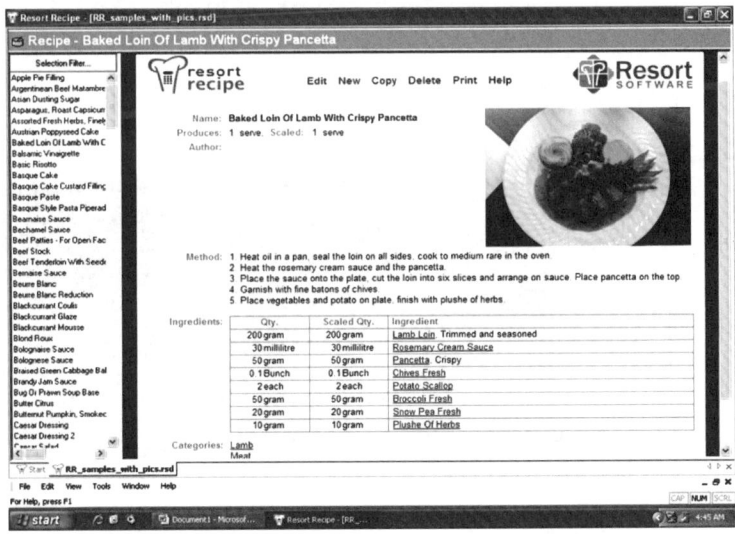

Fig. 9.19   Recipe of baked loin with crispy pancetta

Source: www.resortsoftware.com/downloads/recipe/default.aspx

- The dish belongs to the category lamb. A category can be edited and a new category added (by selecting the New Option), as shown in Fig. 9.20.

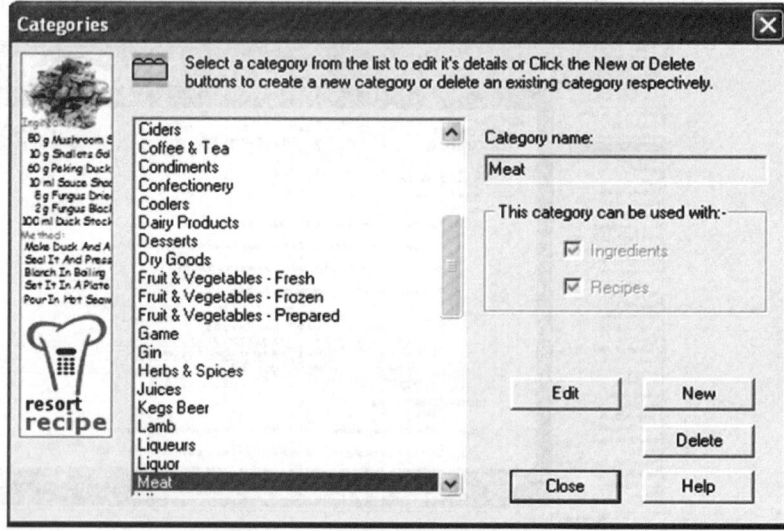

Fig. 9.20 Categories of food items

Source: www.resortsoftware.com/downloads/recipe/default.aspx

- The ingredients to be used in a recipe can also be cross-verified for its quantity and yield. For example, on clicking the ingredient Pancetta, a brief description about purchase specification, usage, yield, and the dish or dishes the ingredient is used for, are displayed on the screen (as shown in Fig. 9.21).

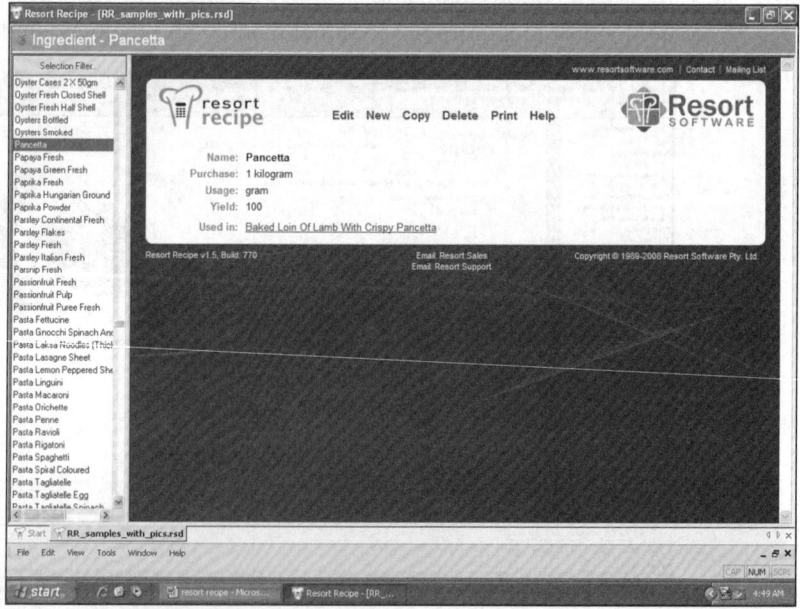

Fig. 9.21 Ingredient pancetta with its use

Source: www.resortsoftware.com/downloads/recipe/default.aspx

- The recipe mentioned uses the loin of a lamb, which belongs to the lamb category (as shown in Fig. 9.22).
- This item belongs to the meat category. By clicking the meat category, we can find the various categories under it, and the dishes that use lamb loin, as shown in Fig. 9.22.

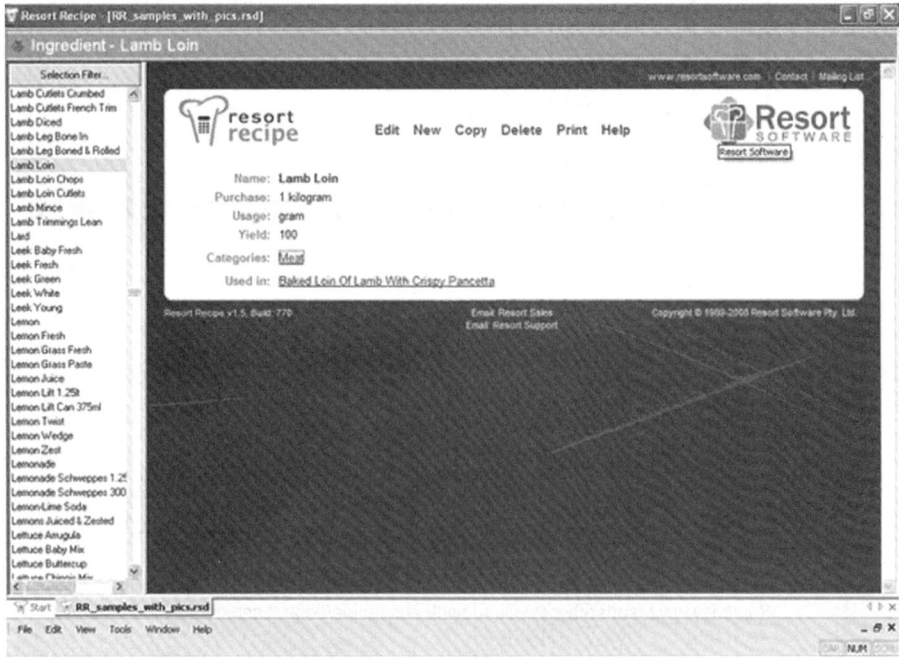

Fig. 9.22   Lamb loin and its category

Source: www.resortsoftware.com/downloads/recipe/default.aspx

- The categories can be edited, deleted, or added as per need. This is shown in Fig. 9.23.

Fig. 9.23   Edit category

Source: www.resortsoftware.com/downloads/recipe/default.aspx

- The ingredients used can also be edited (as shown in Fig. 9.24).

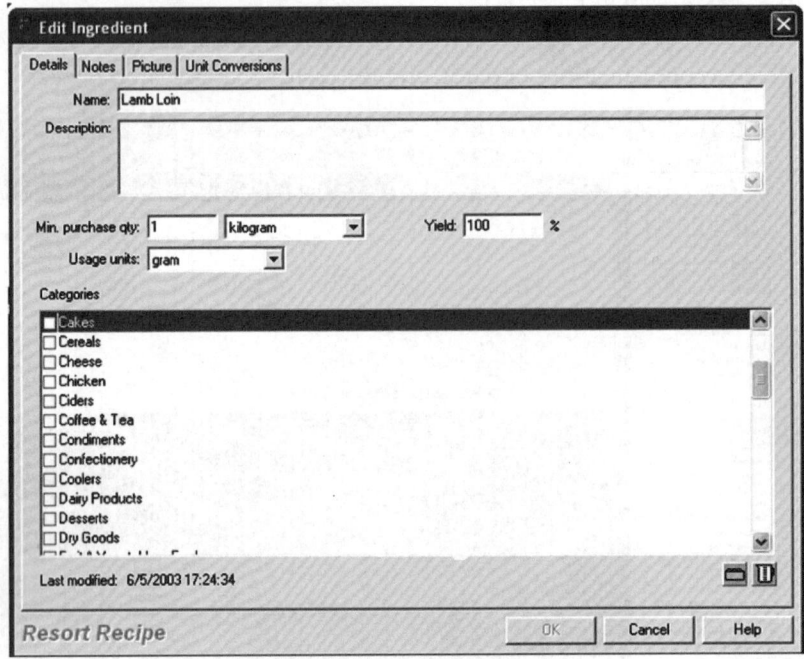

Fig. 9.24 Editing an ingredient

Source: www.resortsoftware.com/downloads/recipe/default.aspx

- The recipe can be copied, printed, or edited. The ingredients and quantity can be altered, or added while the recipe is being copied and then altered. The recipe of the dish can be edited by choosing the Edit option. The following screen appears, as shown in Fig. 9.25.

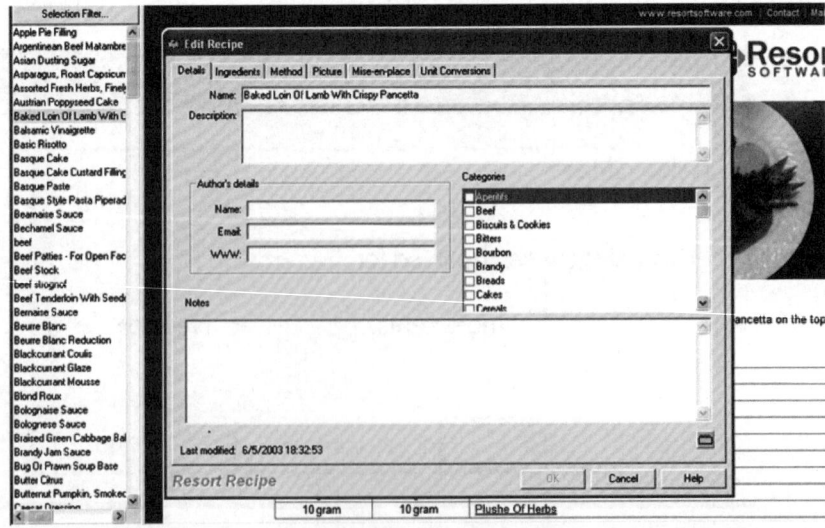

Fig. 9.25 Editing a recipe

Source: www.resortsoftware.com/downloads/recipe/default.aspx

- The ingredients and quantities in the recipe can be altered (as shown in Fig. 9.26).

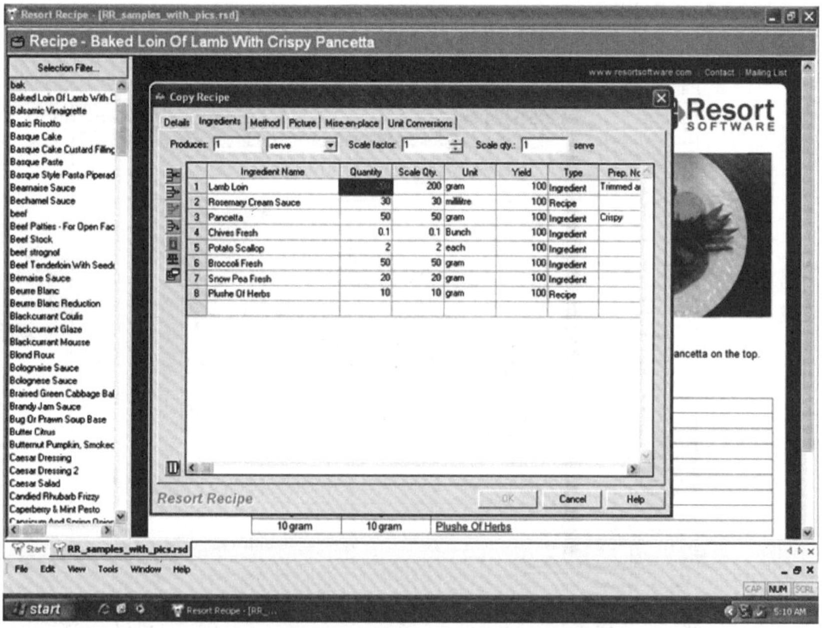

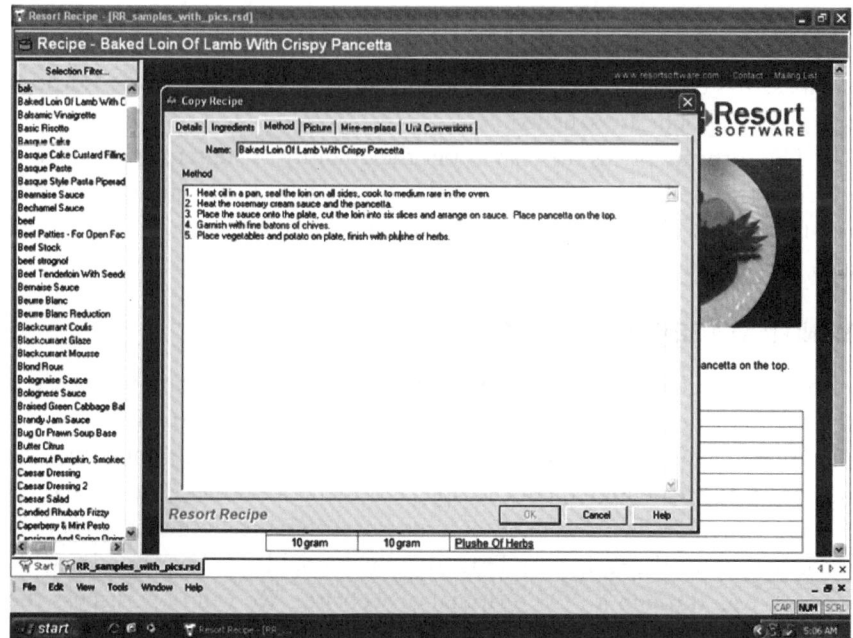

Fig. 9.26   Editing the quantity of an ingredient

Source: www.resortsoftware.com/downloads/recipe/default.aspx

- The method to prepare the dish can also be edited, as shown in Fig. 9.27.

Fig. 9.27   Editing the method of a recipe

Source: www.resortsoftware.com/downloads/recipe/default.aspx

- The image or the photograph can be altered by changing the picture option, as shown in Fig. 9.28.

Fig. 9.28 Editing a photograph

Source: www.resortsoftware.com/downloads/recipe/default.aspx

Some recipe management software mention the nutrition values of each dish. A search option feature makes it possible to search for recipes that primarily revolve around nutritional values. If we have a particular nutritional value in mind to stay fit, the recipe management database has enough recipes that not only answers our concern for nutrition but also ensures that our taste buds are satisfied.

The software prices the dish as per the quantity of ingredients used in preparing it. If any quantity is edited, the price is changed automatically. If any ingredient's price is altered in the master, the prices of all the dishes where the ingredient is used automatically changes, and so also the total dish price.

MENU MANAGEMENT SYSTEM

A menu management system enables a restaurant/hotel operator to price, control, and monitor the entire menu. It provides the operator with a detailed item analysis and insight into what the inventory usage and cost of sales should be. In order to track the menu item costs, it is necessary to create ingredients, recipe, and menu item files. Nutritional values and allergens for the raw ingredients can be entered to automatically generate nutritional and allergen information for the recipes and menus, which could also be printed (e.g., fact sheet, label) or viewed (e.g., through a POS terminal or website). We will now discuss the steps of a menu management system.

Step 1 Create an ingredient file. Every ingredient used in the list of menu items should be inputted into the system. To reduce data entry, menu management systems should have a preloaded database so that it would not take extra time to input every ingredient into the system. Information inputted into the *ingredient file* typically includes the following:

- Ingredient description (e.g., egg)
- Unit description and cost (e.g., case, cost per case, say, is ₹1080)
- Portions per unit (e.g., 360 per case)
- Portion cost (e.g., ₹3 per egg)

The ingredient file must be complete before entering data for recipes, as already mentioned in the recipe management system.

Step 2   Create a recipe file. All recipes are stored in the *recipe file*. A recipe lists the number of ingredients along with preparation procedures (text, audio, photo and/or video) to help maintain consistent food quality.

Recipes can be quickly resized (e.g., increasing portions from 50 to 100). Automatically, the ingredients, quantities, and costs are adjusted.

A menu management system is meant for the creation of sub recipes, that is, recipes placed inside other recipes, to make the construction of complex recipes easier.

The type of information typically inputted into the recipe file includes the following:

- Recipe description
- Cost and quantity of ingredients used
- Serving weight after processing (includes the shrinkage and evaporation factor in a recipe.)
- Recipe or batch cost
- Servings per batch
- Portioning tool
- Serving portion
- Serving portion cost and selling price
- Cost as a percentage of price

Step 3   Create a menu item file. After the recipe file is complete, the final step is to set up the *menu item file*. The kind of information inputted into the menu item file typically includes the following:

- Category description (e.g., breakfast)
- Menu item description (e.g., two eggs, bacon, and potatoes)
- Serving price (e.g., ₹245)
- Accepted food cost percentage (e.g., 21 per cent)
- Actual food cost percentage (e.g., 25.651 per cent)

Step 4   Post the quantities sold and generate the menu analysis reports.

At the end of each day's activities, the quantity of each menu item sold is manually or automatically (if interfaced to the POS system) entered into the menu management system to calculate theoretical usages of all inventoried products and to generate various reports evaluating menu and cost control performances.

A product cost or menu mix report contains the selling price, the ideal cost (quantity of recipe x sold), the food cost percentage, the percentage of total sales, and the gross contribution margin (sales-less food cost) for each menu item. This information is helpful in analysing profitability, food costs, customer preference, menu structure, trends, promotion effectiveness, product performance, and contribution. Menu items on this report can be ranked according to their contribution to profits, enabling management to discern desirable and undesirable menu items quickly, which is also part of menu engineering.

A menu price analysis report shows the impact of price changes on the food cost percentage. It could indicate the food cost percentage for the current menu item prices, the previous menu item prices, and the proposed menu item prices. It would also allow a manager to test speculative menu item prices and to make cost comparisons between various menus.

A theoretical usage report compares ideal usage (the amount that should have been used based on customer sales and the recipe requirements) to actual usage. The exact loss of any food item can then be readily identified.

A perpetual inventory report identifies the theoretical inventory levels based on the initial inventory, purchases, and customer sales. This information is compared to actual inventory counts to compute inventory variances, which are caused by product waste (e.g., burnt kebab), poor controls (employee steals two bottles of vodka), and failure to enter invoices into the menu management system (e.g., report indicates five cases of frozen fish but the actual count is ten cases). Daily reconciliation of the perpetual inventory report for high cost items such as meats, liquor, and seafood, greatly reduces inventory losses.

Sales Analysis

A hotel may have a few F&B sale outlets such as coffee shop, restaurant, banquet, and bar or may have various restaurants, more than one bar, room service, and other service outlets. To monitor the sales of each outlet, the management may require various reports like the comparative sales report to know about the financial status of each outlet in a hotel. First, a sale of the restaurant or an outlet is obtained as a sample sales register, as shown in Fig. 9.29.

Date	Bill No.	Room/Table	Goods Amt	DISC.	Non Taxable	Taxable	Tax Amt	SRV Charge
09/05/05	05897	TB-101	987.00		156.00	831.00	66.48	83.10
09/05/05	05898	TB-103	16.00			16.00	1.28	1.60
09/05/05	05899	TB-106	32.00			32.00	2.56	3.20
09/05/05	05900	TB-101	762.71			762.71	61.02	76.27
09/05/05	05901	TB-305	1444.00			1444.00	115.52	144.40
09/05/05	05902	TB-305	140.00			140.00	11.20	14.00
09/05/05	05903	TB-308	309.00		32.00	277.00	22.16	27.70
09/05/05	05905	TB-102	505.00			505.00	40.40	50.50
09/05/05	05906	TB-202	538.00			538.00	43.04	53.80
Outlet Tot->			4733.71		188.00	4545.71	363.66	454.57
Total-->			4733.71		188.00	4545.71	363.66	454.57

Fig. 9.29 Sample restaurant sales for the day (Copyright protected)

Source: www.datamannet.com/pps/presentation-aatithya.pps

The total sale of the day can be broken down shift-wise, or by meal timings such as breakfast sale, lunch sale, dinner sale, or happy hours sale. The sale of the service outlet can also be obtained according to table numbers, as shown in Fig. 9.30.

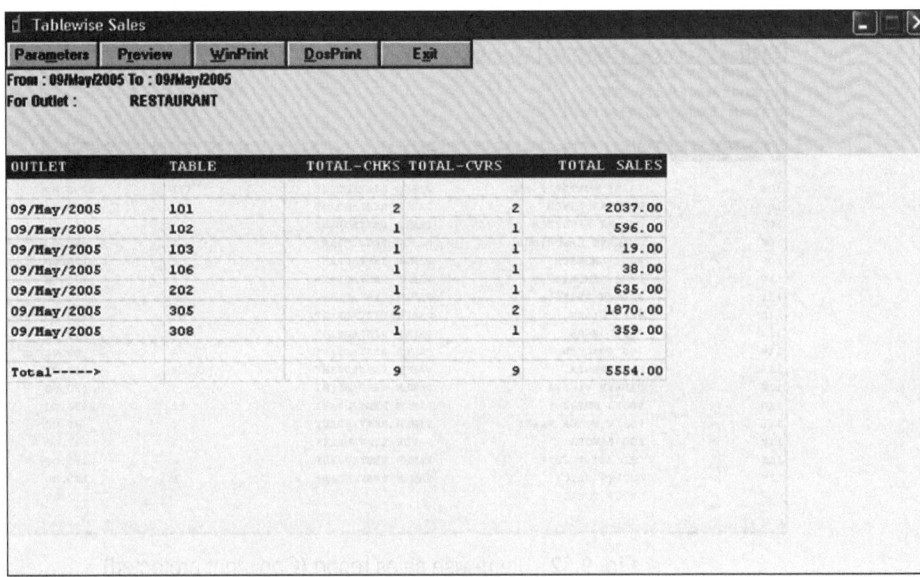

Fig. 9.30   Table-wise sales (Copyright protected)

Source: www.datamannet.com/pps/presentation-aatithya.pps

After a table-wise sales analysis has been done, per cover (person) sale can be obtained from a cover analysis report (as shown in Fig. 9.31).

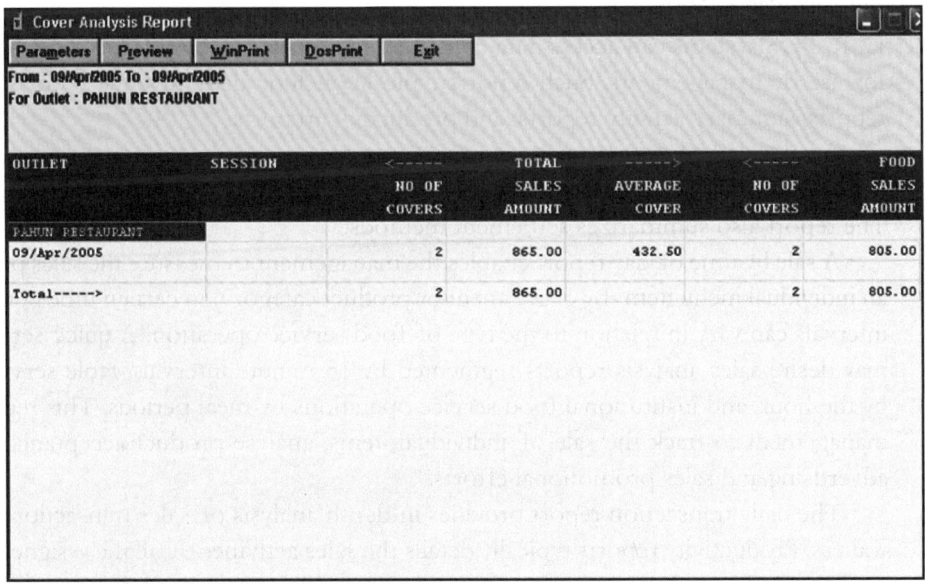

Fig. 9.31   Cover analysis report of a sample restaurant (Copyright protected)

Source: www.datamannet.com/pps/presentation-aatithya.pps

Management may wish to know how each item in the menu is contributing to the total sales of the outlet, how many items in the menu are actually sold, and the quantity of each dish sold (as shown in Fig. 9.32).

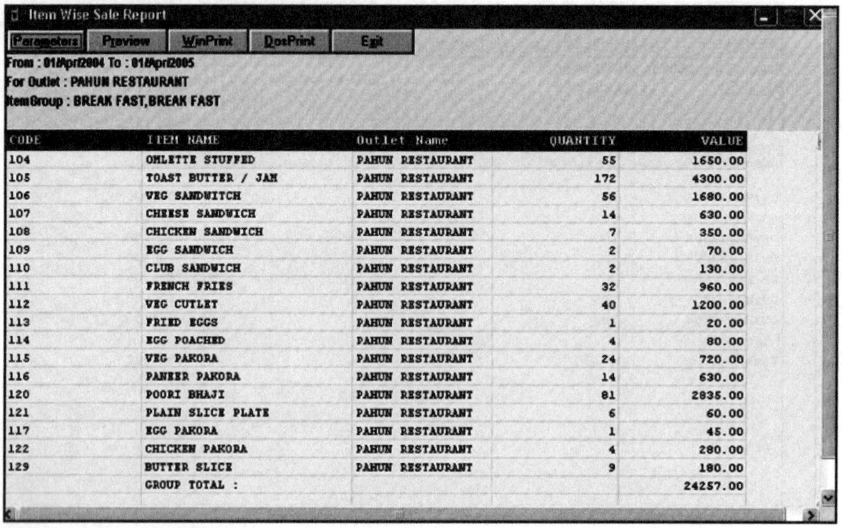
Item Wise Sale Report

Parameters | Preview | WinPrint | DosPrint | Exit

From : 01/April2004 To : 01/April2005
For Outlet : PAHUN RESTAURANT
ItemGroup : BREAK FAST,BREAK FAST

CODE	ITEM NAME	Outlet Name	QUANTITY	VALUE
104	OMLETTE STUFFED	PAHUN RESTAURANT	55	1650.00
105	TOAST BUTTER / JAM	PAHUN RESTAURANT	172	4300.00
106	VEG SANDWITCH	PAHUN RESTAURANT	56	1680.00
107	CHEESE SANDWICH	PAHUN RESTAURANT	14	630.00
108	CHICKEN SANDWICH	PAHUN RESTAURANT	7	350.00
109	EGG SANDWICH	PAHUN RESTAURANT	2	70.00
110	CLUB SANDWICH	PAHUN RESTAURANT	2	130.00
111	FRENCH FRIES	PAHUN RESTAURANT	32	960.00
112	VEG CUTLET	PAHUN RESTAURANT	40	1200.00
113	FRIED EGGS	PAHUN RESTAURANT	1	20.00
114	EGG POACHED	PAHUN RESTAURANT	4	80.00
115	VEG PAKORA	PAHUN RESTAURANT	24	720.00
116	PANEER PAKORA	PAHUN RESTAURANT	14	630.00
120	POORI BHAJI	PAHUN RESTAURANT	81	2835.00
121	PLAIN SLICE PLATE	PAHUN RESTAURANT	6	60.00
117	EGG PAKORA	PAHUN RESTAURANT	1	45.00
122	CHICKEN PAKORA	PAHUN RESTAURANT	4	280.00
129	BUTTER SLICE	PAHUN RESTAURANT	9	180.00
	GROUP TOTAL :			24257.00

Fig. 9.32 Item-wise sales report (Copyright protected)

Source: www.datamannet.com/pps/presentation-aatithya.pps

The statistics obtained from the item-wise sales report helps the chef, restaurant manager, and purchase manager to plan accordingly. The report may also help in menu card planning to include new items in the menu card and remove those items that are only sparingly sold.

Consolidated Reports

ECR/POS systems can access data contained in several files to produce consolidated reports for use by managements. Such reports typically include daily revenue reports, sales analysis reports, summary activity reports, and productivity reports.

A sales and payment summary report provides managers with a complete statement of daily or monthly sales (shift-wise or broken down according to food and beverage categories). The report also summarizes settlement methods.

A sale by time of day report enables the management to measure the sales performance of an individual menu item (by department or product category) in certain time periods. The time intervals can vary in relation to the type of food service operation. A quick-service restaurant may desire sales analysis reports segmented by 15 minute intervals, table service restaurants by the hour, and institutional food service operations by meal periods. This report allows the management to track the sale of individual items, analyse product acceptance, and monitor advertising and sales promotional efforts.

The daily transaction report provides in-depth analysis of sales transactions by individual waiters. Productivity reports typically details the sales activities of all the assigned waiters' sales records. Daily productivity reports can be generated for each waiter and cashier in terms of guest count, total sales, and average sales. In addition, a weekly productivity report can also be generated, describing the average sales per guest for each waiter.

Automated Beverage Control Systems

Automated beverage systems reduce much of the time-consuming management tasks associated with controlling beverage operations. While automated beverage systems vary, most systems can dispense drinks according to the operation's standard drink recipes and count the number of drinks poured.

Automated beverage systems can be programmed to dispense both alcoholic and non-alcoholic drinks of different portions. They can also generate expected sales information based on different pricing periods as defined by the management. With many systems operating the stations at which drinks are prepared, it can be connected to a guest cheque printer, which would record every sale as and when the drinks are dispensed. As a control technique, some systems also require that a guest bill be inserted into the printer before the drink is dispensed. The equipment should be connected to the bar's cash register so that it would automatically record all the sales generated through the automated equipment.

In one type of automated beverage system, liquor is stored at the bar and price-coded pourers (special nozzles) are inserted into each bottle. These pourers cannot dispense liquor without a special activator ring. The bartender slips the neck of a liquor bottle (with the price-coded pourer already inserted) into the ring and then prepares the drink with a conventional hand-pouring motion. A cord connects the activator ring to a master control panel, which records the number of drinks poured at each price level. Figure 9.33 shows a bottle with an activator ring.

The master control panel can also be connected to a POS system that records the sale. Some of the master control panels are equipped with printers and can produce sales reports for each station. The reports indicate the number of drinks poured at different price levels by the bartender and hence the total expected income from each station.

In another type of automated beverage system, liquor is stored in racks in a locked storage room. The bartender prepares the drink by pushing an appropriate key on the keyboard. The liquor and the necessary mixes travel to the dispensing device at the bar through a separate plastic tubing. The system then pours the drink as and when the bartender holds the glass (with ice) under the dispensing device. The drink is then finally garnished and served to the guest. Figure 9.34 shows liquor stacked in stores attached to an automatic beverage system.

Fig. 9.33 Bottle with activator ring
Source: www.easybar.com/oldsite/easy_pour.html

The keyboard-operated automated beverage control systems can employ different types of sensing devices, which would increase the operational controls and also maintain data integrity within the system. Three common sensing devices that are commonly used are glass sensors, guest cheque sensors, and empty bottle sensors. A glass sensor is an electronic mechanism located in a bar dispensing unit, which does not permit the liquid to flow from the dispensing unit till there is a glass in place to hold the liquid below the dispensing head. This helps in reducing wastages as no liquid can be dispensed without a glass below the sensor. The guest cheque sensor, on the other hand, prevents the system

Fig. 9.34   Liquor stacked in stores

Source: www.easybar.com/oldsite/liquor_room_setup.html

from fulfilling beverage orders unless they are first recorded on the guest cheque. As and when a waiter places a beverage order whose ingredients are close to becoming out-of-stock, an empty bottle sensor relays the signal to the order-entry device about the bottle quantity status so that it can be replaced in time.

Automated beverage control systems can also enhance production and service capabilities while improving accounting and operational controls. Sophisticated systems are capable or recording data input through the order-entry devices, transport beverage ingredients through a controlled delivery network, dispense ingredients for the ordered items, and also track important service and sales dates that can be used to produce various reports for the management. The next section discusses the basic components of an automated beverage control system: order-entry devices, delivery networks, and dispensing units.

Order-entry Devices

In an automated beverage control system, the primary function of an order-entry device is to initiate activities involved with recording, producing, and pricing beverage items requested by the guests. There are two basic order-entry devices—a group of preset buttons located on a dispensing unit and also the keyboard unit—that function as pre-check terminals.

A group of preset buttons on the dispensing unit is the most popular order-entry device. These devices can result in lower system costs since the dispensing unit serves both as an order taker and a delivery unit. However, since many dispensing units can support a maximum of only 16 preset buttons, the number of beverage items under the control of the automated beverage system is limited.

Keyboard units function like the pre-check terminals with the beverage dispensing performed by a separate piece of hardware. Since they support a full range of keys (preset keys, price look-up keys, and modifier keys), keyboard units place a large number of beverage items under the control of the automated system.

Delivery Networks

An automated beverage control system relies on a delivery network to transport the beverage item ingredients from the storage area to the dispensing unit. The delivery network should be a closed system capable of regulating temperature and pressure conditions at various locations and stages of delivery. To maintain proper temperature conditions, the delivery network typically employs a cooling sub-system, which controls mechanisms such as cold plates and cold storage rooms. Usually beer is dispensed from the tap, passes through the chilling station after mixing with carbon dioxide and compressed air, and is then delivered through pipelines to the bar area.

Most of the systems are capable of delivering the beverage ingredients by controlling pressure sources such as gravity, compressed air, carbon dioxide, and nitrous oxide. Gravity and compressed air are used for delivering liquor, nitrogen or nitrous oxide for wine, compressed air for beer, compressed air for perishables, and a carbon dioxide regulator for the post-mixes. A post-mix soft drink dispenser places the syrup and the carbonated water together at the dispenser instead of storing, transporting, and then distributing the soft drink as a finished product.

A particular pressure source selected to transport a specific ingredient is a function that also affects the taste and wholesomeness of the finished beverage. For example, if carbon dioxide was attached to a wine dispenser, then the wine would be carbonated and hence spoiled. Similarly, if compressed air was attached to a post-mix soft drink dispenser, the finished beverage item would not have any carbonation. Pressure sources not only affect the quality of the finished beverage items, but may also affect the timing, flow of mixture, portion size, and the desired foaming.

Almost any brand of liquor and accompanying liquor ingredients can be stored, transported, and dispensed by an automated beverage control system. The quantities of the liquor dispensed can be controlled with remarkable accuracy. Typically, systems can be calibrated to maintain portions ranging from one-half fluid ounce to three and one-half fluid ounces.

Dispensing Units

Once the beverage item or ingredients are removed from storage and transported by the delivery network to production areas, they are ready to be dispensed to the guest. Automated beverage control systems can be configured with a variety of dispensing units. Common dispensing units include the following:

- Touch-bar faucet
- Hose and gun
- Console faucet
- Mini-tower pedestal
- Bundled tower

A touch-bar faucet can be located under the bar, behind the bar, on top of an ice machine, or on a pedestal stand. Touch-bar faucets are dedicated to only a single beverage type and are preset for only a particular quantity of output per push on the bar lever. A double shot of whisky or beer would therefore require the bartender to push the same lever twice.

The hose and gun device is another more popular dispensing unit. The control buttons on the handle of the gun can be connected to various liquors, carbonated beverages, water, and wine tanks. Dispensers can be installed anywhere along the bar and can be frequently included as standard equipment at portable and service bars. Pressing the control button leads to a pre-measured fixed flow of the desired beverage. The number of beverage items under the control of a hose and gun dispensing unit is limited to the number of control buttons the device can support. Some of the newer units offer the bartender a maximum of 16 buttons.

Console faucet dispensing units are similar to touch-bar faucet devices, which can be located at almost any part of the bar area. In addition, these units may be located up to 300 feet from the beverage storage areas. Touch-bar faucet devices, unlike the console faucet units, can dispense many beverages in a number of quantities. Using the buttons located above the faucet unit, a bartender can trigger up to three different portion sizes from the same faucet head. An optional feature of this kind of dispensing unit is a double hose faucet unit, which has the capacity of transporting large quantities of liquids in a short span of time.

The mini-tower pedestal dispensing unit combines the button selection technique of hose and gun devices with the portion size capabilities of a console faucet unit. In addition, the mini-tower concept offers increased control of bar operations. For a beverage to be dispensed, the mini-tower unit does not require a button to be pressed; the glass sensing device requires a glass to be placed directly below the dispensing head. The automated dispensing unit is popular for dispensing beverage items that need no additional ingredients before service, such as wine, beer, and all branded liquors. A mini-tower unit can also be located on a wall, ice machine, or pedestal base in the bar area.

The most sophisticated and flexible dispensing unit is the bundled tower unit, also referred to as a tube tower unit. A bundled tower unit has been designed to dispense a variety of beverage items. Beverage orders can also be entered on a separate piece of hardware, not on the tower unit. Bundled tower units can support in excess of 110 beverage products and also contain a glass-sensing element. Each type of liquor has its own line to the tower unit, and a variety of pressurized systems can be used to enhance delivery from the storage areas. While other units sequentially dispense all the ingredients required for a specific beverage item, bar servers merely make the finished product look appealing. The dispensing unit can be located at a maximum distance of 300 feet from the beverage storage areas. Even a cocktail dispensing system with pre-programmed ingredients and quantities is available (as shown in Fig. 9.35).

Fig. 9.35   Cocktail dispensing system
Source: www.easybar.com/oldsite/cocktail_station.html

SUMMARY

This chapter focuses on service-oriented applications of a computer-based restaurant management system. Important hardware components of an ECR/POS system have also been examined. The components include cashier terminals, pre-check terminals, keyboards, display screens, printers, kitchen monitors, and printer controllers.

Keyboards have been discussed in relation to design, menu boards, and types of keys. The differences between on-board printers and remote printers have been shown; several kinds of ECR/POS printing devices have also been explained. These devices include guest cheque printers, receipt printers, work station printers, and journal printers.

The chapter talks about ECR/POS software and also recognizes the different kinds of data obtained from them. The data maintained by ECR/POS terminals and the formatted reports can be printed on a narrow register tape. Several important management reports have also been described.

The chapter also throws light on advanced input devices such as touch screen, barcode, and wireless server terminals. In addition, magnetic strip reader devices have been discussed in relation to rapid data entry and settlement processing.

Recipe management systems, which help in preparing and maintaining standardized recipes with ingredients, quantity, method, and pictures, have been described.

The chapter explains the menu management system in F&B management, which helps an operator to price, control, and monitor the entire menu. It also details the sales analysis on a particular day, table-wise, and even item-wise.

The chapter ends by examining the features of automated beverage control systems such as order-entry devices, delivery networks, and various dispensing units.

KEY TERMS

Automatic slip feed   It is a function that prevents overprinting of items and amounts on guest cheques, operates restaurant costs, and controls and monitors the entire menu.

Barcode terminal   This terminal is used for entering menu dishes that are barcoded. The waiters use hand-held, pen-like barcode readers to enter orders at service station terminals from a laminated barcoded menu.

Customer display unit   This is a display unit that rests on the top, inside, or alongside an ECR/POS. It is usually used to display items and rates for customers.

Daily transaction report   It is a report of the total transactions made by an outlet in a single day.

Delivery network   This system is used in automated beverage management systems to transport beverage ingredients from storage areas to dispensing units. It is a closed system capable of regulating temperature and pressure conditions at different locations and stages of delivery.

Empty bottle sensor   This is a sensing device that relays signals to the order entry device to show that the beverage in the system is going out of stock.

Function keys   These keys are used for particular functions and are usually important for error corrections such as clear and void or for discount.

Glass sensor   It is an electronic mechanism located in a bar dispensing unit that does not permit liquid to flow from the dispensing unit unless there is a glass in place to hold the liquid.

Guest cheque sensor   It is a sensor that prevents the system from fulfilling beverage orders unless they are first recorded on a guest cheque.

Hard keys   These are keys on the keyboards that are dedicated to specific functions programmed by the manufacturers.

Hand-held terminals   These units having low-frequency FM radio transmitters and receivers are used by waiters to enter orders from the guest's table and are sent to radio base stations where a modem converts the analog signals into digital signals, which are cabled to the restaurant's computer system processing unit.

Journal printers   These are remote printers that produce a continuous detailed record of all transactions entered.

Kitchen display   This display unit found in a kitchen is used for viewing orders.

Modifier keys   These keys are used to alter prices according to various portion sizes.

Menu mix report   This report contains the selling price, the ideal cost (quantity of recipe x sold), the food cost percentage, the percentage of total sales, and the gross contribution margin for each menu item.

Menu price analysis report   This shows the impact of price changes on the food cost percentage.

Preset keys   These keys are programmed to maintain the price, descriptor, department, tax, and inventory status for a limited number of menu itmes.

Price look-up keys   These keys are used to view the price of a dish

Printer controller It coordinates communications between cashier or pre-check terminals and work station printers or kitchen monitors while ensuring that waiters enter their orders only once.

Perpetual inventory report This report identifies theoretical inventory levels based on the beginning inventory, purchases, and customer sales.

Receipt printer This is a printing device that produces hard copy of orders on a narrow registered tape.

Settlement keys These keys record the methods by which the guest account is settled (i.e., by cash, credit card, debit card, or other modes).

Soft keys These keys on the keyboard can be programmed by users to meet the specific needs of a particular establishment.

REFERENCES

http://ssmg.in/smart_card_software.html&docid, last accessed on 18 April 2012.

http://www.beveragecontrol.net/Wunder-Bar-Flexible-Hose-Post-Mix-Bar-Guns.asp&docid, last accessed on 10 January 2012.

http://www.bridgat.com/msr605_usb_magnetic_stripe_reader_writer_encoder-o219786.html&docid, last accessed on 18 April 2012.

http://www.duerr-technik.eu/oil-free-compressors-beverage.html&docid, last accessed on 24 April 2012.

http://www.easybar.com/oldsite/cocktail_station.html, last accessed on 24 April 2012.

http://www.easybar.com/oldsite/easy_pour.html, last accessed on 24 April 2012.

http://www.easybar.com/oldsite/liquor_room_setup.html, last accessed on 24 April 2012.

http://www.easytouchpos.co.uk/data/_uploaded/image/VFD_Customer_Display.jpg, last accessed on 15 February 2013.

http://www.escoservices.com/Merchant2/graphics/000000 01/9450074xl.jpg, last accessed on 13 February 2013.

http://www.escoservices.com/Merchant2/merchant.mvc, last accessed on 10 January 2012.

http://www.fileguru.com/Plexis-POS/screenshot&docided, last accessed on 19 April 2012.

http://www.hellotrade.com/geller-business-machines/labware-thermal-receipt-and-guest-check-printer.html&usg, last accessed on 15 February 2013.

http://www.hotelogix.com/videotut/flash/POS/How_to_make_a_Sale.html, last accessed on 6 January 2012.

http://www.lucaspos.com/Hardware.aspx&docid, last accessed on 18 April 2012.

http://www.manitowocfsusa.com/, last accessed on 24 April 2012.

http://www.resortsoftware.com/downloads/recipe/default.aspx, last accessed on 4 November 2011.

http://www.rkeeper.com/index/cont.meniu/5&docid, last accessed on 18 April 2012.

http://www.scribd.com/doc/29919138/Hotel-Property-Management-Systems,last accessed on 4 October 2011.

http://www.sodabarsystem.com/index_Soda_Dispenser_Wunder_Bar_Mini_Tower_IC.htm&docid, last accessed on 10 January 2012.

http://www.sumudra.com/recipemanagement.html, last accessed on 3 March 2012.

http://www.totalpos.co.nz/Casio-QT-6000.html&docid, last accessed on 18 April 2012.

https://cenium.com/factsheets/CeniumEPOS.pdf, last accessed on 19 April 2012.

https://www.kendallhunt.com/samples/4051.pdf, last accessed on 16 April 2012.

www.datamannet.com/pps/presentation-aatithya.pps, last accessed on 13 February 2013.

www.idsnext.com, last accessed on 15 February 2013.

EXERCISE

Concept Review Questions

1. How do preset keys differ from PLU keys?
2. What functions do modifier and numeric keys perform?
3. How can a customer display unit on an ECR/POS register enhance a management's internal control system?
4. What are the two important features of guest cheque printers?
5. How are guest cheques opened and closed within an ECR/POS system?
6. Why would managers prefer touch screen, barcode, or wireless terminals to conventional keyboard order-entry devices?
7. What kinds of sensor devices do some types of automated beverage systems have?
8. What are the basic components of an automated beverage control system?
9. How does recipe management software differ from menu management software?

10. How is a nutritional value calculator interfaced with recipe management software?

11. What are the precautions taken to lay out the delivery networks in an automated beverage management system?

Multiple Choice Questions

1. Price look-up keys (PLU) are used for
   - (a) looking at price
   - (b) selecting price
   - (c) modifying price
   - (d) none of these

2. Preset keys are used to maintain
   - (a) the description of the item
   - (b) the retrieving price of a dish
   - (c) both (a) and (b)
   - (d) none of these

3. Function keys are used for
   - (a) clearing
   - (b) voiding
   - (c) discounting
   - (d) all of these

4. A receipt printer helps in producing a hard copy on
   - (a) a registered tape
   - (b) paper
   - (c) both (a) and (b)
   - (d) none of these

5. A void report is generated to
   - (a) avoid pilferage
   - (b) find reasons for void
   - (c) guests over payment
   - (d) all of these

6. Touch screen technology helps
   - (a) a guest choose and order
   - (b) a waiter choose and order
   - (c) an order to reach a kitchen quickly
   - (d) all of these

7. A recipe management software helps to
   - (a) edit ingredients
   - (b) record the preparation of a dish
   - (c) add new ingredients
   - (d) all of these

8. A menu mix report contains
   - (a) the selling price
   - (b) the food cost percentage
   - (c) both (a) and (b)
   - (d) none of these

9. A sales analysis report helps in the analysis of sales in a restaurant on
   - (a) an hourly basis
   - (b) a shift basis
   - (c) a daily basis
   - (d) all of these

10. An automatic beverage management system can be programmed to deliver the following beverages:
    - (a) alcoholic beverages
    - (b) different portions of beverages
    - (c) non–alcoholic beverages
    - (d) all of these

Project Work

In a recipe management software, add a new recipe, ingredients, quantity, method of preparation, and the price for serving four portions.

In a menu management software, prepare a menu, including recipes and sub-recipes.

CASE STUDY 1

Arshad Warsi checked into Hotel Deep International with his family. He was a regular customer at the hotel. However, Sanjay Gomes, the executive chef, was not very pleased with Arshad's arrival. During his last visit, Arshad had requested that the calorific values of all the dishes be mentioned. He had also requested that no food be above 250 calories. Though they loved the food, Arshad and his family were obese and were hence conscious of their food intake. Calculating the calorific values of all the dishes and preparing dishes that are below a particular amount of calories is a tedious process. Under these circumstances, Chef Gomes wished he had a nutritional value calculating software, which would have made his job easier.

(a) How is calorific value of a dish calculated?
(b) How can nutritional value of a dish be calculated with the help of a recipe management software?
(c) What are the benefits of using a nutritional recipe management software?

CASE STUDY 2

Mukesh Negi, the food and beverage manager of Hotel Holiday Retreat, wanted to bring in a new menu card for the coffee shop and room service. Along with him, Coffee Shop Manager Sushil Sharma, Room Service Manager Niranjan Sinha, and Executive Chef Srinivasan Iyer were talking about bringing a change in the menu. The menu had to be altered as the sales from both the outlets had decreased in the last six months (all the other factors for reduction in sales had remained unchanged).

Negi asked Sushil and Niranjan about the sale of each item in the present menu and their contribution. Though both the managers had a brief idea about this from the kitchen order tickets (KOTs) collected in a day, the exact figures were not available. Since Sushil and Niranjan neither had any knowledge regarding menu engineering nor a dish-wise sales report, coming to a conclusion about a particular dish to be removed was not feasible.

(a) What went wrong in this case?
(b) What is menu engineering?
(c) How does menu engineering help in planning or setting up a menu?

CASE STUDY 3

Ajit Chadha, the restaurant manager of 'The Bistro' was suspicious that some waiters in the restaurant were not following proper procedures, as a result of which the restaurant was losing revenue. He first thought that the kitchen staff were responsible for the loss of revenue but when the kitchen's opening and closing inventories were tallied for the day, no considerable differences were found.

Ajit was now sure that the pilferage was by the service staff only. One day when Ajit was on rounds, he saw a waiter, Vikramjit, consuming chicken kebab prepared for a guest. When Ajit enquired about this, he was told by Vikramjit that it was a piece left by the guest.

Ajit felt a little suspicious about what Vikramjit had said, and so went to check the guest bill. To his surprise, he found that no guest bill had chicken kebab nor was any bill pending. When Ajit enquired from the chef, he was informed that chicken kebab had been served about 30 minutes back. The kitchen had indeed prepared the dish but it was neither sold to the guest nor returned to the kitchen.

(a) What went wrong in this case?
(b) What procedure must be taken to prevent pilferage?
(c) Should the kitchen be informed about void items? Justify your answer.
(d) How does a PMS help in maintaining better control of void/cancelled bills?

Answers to Multiple Choice Questions

| 1. (a) | 2. (c) | 3. (a) | 4. (a) | 5. (d) | 6. (d) | 7. (d) | 8. (c) | 9. (d) | 10. (d) |

CHAPTER 10
Property Management System Interface

LEARNING OBJECTIVES

After reading this chapter, you will be able to understand the following:

- Features of point-of-sale systems and how they function
- Call accounting systems
- Types of electronic locking systems
- Need for energy management systems
- Features of auxiliary guest services
- Guest-operated devices
- Selection of a property management system

POINT-OF-SALE SYSTEMS

A point-of-sale (POS) system consists of a number of terminals, which interface with a remote central processing unit (CPU). A POS terminal has its own input/output component and can even possess a small storage (memory) capacity, but may not contain a CPU. For a POS transaction to be processed, the terminal must be interfaced with a CPU located outside the terminal board.

EXHIBIT 10.1   Technology Increases Efficiency

Pearl Residency had always been a busy hotel. Its proximity to the railway station had made it a transit hotel. The hotel always had problems with getting the room keys back from the guests as they had a tendency to take it along with them, while in a hurry to leave the hotel. The hotel had to bring in a locksmith to change the lock, which was a time-consuming and expensive affair. In addition, for security and privacy reasons, the room could not be let out till the lock had been replaced. Sometimes the guests would keep all the electrical appliances in the room, including the air conditioner, switched on. This resulted in increased electricity bills and loss of renewable energy.

After observing this problem, the management of Pearl Residency came up with a plan to implement an energy management system and an electronic door-locking system. Though the installation of an electronic door-locking system was expensive, the maintenance and operation costs were relatively cheaper when compared to traditional brass locks. While the initial costs are high, an energy management system helps in reducing electricity charges since it switches off all the appliances with the help of sensors, if an individual is not present in the room.

Though a POS system's processing unit is a stand-alone device, a connection to an electronic cash register (ECR) is also possible.

An ECR is defined as an independent, stand-alone computer system. The ECR frames have all the necessary components of a computer system: an input/output device, a CPU, and storage (memory) capacity.

In the newer POS system designs, there is a microprocessor at each terminal's location. The microprocessors are then networked to form a complete POS system, which usually functions without a large, remote CPU. These systems are said to be micro-based POS systems.

When the main processor of a POS system is interfaced with the property management system (PMS), data can be directly transferred from the POS system to various front office and back office PMS modules for further processing. The time required to post the sales to a guest folio is reduced, and the number of times that the sets of data have to be managed is minimized. Relaying the data collected by the POS terminals to the PMS helps to considerably reduce posting errors and also decreases the possibility of any late charges.

For example, a large resort or a hotel could place POS terminals at every revenue collection area, as mentioned here:

- Restaurants
- Bar and lounge areas
- Room service outlets
- Gift shops
- Cake shop
- Swimming pool
- Spa

POS Postings and Account Entries

Account entries can be made from terminals at the front desk or from any remote POS terminal that interfaces with the guest accounting module of the PMS. The accounting entries can also be made internally, that is, from within the guest accounting module itself (as explained in Chapter 8). For example, when the system is in the 'update' process, the room charges and taxes can be automatically posted to all the active guest folios. Although guest accounting modules vary in the specification of their operations, most modules rely on specific data entry requirements so that the amounts are properly posted to the designated folios.

If an in-house guest makes purchases during his/her stay at the hotel, the guest may be asked to present the room key as verification that a valid posting status exists for the guest folio. Some types of electronic locking systems (ELS), which can be interfaced with the PMS and guest accounting module, depend upon the insertion of plastic electronic keycards (discussed in the section on electronic locking system) to authorize the posting of charges from remote POS terminals. If, for any reason, a guest presents a keycard for an unoccupied room, there is an account with a no-post status, or a guest account whose bills have been cleared (settled), the system does not permit the cashier to post the amount. Sometimes a hotel may ask the guest to enter a guest identification code (usually the first few letters of his/her last name), which could also provide further evidence to the hotel that the person being charged is an authorized guest of the hotel.

Concerns for Management

A POS/PMS interface offers lodging properties significant advantages, but there are also some important concerns to be addressed. The following are the problems that could arise:

- Data transferred from the POS system may not meet the specific needs of the PMS.
- Some POS system data could be lost during the routine updation of the PMS.
- Certain limitations of an interface technology could interfere with effective system operations.

The quantity and quality of data that is communicated from a POS system to the PMS usually varies in relation to a particular type of POS system and the PMS design that is employed by the property. There could be some problems when the type of data needed by the front office or back office PMS modules cannot be collected or transferred from the POS system. For example, a POS system cannot do the following:

- Separate the amount for food and beverage from the total amount on a guest bill
- Transfer data relating to special hotel meals and promotions
- Check taxes, tips, and service charges

Management officials would also have to address the following questions:

- Would the individual transactions or the consolidated transactions be transmitted?
- Would the data be transmitted as and when it has been collected or at a later time?
- How much data should be stored in the PMS files and how much data should be retained by the POS system memory?
- How and when would the settlement affect the stored transaction data?
- What audit procedures should be followed to ensure proper posting and monitoring of the transactions?

The PMS in all hotels might undergo a system update routine (an automated version of the traditional night audit functions). It usually occurs sometime during the evening hours. While the system is being updated, the POS interface might be inoperable. The interruption of data flow along the interface channel might lead to some transaction records being lost or blocks at either the POS or PMS end. Hence, proper care should be taken to schedule the PMS update when the food and beverage outlets and other revenue-producing centres are closed or during low sale hours. If this cannot be done, then timed non-automated/manual procedures have to be implemented at all revenue outlets till all of the PMS is updated.

Before actually interfacing a POS system to a PMS, the management should first try to resolve problems (if any, related to the interface technology) since a POS system might sometimes be dependent upon a set of applications software, which may not be related to the requirement of the hotel's PMS. In such a condition, the primary applications software of the POS system should first be enhanced before the system is interfaced with the PMS.

CALL ACCOUNTING SYSTEMS

It has now been legalized for lodging properties to resell telephone services to the guests in the hotel. This resale capability has enabled the hotel's telephone department, traditionally a loss-making department, to become a potential profit-making department. A call accounting system (CAS) enhances the management's control over expenses relating to local and long distance telephone services.

A CAS can operate as a stand-alone system or be interfaced with a hotel's PMS. Usually, a CAS is able to handle direct-distance dialing, distribute various calls through the least-cost routing network (a hotel may have various service providers whose rates could be different), and price each outgoing call. If the CAS is interfaced with the guest accounting module of a PMS, the telephone charges can be posted automatically to the proper guest folio.

These systems usually reduce working space, and maintenance and labour costs, which are associated with the telephone systems. The hardware of CAS occupies lower space and requires lesser maintenance, when compared to traditional switchboard equipment. Labour costs have decreased considerably since a telephone operator is not involved in the CAS call placement and distribution functions. The automatic pricing of the calls eliminates the need for manually calculating charges and posting them to the guest's accounts.

Some calls, which are usually direct-distance-dialled, are channelled through the call accounting system; outgoing calls, which required an operator's assistance, are channelled through a hotel billing information centre (HOBIC). A CAS includes many features as mentioned here:

- Automatic identification of outward dialing (AIOD)
- Automatic route selection (ARS)
- Least-cost routing (LCR)
- HOBIC system interface

CAS are usually designed to monitor the non-guest (administrative) and guest telephone calls, both incoming and outgoing. Since all the telephone extensions are interconnected to the hotel's switchboard, it serves as a primary control device for the CAS. The switchboard might also have an optional station message detail record (SMDR), which is accountable for charting and monitoring the telephone traffic.

HOBIC System Interface

Some hotels use the HOBIC system to provide telephone services to the guests. The system is still in use today and often serves as a backup system for properties that have already installed CAS.

The HOBIC system is a service that is usually supplied by a telephone company and records the time and charges when a guest makes a long-distance call. The guest calls are usually placed on special telephone lines called HOBIC lines. When the guest makes a call on these lines, the telephone company operator asks the guest for his or her room number. When the operator receives the room number, he/she allows the call to be made. After completion of the call, the hotel receives details pertaining to the duration and charge of the call, either from an operator calling the property or from the telephone company's transmission received by the front desk's teletype machine.

The HOBIC system is often interfaced with CAS to perform the following functions:

- Supervise all operator-assisted calls
- Process overflow of telephone services from the CAS
- Serve as a fail-safe telephone service

Sometimes it may so happen that a guest makes a direct-dial call but all the available CAS lines are found to be busy; the call is then automatically transferred to the HOBIC lines for completion. Though a HOBIC system might be used to process an overflow of calls, the calls are usually sent back to the CAS for final pricing, recording, and reporting. If, for unavoidable circumstances, the CAS is inoperable, then the HOBIC system serves as an alternative for processing calls in the same manner as seen in lodging properties without CAS. This also ensures that a telephone service is continuously available for the guests. HOBIC is a generic term that is used in the telecom industry to define, as a standard, the format of the data sent to the PMS by HOBIC. The general format is shown in Fig. 10.1.

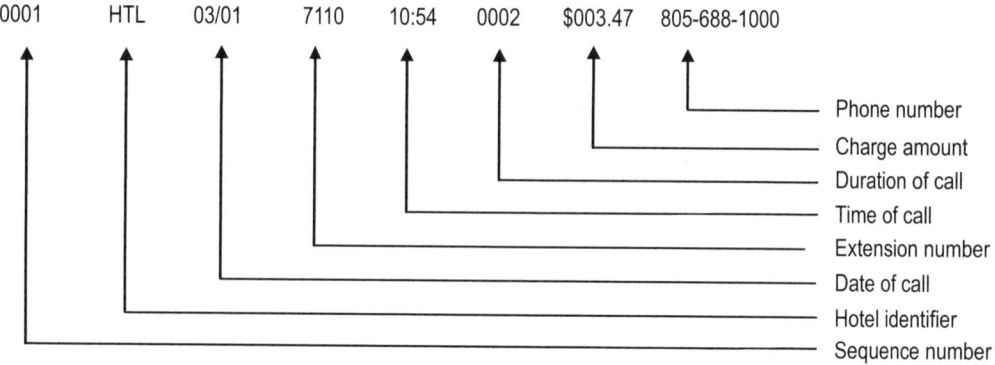

Fig. 10.1 Call accounting details by HOBIC

Features of Call Accounting Systems

The functions of a CAS include the following:

- Call placement
- Call distribution
- Call routing
- Call rating
- Call record

Such systems have now simplified the sequences involved in call placements. A guest can directly make long-distance calls, thus eliminating an operator's intervention. In the HOBIC system, an operator used to earlier intercept outgoing calls to identify a guest's room number. The AIOD feature of the CAS now immediately identifies the extension of the room number from where the call is made.

As and when an outgoing call is made, the call distribution equipment of the CAS is engaged. The ways and means by which a specific call is routed is essential for determining its cost. In a passive CAS, there are actually no options available to the call distributing centre. However, an active CAS employs an ARS switch and also possesses an LCR device. The ARS feature has now become an essential CAS component and is capable of connecting with a variety of common service providers. A service provider is any recognized entity, which can transmit messages or other communications for general use, at mutually acceptable rates. The LCR device directs calls over the least-cost available line, regardless of the service provider. When the least-cost line is busy, the LCR drives the system to find out the next least expensive line. This can go on until the call is actually made, at high speed and precision.

The manner in which a call is rated will vary in relation to different vendors, equipment packages, and electronic switches. A station message detail record (SMDR) might be engaged to chart and monitor telephone traffic. The data obtained from the SMDR is used to rate calls. Some systems even use base calls on a ring-back mechanism, whereas others incorporate a timeout feature. In a ring-back mechanism, the guest is charged only if the call has been answered by the guest. In a timeout feature, callers begin to pay for the calls after a predetermined amount of placement time (the time interval since placing a call). After the call has been rated, it is then entered into a call record file.

The call record file monitors details regarding the calls processed by the CAS. This file includes the following details:

- Date
- Guest room's extension number
- Telephone number that a guest has dialed
- Time when the call was made
- Duration of the call
- Cost of the call
- Tax and service charges, if any

While most of the call rating systems calculate the cost and tax involved in a call and automatically post the necessary data to the appropriate call record files, other systems that price and tax calls might also require the data to be manually posted to the call record files. The call record is a hard copy document containing essential transactional support data for individually placed telephone calls. The call records could be referred to a guest folio for resolving guest discrepancies, if any, relating to telephone charges.

Usually, call records are automatically logged in a call traffic transaction file. This file usually maintains data that is necessary for the management to generate reports. Typically, records are organized, according to either the time of call placement (chronological file) or the room extension number (sorted file). The report details many requirements of the management.

Advantages of CAS/PMS Interface

A CAS/PMS interface offers lodging properties a number of significant advantages, as mentioned here:

- Improved guest services leading to more guest satisfaction
- Better communications networking for the guest
- Improvized call pricing methods
- Minimized telephone expenses

- Automatic posting of call charges to the guest folios
- Automatic call detail records
- Detailed daily reports of various telephone transactions

As the CAS reduces operator intervention, the hotel's telephone department can now save on both time and labour. It also helps in eliminating telephone meter readings and thus reduces guest telephone charge discrepancies, which could contribute to faster checkout and more efficient front-desk operation, leading to more valuable time being spent with guests.

Emergency backup procedures could be a major management concern in relation to a CAS and its interface with the PMS. The CAS/PMS interface is usually backed by the HOBIC call system. The CAS would however require an energy backup and access to uninterrupted power supply to work smoothly.

The storage capacity of the CAS is also very important. Before a CAS is purchased and installed in a hotel, the management must ensure that the telephone traffic across the property has been properly determined so that the proposed CAS would have adequate storage capacity for processing and storing the telephone data. The management of the hotel would also like to ensure that the proposed system is able to distinguish administrative (non-guest) calls from guest calls. Other important concerns focus on systems maintenance, services, and after-sale services.

There are other new methods of making calls, which are faster, cheaper, and efficient; one of these is voice over Internet protocol (VoIP).

VOICE OVER INTERNET PROTOCOL

VoIP is a method of taking analog audio signals (the kind we hear when we talk on the phone) and turning them into digital data, which can be transmitted over the Internet. VoIP can then use a standard Internet connection to place free phone calls. The advantage of this is that by using some of the free VoIP software, which is available to make Internet phone calls, we would actually bypass the telephone company (and its charges) entirely.

The most important feature of VoIP is that there are many ways of placing a call. The following are the three different ways in which a VoIP service is used today.

Analog telephone adaptor   The simplest and most common way is by using a device called an analog telephone adaptor (ATA). The ATA allows us to connect a standard phone to our computer or Internet connection for use with a VoIP. The ATA is an analog-to-digital converter. It takes the analog signal from the phone and converts it into digital data for transmission over the Internet.

IP phones   These specialized phones look just like normal phones with a handset, cradle, and buttons. However, instead of having standard phone connectors, IP phones have an Ethernet connector. IP phones directly connect to a router and have all the necessary hardware and software onboard to handle the IP call. Wi-fi phones allow subscribing callers to make VoIP calls from any Wi-fi hot spot.

Computer-to-computer   It is the easiest way of using a VoIP. We don't even have to pay for the long-distance calls. There are several companies offering free or very low-cost software that can be used for this type of VoIP. All we need is the software, a microphone, speakers, a sound card, and an Internet connection, preferably a fast one that we can get through a cable or DSL modem. Other than the normal monthly ISP fee, there is usually no charge for computer-to-computer calls, irrespective of the distance. Figure 10.2 shows a diagrammatic representation of how VoIP operates.

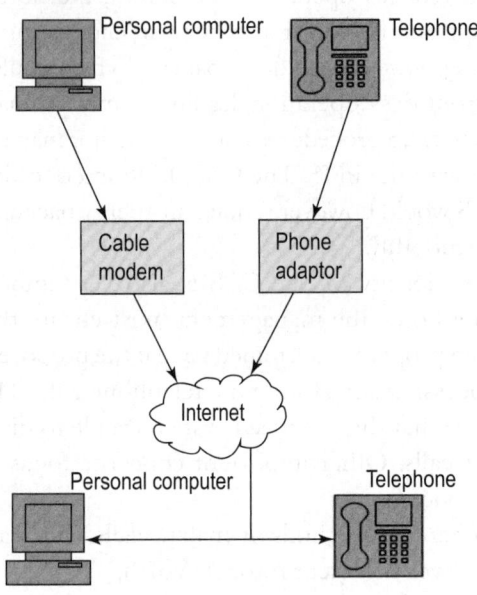

Fig. 10.2 VoIP operation

Source: http://transition.fcc.gov/voip/

ELECTRONIC LOCKING SYSTEM

An ELS has now replaced traditional brass keys and mechanical locks with sophisticated computer-based guest room access devices. Installation of electronic locks on guest room doors with traditional locks can either be an easy job or a complex task with major reconstruction. Some systems might only require drilling of a small hole for the wires to pass through from the outside to the inside portion of the lock and retain the lock hardware as part of the new lock, whereas some systems could require new locking hardware or even a new door.

There are presently a variety of electronic locking systems available in the hotel industry. These include hard-wired, micro-fitted, radio frequency identification (RFID), and biometric locking systems.

Hard-wired Systems

Hard-wired electronic locking systems usually operate through a centralized master code console, which is interfaced with every controlled guestroom door. The console can also be a slotted switchboard, which is centrally located at the front desk. In this type of hardwired system, a front-desk employee, during the check-in procedure, inserts a formerly encoded keycard into the proper room location slot on the console. The console then immediately transmits the keycard's code to the remote guest room door's lock.

As the guest leaves the front desk after completion of check-in, the keycard that he or she has been issued is the only functional guest room key. Keycards that were issued to guests who had previously occupied the same room would have already become invalid.

For this, all the controlled guest/non-guest doors should be cabled to the master console. A hard-wired system is challenging, considering its expensive design, but improves security. Before such a system is installed, the management should also identify emergency energy backup sources since these systems require continuous source of electricity. Hard-wired locking systems use normal alternating current as their primary energy source with batteries as an emergency backup.

Micro-fitted Systems

Micro-fitted electronic locking systems operate on an independently configured stand-alone unit, thereby avoiding complex dedicated circuits that are required for hard-wired locking systems. Every door in this system has its own microprocessor, which has a unique, predetermined sequence of codes. The master console at the front desk usually contains a record of all such code sequences for each and every guest room's door. With the help of a micro-fitted locking system, the front-desk employee completes the guest check-in process by encoding the keycard with the preset sequence of codes already assigned for a particular room.

In a hard-wired system, the codes are directly communicated from the master code console to the controlled doors, which are connected to the console. A micro-fitted system does not require connectivity between the console and the door unit. The front-desk console and the microprocessors of the controlled doors are separate units. The connection between the microprocessors and the controlled doors is the preset sequence of codes, which means that the front-desk console should be programmed simultaneously with the door unit so that the codes in the microprocessor and the console are the same.

Since the console and the doors are separate entities, it may create inconvenience for some guests. For example, a family of four comes to the hotel and requests for two rooms; the parents plan to stay in one of the rooms, whereas their children would be put up in the next room. On reaching the assigned rooms, the family enters the first room and rests there, and does not use the second room. Hence, the second designated room key is not used. The next morning, the family checks out from both the rooms. Here, the locking mechanism in the second room's door, which was not operated, would not have advanced to the next code in the preset sequence because the keycard would not have been used. However, the master console at the front desk would have automatically advanced to the next code in the sequence as and when a new guest checks into the same room because it assumes that the last issued keycard had already been used and there would be a mismatch between the codes. In such a situation, the new guest (receiving the next keycard) would find that the keycard he or she has been issued fails to operate the lock, and is hence unable to open the door. Then, the front-desk employee would have to use a specially designed keycard to reprogramme the door's microprocessor so that the current code synchronizes with the front-desk control console. Keeping in mind the drawbacks of a micro-fitted system, newer technologies have come into the market, which are discussed later in this chapter.

An important energy feature of micro-fitted electronic locking systems is that the microchips in the doors are powered by the battery packs and therefore do not require wiring to an external power source, which is however required for hard-wired systems. Some systems also use penlight-sized batteries, some use D-size cells, whereas others use even larger battery units, which also give an inside view of a door locking system.

Features of Electronic Locking Systems

Electronic locking systems can produce several levels of master keys. The most common systems are configured to provide distinct levels of security as per need. Levels may be established for the following: housekeeping personnel, security officials, property management officials, management.

Some of the electronic locking system designs also provide a 'do not disturb' option for the guests. This option usually employs an indicator, which displays a notice that the guest wants privacy. If the guest later wants the room to be made ready, the indicator is changed.

The notice is often given by a flashing red light located within the locking mechanism. This indicator can also be triggered when the room attendant inserts a keycard into the locking mechanism to prepare the room.

A safety inbuilt feature of some of the electronic locking systems prevents the door from opening when the keycard is present in the lock. This would prevent the guest from entering the room if he/she forgets to take the keycard from the lock. While some of the systems permit entry into the room even if the keycard is attached to the lock, they also track the duration for which the keycard is in the door. If the keycard remains in the locking mechanism beyond a fixed time interval, the system would destroy the keycard by scrambling the card's code. The reason for scrambling is to provide the guest with better security. A keycard that remains in the lock might be taken by someone other than the room's occupant and might be misused. To avoid such repeated problems, the hotel staff should inform the guest during check-in that the failure to promptly remove the keycard will make it invalid.

Electronic locking systems are an essential hotel feature as self-service terminals, which enable unassisted guest check-in and checkout, are seen throughout the hotel industry. (Self-service terminals have been discussed later in the chapter.) There are newer forms of electronic locking systems that do not require guests to use keys or cards (e.g., RFID and biometrics) and are better than hard-wired or micro-fitted systems.

Radio-frequency identification Radio-frequency identification (RFID) is an inexpensive technology that enables wireless data transmission. It is a digital door-locking system implemented and governed by an RFID reader, which authenticates and validates the user and then opens the door automatically. It also maintains a record of the check-in and checkout of the user and authenticates a guest before he/she enters the room. The system also includes a door-locking system, which opens the door as and when the user brings the tag in contact with the reader; the user information must match with the information stored in the database.

This system works by storing all necessary information about the user. A new user is first registered with the system, and then the corresponding information is burned on the RFID tag. When a registered user comes to the entry point and puts the tag into the reader, the system checks whether it is a registered user or an imposter. If the user is registered, then the tag information is matched with the user information stored in the system. The door is opened after successful authentication and closes after a specific time interval.

Biometric locks   These are fingerprint scanning locks that can be installed on the doors in a hotel. These work by scanning the finger and saving the fingerprint. When an individual with the specific fingerprint comes in contact with the lock, the lock opens up.

Electronic Locking System Reports

The most important advantages of an ELS is that the management can easily find out which keycard has been used to open which door, by date and time. Communicating the ELS capability to hotel staff and guests helps in reducing confusion.

An ELS typically maintains an audit trail of all the activities involving the use of system-issued keycards. Some systems also print out reports detailing the activities in a chronological order. A system that records events as and when they occur generally does so due to its limited memory, and not because the resulting printouts are fundamentally more useful or effective. These systems record and store activity data, which can be formatted to provide printed reports on demand.

ENERGY MANAGEMENT SYSTEMS

Heating, lighting, ventilating, and air conditioning equipment are basic essentials for a hotel's existence. The greater the efficiency of the equipment, the better would the hotel be able to serve the requirements of the guests. An energy management system (EMS) can conserve energy, hold and maintain energy costs, and incorporate operational controls over the guest rooms and public areas. An important feature of the system is the ability to minimize the building's energy requirements while not considerably affecting the hotel's comfort conditions.

An EMS is a computer-based control system that is specially designed to automatically manage the operation of all mechanical equipment in a property. The programming of this system allows the management to determine when a particular equipment has to be turned on, off, or regulated. For example, if a hotel's conference room has to be used from 9 a.m. to 5 p.m., the computer can be programmed accordingly so as to automatically conserve energy during the hours the room will not be functional, but ensuring that by 9 a.m., the room reaches the desired comfort level for the guests. This programming technique can also be applied to other types of equipment affecting the numerous spaces throughout the property such as the lobby area and corridors after midnight (when guest activity is considerably reduced).

Although the actual operating features of an EMS vary, common energy control designs include the following:

- Demand control
- Duty cycling
- Room occupancy sensors

Demand control   The demand control mechanism maintains usage levels below a given limit by shedding the energy loads in an organized manner. Equipment units assigned for the demand control programmes are those that can be turned off for varying periods without adversely affecting the surrounding comfort conditions. Lights in the public area and the ventilation system can be suitably reduced during non-peak hours, especially during the night.

Duty cycling   The process of duty cycling involves serially turning off all equipment for a period of time every day. Heating, ventilating, and air conditioning systems are usually duty-cycled so as to reduce energy consumption while maintaining the comfort conditions of the guests. However, duty cycling is not normally applied to large motors, which cannot be stopped and started on a frequent basis.

Room occupancy sensors   These sensors use either infrared light or ultrasonic waves to register the physical occupancy or presence of an individual in the room. As and when the guest enters a monitored space, sensors turn on devices such as lights, air conditioning equipment, and heating equipment that are under its control. When the guest leaves the monitored room, the sensor reacts after a preset time frame and turns off the lights and/or automatically resets the temperature of the room.

An EMS/PMS interface may offer huge opportunity for the energy control of a hotel. For example, let us assume that on a particular night, 40 per cent occupancy is forecasted for a 100-room hotel property. Reducing the hotel's energy consumption on that night becomes an important factor in deciding which rooms the hotel should sell. The hotel would like to assign guests to only the lower floors of the property, which could significantly lead to reduction of energy demands (in comparison to rooms on the upper floors). By interfacing an EMS with the front office deparment's room management module, it is also possible to control room allotments by the front desk staff and achieve the much desired energy cost savings. In many such incidents, the energy savings are also recorded by an in-house microcomputer in specially created electronic spreadsheets.

Comfort conditions in the guest rooms, meeting and conference rooms, public spaces, administrative offices, and other areas can be controlled through the system console. The energy management system provides heat, ventilation, and air conditioning (HVAC) levels at various remote locations and also displays its readings on the console screen

An energy management system's energy controls are virtually useless if the hotel is operating a system that is neither properly designed nor adequately maintained.

AUXILIARY GUEST SERVICES

Automation has now simplified many auxiliary guest services such as the placement of wake-up calls and the delivery of messages to the guests. These functions can often be performed by devices that are sold as stand-alone systems and which can also be interfaced to the room management module of a PMS.

The main reason for interfacing the auxiliary guest services to a PMS is the complete coordination and tracking of guest-related functions. Though the automated wake-up call devices are often best operated as stand-alone independent units, it could become beneficial if it is interfaced with the guest messaging system and the PMS. The ability to alert guests about the messages waiting for them depends upon the accessibility of the PMS mechanism, which links the guest room telephones and in-room televisions.

An automated wake-up system allows front-desk employees to input the guest's room number and the requested time for a wake-up call. At the precise time, the system automatically rings the room and reverts at fixed intervals till the guest answers the telephone. If the guest does not respond even on the third or fourth attempt (the way the system is usually programmed), the system stops calling and makes note of the guest's inability to respond. When a guest answers the call, the system plays a prerecorded morning greeting and then disconnects. In some hotels, a register of all such answered calls are often stored for the entire day within the system for reference. The system's ability to keep calling the guest till he/she actually answers is beneficial for the organization as the switchboard operator need not waste time in calling guest rooms.

Electronic message-waiting systems are now designed in such a way, so that a guest knows about a message waiting at the front desk. The message-waiting devices come with an indicator light, which either flashes on the telephone or television screen in the guest's room. This system is also now capable of displaying messages on the television screen in the guest's room.

Nowadays, hotels use voicemails. These devices record telephone messages of the guests. A caller has to just leave a message for a guest over the phone; the message is recorded and can later be accessed by the guests. To get back the message, the guest dials a specific telephone number, which is connected to a voice mailbox, where he/she can listen to the message delivered in the caller's voice.

GUEST-OPERATED DEVICES

Guest-operated devices can either be located in a public area of the hotel or in individual guest rooms. In-room guest-operated devices are very user-friendly. The devices usually provide concierge-level services within the convenience of a room. The following guest-operated devices are discussed here:

- Self check-in/checkout systems
- In-room movie systems
- In-room beverage service systems
- Information service systems

Fig. 10.3   Auto check-in machine

Source: http://www.autocheckin.com/interactive_slideshow_b.php

Self Check-in/Checkout Systems

Self check-in/checkout terminals are typically located in the lobby or at the entrance of a fully automated hotel. The designs of the terminals vary as some might resemble an automatic bank teller machine (ATM), whereas others may have a unique design and might also possess both video and audio capabilities. Figure 10.3 shows an auto check-in machine.

This machine is mainly used for guests who arrive at the hotel with prior reservation. If an advance reservation has already been made by the guest, the system would ask for the reservation identity to be entered (as shown in Fig. 10.4).

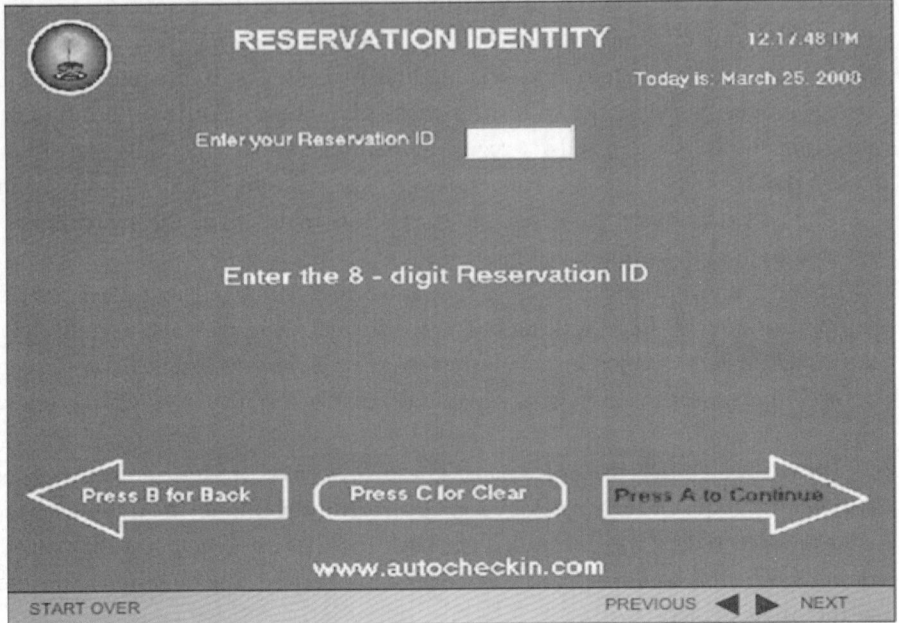

Fig. 10.4 Addition of reservation ID

Source: http://www.autocheckin.com/interactive_slideshow_b.php

When the reservation identity entered by the guest matches with the ID in the system, the system queries the guest regarding the duration of stay and the number of guests (to ascertain room occupancy requirement), as shown in Fig. 10.5.

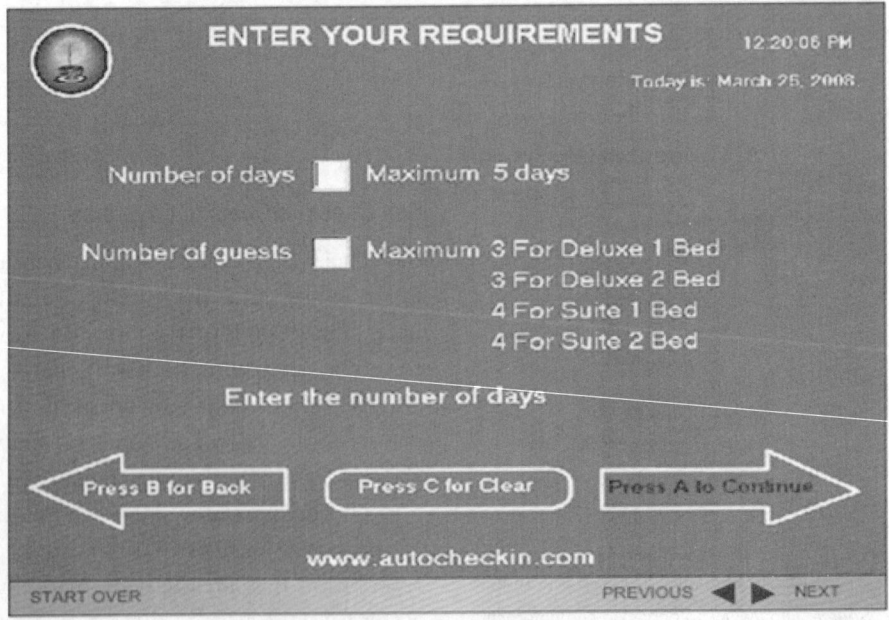

Fig. 10.5 Room types with capacity

Source: http://www.autocheckin.com/interactive_slideshow_b.php

After opting for the number of days of stay and number of guests, the next screen shows the different types of room available, some of the key features it has, and the rate of the room.

Based on the guest's choice and availability of rooms, the booking is confirmed with the type of room, number of guests, and the room rent. The guest is next asked to enter his/her credit card in a predetermined slot of the machine for payment. After the payment has been made, the machine generates a receipt for the payments and also delivers a keycard for the guest.

Auto check-in machines or terminals are also capable of handling self checkout procedures. The guest might use the same credit card (used at check-in) to access the guest folio and check its contents. As the guest completes checkout, the system automatically posts the account balance to the guest's credit card for billing and delivers the bill (with the break-up) to the guest.

Self check-in/checkout systems are now also available even for small properties. These systems allow the manager of a small property to benefit from technological advances, which in the past would have only been available for owners of large properties. The system can register guests, assign rooms, process credit cards or cash transactions, provide room keys or keycards, and also print out receipts for guests.

For the convenience of the guests, many self check-in systems have step-by-step instructions printed on the machine or beside it. For security reasons, the only way to access the machine's contents (cash) is from the rear of the machine, which might either open into the manager's office or a secured location. In addition, as a precaution, the system does not give out any cash. If a guest arrives late at the hotel and some amount is found to be due from the hotel's end, the guest is requested to collect the amount from the front desk next morning. When a guest pays by credit card, the credit authorization is secured by a telecommunications facility. If, for any reason, a guest's credit card is declined, instructions would be given requesting him/her to use another card or pay by cash.

The present technological advances allow guests to review the guest folio and check out from the room itself. For these, hotel systems might use in-room computer terminals, and guestroom telephones and television sets to display the guest folios. In an establishment, if the in-room computers are interfaced with a PMS and guest accounting module, the folio data can be easily viewed and the accounts settled too. In some hotels, printers are attached to the television and a guest can take a printout if desired.

Whatever the kind of guest-operated devices used in a hotel, the self check-in/checkout terminals and in-room computer interfaces have now considerably reduced the time used for processing guest registrations, check-ins, and checkouts. With advancements in technology, some automated terminals also have an inbuilt video facility, which allows the guest a view of the hotel features and rooms. Automated check-in and checkout devices can also, to some extent, make the employees of the front office spend some more time with those guests who require more attention.

In-room Movie Systems

In-room movie systems can either be interfaced with a hotel's PMS or can function as independent or stand-alone systems.

An in-room movie system is interfaced with the PMS to provide this facility in individual guest rooms through a dedicated pay channel on a television set. The interface also has a timing device; as and when a guest tunes in to the channel for a prefixed amount of time (usually several minutes), the device is activated and a charge is posted to the appropriate guest folio.

Since its launch, many a time, disputes have arisen due to the charges posted in connection with the in-room movie system. A guest might unknowingly turn on the television set for entertainment, and later find out at checkout that he/she had tuned in to a pay channel. To avoid this kind of situation, it is best for the hotel to incorporate a pay television preview channel, which might allow the guest to view a program for a set time, after which the guest would have to pay for it. For the transition, the guest may have to personally shift the channel to a pay channel (In this case, the guest cannot later deny the charges made). At the time of check-in too, the front-desk staff can explain the system and its charges.

In stand-alone, in-room movie systems, it is required of the guest to dial an in-house service number and request for the pay channel to be switched on. The operator, who then actually switches the program, is also responsible for posting the charge to the designated guest folio. The telephone system may not be a fully automatic option, but might be helpful if guests are not accustomed to a PMS interface. Hence, guest disputes during billing can be reduced.

In-room Beverage Service Systems

In-room beverage service systems have the capability of monitoring the sales made during a designated time span and also decide upon replacement of the inventory. The two most popular in-room beverage service systems are non-automated minibars and microprocessor-based vending machines.

Non-automated Minibars

A non-automated minibar usually stocks items (for dry or cold storage) within a guest's room. The bar's starting inventory is noted, either by housekeeping attendants (as part of routine schedules) or by specially chosen room service employees, before a guest checks in. If there is any consumption by the guest, an employee connects with the dedicated bar computer using a touch-tone telephone in the guestroom, then enters the product code and the numbers of items consumed by the guest, and updates the status. The bar system's CPU then relays the consumption made to the guest room information, and thereby the charges for the consumed items are posted to the guest folio. If there are any more requirements, the stock is replenished by the employee.

Though these types of non-automated bar systems are extremely convenient for the guests, they might be revenue losers for the hotel. Since a dedicated set of employees have to be involved in taking the physical inventory of each in-room bar, the labour cost would be high. The employee might sometimes also forget to report the consumption. In addition, the guest could also consume some beverage just before leaving the premises, which may not be immediately posted to the guest account, thus leading to loss of revenue for the organization.

Microprocessor-based Vending Machines

Microprocessor-based vending machines contain all the beverage items in a transparent closed compartment. The compartment doors are usually equipped with fiber optic sensors that record any removal of the stored beverage in the compartment. When a guest removes some beverages from the compartment, the sensors activate the relay about the transaction to the microprocessor, which then records it.

All the rooms having microprocessors are then typically cabled to a CPU, which stores the recorded transactions. The CPU then passes the transactions to prepare the account entries, and thereby relays them to both the PMS and the guest accounting module for guest folio posting. The bar system's central processing system also maintains an inventory replacement data, which directs restocking of the vending units.

A microprocessor-based vending system also avoids some of the problems that are associated with non-automated bars. For example, hotel managers could use a remote central console to lock in-room vending units when the rooms are not allotted to guests to avoid pilferage. Some systems also allow guests to lock their in-room bar units with their room keys when they are leaving their rooms. Since the system is interfaced with the PMS, the delayed posting of charges is avoided. A microprocessor-based vending system maintains a continuous inventory record, hence the labour costs associated with maintaining the inventory record manually, is reduced.

Some microprocessor-based beverage vending systems, which do not have a see-through facility, have infrared sensors below the bottle or can.

As and when a guest chooses a product in the bar, the infrared beam is broken, thus starting a timer set by the hotel. As timeout is reached, the sale is said to be completed; if not, the movement is stored. This timeout is allowed so that a guest can view and then make a final decision regarding purchasing a product.

Guest Information Services

Many huge shopping malls have installed information terminals for the guest to locate a particular sale outlet; many hotels too have followed this. These automated guest information services are usually kept in the public area of a hotel, thus allowing the guests to enquire about both in-house events and local activities across the city. Guests can also obtain a printout if the terminal is connected to a printer (detailing the events of a particular day).

Guest information systems, also called in-room electronic services, are important guest amenities, which have now transformed rooms. These systems can connect to the cable broadcast systems, several news services, transportation schedules, restaurants, various room service menus, and might also access external computer systems via the Internet. The in-room computers are able to link with external computer information services and the guests may have access to the following:

- Airline schedules
- News and sports updates
- Local restaurant guides
- Shopping catalogs
- Entertainment guides
- Historical places
- Stock market reports

If all the guestroom terminals are directly interfaced with a PMS, a great deal of system processing might be required at a particular time for responding to all the activities coming from the in-room terminals. To avoid such a situation, the PMS and in-room computers might be connected to a remote CPU, which is then interfaced with the central processing unit of the PMS. Since only a single connection must be made to the PMS, this configuration simplifies the guest information services interface.

In addition to the PMS interface, in-room guest information terminals might also be connected to a hotel's cable television. These connections enable the property to ensure that convention attendees are informed of various events and functions in and around the city, provide tourists with information about local tourist attractions, and inform business travellers about various add-on services provided by the property.

SELECTING AND IMPLEMENTING COMPUTER SYSTEMS

Whether a hotel selects its own PMS or is advised by the franchisor, the quality of the PMS is determined by examining the needs of the hotel. Since the PMS to be implemented requires hardware and software, the needs of the hotel have to be taken into consideration before a PMS is installed. A PMS should have the following characteristics:

Dependable   The cost of a PMS is of lesser importance than its reliability, as this is the most important concern. The PMS is a computer system and can sometimes become inoperable (due to system crash). Since the PMS is the heart of a hotel, if it fails during peak hours, it creates major problems for employees and also causes dissatisfaction among guests due to the delay. A PMS can fail because of hardware, software, or link problems. A PMS should always be linked to a battery or generator backup, which is capable of operating the system if the hotel loses power due to circumstances beyond the control of the hotel. Backup of the system should be maintained so that if the computer or server has problems, the stored data can be retrieved.

Moderate operation cost   The hardware used for a PMS should be easily replaceable. The cost for replacement should not be very high and the component must be easily available. The cost of other operating supplies such as paper and cartridges should also be considered. If a particular system requires special ink cartridges, special types of paper, or special hardware (printer) that is expensive, such a system should be avoided.

Easily installable   A hotel operates 24 hours a day for 365 days. Hence, disruption of operations during installation of a PMS may affect the smooth running of the hotel. Therefore, before a PMS is selected, the management must consider the amount of time it will take to install the new system. In addition, the hotel's existing data also has to be transferred to the new system, which is important for management decisions.

User-friendly   New employees must learn to use the PMS. Even existing employees should easily adapt to the new system. Keeping this in mind, the PMS should be an easy-to-learn system, reducing training cost, and thereby increasing the pace and efficiency of employees.

Interfaced integration   The front-office manager needs information from a variety of sources to profitably manage the front office. For example, a hotel may have various service outlets, restaurants, bars, and gift shops. All the charges of the guest have to be posted to the guest folio. In any PMS, the charges can be posted manually. In a large hotel where the number of rooms is too many, manual posting of each transaction for each guest would be time-consuming and subject to data-entry error. These risks are unnecessary, considering the advancement in technology.

An interfaced information system is beneficial. A PMS may not always be capable of interfacing with all other information-generating subsystems in the hotel. Taking this into account, the management must ensure that a PMS is interfaced with the hotel's critical, pre-existing, information-generating system. The best PMS is one that can interface with the hotel industry's best and most popular ancillary hardware and software products.

Upgrading and updating capability   Hardware and software upgrades are routinely issued by the manufacturers and most of these significantly improve the effectiveness of a PMS. Difficulty may arise when a system's hardware component is not adequate to effectively operate the new software system. PMS software upgrade should be easy to install and minimize system upgradation time. A PMS manufacturer issues newer versions of the system's software on a regular and frequent basis, in keeping up with the hotel industry's changing demands for information.

Maintenance requirements   A PMS, like all other hotel equipment, must be properly maintained to operate efficiently. The cost of effective preventive maintenance programmes for each PMS will vary. The management must determine and understand the time and money required for maximizing system efficiencies by providing routine ongoing cleaning and maintenance.

Availability of quality support service   A PMS is a computerized system that is subject to hardware malfunctioning, software glitches, and potentially damaging intrusive viruses in any computer linked to the outside environment via the Internet. Since many other computerized systems interface with the PMS, difficulties may arise in one or more of these interfaces. Support service personnel should be easily accessible by e-mail or phone. Access to support services must also be offered on a 24×7 basis, including all holidays.

SUMMARY

This chapter examines independent, stand-alone computerized systems, which may interface with a PMS. These PMS interfaces are point-of-sale systems, call accounting systems, electronic locking systems, energy management systems, auxiliary guest services, and guest-operated devices.

A point-of-sale system is made up of a number of point-of-sale terminals, which typically interface with a remote central processing unit. The processing unit interface is usually a stand-alone CPU or a host electronic cash register. When the remote CPU or the host electronic cash register is interfaced with a PMS, data can be transferred from the POS system to various front office and back office PMS modules for further processing.

A call accounting system is capable of directly handling long-distance dialling, distributing calls by a least-cost routing network, and also pricing outgoing calls. When a call accounting system is interfaced with the PMS and front office guest accounting module, telephone charges can be directly posted to the guest folio.

The chapter also examines the types of electronic locking systems: hard-wired systems, micro-fitted systems, RFID, and biometric locks. Hard-wired systems employ a key code console or master control board that determines a door's lock combination and relays that code to the door. Micro-fitted systems rely upon a predetermined sequence of code numbers residing in each individual door lock. All codes and code sequences are stored in a central console at the front desk. The insertion of a keycard into a microprocessor lock advances the stored sequence to the next number and makes all previous codes invalid. The audit trail obtained by the electronic locking system forms the basis of security reports for use by the management. Radio-frequency identification is a digital door-locking system that authenticates and validates the user and then opens the door automatically. It is inexpensive and uses wireless technology. Biometric locks are fingerprint scanning locks that can be installed in the doors of a hotel's rooms. It scans a fingerprint and saves it; the guest fingerprint is tallied with the stored one, and then the door is opened.

A computerized energy management system is designed to automatically manage the operation of some types of equipment responsible for maintaining comfort levels throughout a hotel. The programming of these systems enables the management to determine when a particular equipment is to be turned on, off, or regulated.

Guest auxiliary services include the automation of wake-up calls, electronic message systems, and the use of voicemail. The recent introduction of advanced technology to auxiliary services has enhanced the hotel's productivity, especially with regard to guest support services.

The chapter also discusses guest-operated devices, for example, self check-in/checkout technology, which allows a preregistered guest to check-in without a manned front office and delivers a keycard after the registration and necessary payment is carried out. An in-room movie system allows guests to choose movies, sitting in the comfort of their room. Since such a system is interfaced with a PMS, the charges are posted to the guest account. The chapter also discusses in-room beverage service systems and guest information services. It also talks about the criteria for selecting and implementing a PMS for a hotel, taking into consideration the present needs of a hotel and those of the future.

KEY TERMS

Automatic identification of outward dialing (AIOD)   A terminal equipment that can transmit data from a private branch exchange (PBX) so that a vendor can provide a detailed monthly bill identifying long distance calling usage by individual PBX stations.

Biometric locks   They are fingerprint scanning locks that are installed within the doors of hotel rooms. They scan the fingerprint and save it.

Call accounting system (CAS)   It is a stand-alone system that interfaces with a hotel's PMS and is capable of handling direct distance dialing and prices outgoing calls.

Demand control   This is a system that maintains usage levels by shedding energy loads in an orderly fashion.

Electronic cash register (ECR)   It is an independent stand-alone computer system possessing an input/output device, a central processing unit, and storage capacity.

Energy management system   It is a computer-based system that is designed to automatically manage the operation of mechanical equipment within a property. Such a system conserves and contains the energy cost of a building.

Guest operated devices   These are user-friendly devices that are located in a public area of a hotel or in private guest rooms.

Hard-wired system   This is a system that operates through a centralized master console and is interfaced to control all guest room doors.

Hotel billing information centre (HOBIC)   This system is a service provided by a telephone company, which records the time and charges when a guest makes a long-distance call.

In-room beverage service system   This service system is capable of monitoring sale transactions and determining inventory replenishments.

In-room movie system   This system interfaces with a hotel's PMS and is used for providing a guest with entertainment through a dedicated television pay channel or movie, in the room.

Micro-based POS system   This point-of-sale system contains a microprocessor in each terminal and functions without a large, remote central processing unit.

Micro-fitted system   This consists of individually configured stand-alone units, which contain a unique, predetermined sequence of codes.

Microprocessor-based vending machine   This machine contains beverage items in see-through compartments, whose doors are equipped with fiber optic sensors that record removal of stored product.

POS terminal   A point-of-sale terminal is a terminal that has its own input/output component and can also possess a small storage (memory) capacity, but usually does not contain a central processing unit.

Radio-frequency identification (RFID)   It is a digital door-locking system, which authenticates and validates the user and then opens the door automatically.

Room occupancy sensors   These sensors use infrared light or ultrasonic waves to register physical occupancy of a room and turn on devices that are under their control.

Self check-in/checkout system   This refers to a terminal that does not require manual handling (e.g., automatic bank teller machine), possesses both audio and video abilities, and initiates self-registration and checkout processes by using a credit card.

Station message detail record (SMDR)   It is an optional section within a switchboard for charting and monitoring the telephone traffic in a hotel.

Voice over Internet protocol (VoIP)   It is a method by which analog signals are turned into digital signals (and vice versa) and transmitted over the Internet.

REFERENCE

Woods, Robert , Jack Ninemeir, David Hayes, and Michele Austin, *Professional Front Office Management*, First edition, Pearson Education, New Delhi, 2008.

Web References

http://communication.howstuffworks.com/ip-telephony.html, last accessed on 6 January 2012.

http://support.resortdata.com/Customers/Knowledge/KB-Interfaces/KIN0035.htm, last accessed on 2 January 2012.

http://transition.fcc.gov/voip/, last accessed on 6 January 2012.

http://www.autocheckin.com/interactive_slideshow_b.php, last accessed on 28 April 2012.

http://www.scribd.com/doc/29919138/Hotel-Property-Management-Systems, last accessed on 4 October 2011.

http://www.securitylockkey.com.au/content_images/ibt201003015-01_072dpi.jpg&w, last accessed on 6 January 2012.

www.uptuplus.com/index.php?/downloads/...vending-machine.html, last accessed on 10 February 2012.

EXERCISES

Concept Review Questions

1. How do micro-based POS systems differ from other POS systems?

2. What are the typical POS data entry requirements for posting charges to the appropriate guest folios?

3. What are some of the concerns that a management should address in relation to interfacing a POS system with an integrated PMS system?

4. What are the advantages of a call accounting system when compared to the HOBIC system?

5. What kinds of data are maintained by a CAS call record file? Explain how this data may be useful to a management.

6. What are the major differences between hard-wired and micro-fitted electronic locking systems? Identify the advantages and disadvantages of each system.

7. What are the three energy control strategies that may be used by an energy management system?

8. How can lodging properties benefit from automated self check-in/checkout systems?

9. What are the two in-room beverage service systems?

10. What external information services may guests be able to access from in-room computer terminals?

11. What are RFID and biometric door-locking systems?

12. What are the criteria to be considered while selecting a PMS? How will a person implement a PMS after its selection?

13. What are the ways by which demand control can be maintained in hotels?

14. How is RFID technology used? How is it better than a wired system?

15. What are the points to be considered for a self check-in/checkout system?

Multiple Choice Questions

1. The number and location of POS terminals throughout a property is a function of a variety of factors such as
   (a) size and type of operation
   (b) physical design limitations
   (c) communication requirements
   (d) all of these

2. Which of the following features does an electronic cash register have?
   (a) Dependant
   (b) No input/output device
   (c) No memory capacity
   (d) None of these

3. A call accounting system includes which of the following features?
   (a) Handles direct-distance dialing
   (b) Prices outgoing calls
   (c) Distributes calls through the least-cost routing network
   (d) All of these

4. HOBIC
   (a) supervises all operator-assisted calls
   (b) is often interfaced with CAS
   (c) both (a) and (b)
   (d) none of these

5. The functions of CAS include
   (a) call placement
   (b) call routing
   (c) call distribution
   (d) all of these

6. Electronic locking systems are usually
   (a) hard-wired
   (b) micro-fitted
   (c) both (a) and (b)
   (d) none of these

7. HVAC controls
   (a) heating
   (b) cleaning
   (c) cooking
   (d) none of these

8. Duty cycling helps in
   (a) reducing energy consumption
   (b) enhancing guest comfort
   (c) reducing man hours
   (d) none of these

9. Room occupancy sensors use
   (a) infrared light
   (b) ultraviolet light
   (c) microwaves
   (d) none of these

10. These are required for self check-in terminals.
   (a) Advance reservation
   (b) Valid credit card
   (c) Both (a) and (b)
   (d) None of these

Project Work

1. What are the points you will consider before implementing an energy management system in a hotel?

2. If you have been asked to select and implement a PMS, what steps will you take for the same?

CASE STUDY 1

Front office manager Vivek Patwardhan of Hotel Skylark was in the lobby interacting with some foreign guests. One of the guests, Utsav Roychoudhury, staying in room number 328, was checking out from the hotel. When the final bill was presented to him, he was perplexed. A ₹1246 ISD call made to London and a movie charge of ₹300 had been posted to his account.

Utsav told the cashier that he had never made any ISD calls from his room. Regarding the movie, Utsav said that he had enquired about the in-room movie facility and had asked for a few Bollywood blockbusters. As the movie was available with the hotel, he had said that he would view it later in the evening; however, he forgot to watch it.

The hotel had an automatic call accounting system and an in-room movie system. Hence, it was almost impossible that a guest would be charged without using the services. The call accounting system showed details regarding the call—the number to which the call was made, the time and date the phone call was made, the duration of the call, and the room number from which the call was made. The audit trail report showed that in both instances, Utsav was in his room only.

Utsav was not convinced with the front-office cashier and wanted to speak to the front-office manager about his bill.

(a) If you were Vivek, how would you have handled the situation?
(b) Is Utsav right about overcharging? Justify your answer.
(c) How does a call accounting system function as a foolproof mechanism for guest billing?
(d) How is an in-room movie system interfaced with a POS system for better billing?

CASE STUDY 2

Hotel City Palace is situated in the heart of the city and is famous for its hospitality, guest services, and use of information technology for better guest comfort. Hence, Sheila Shinde, a business woman, booked a room in the hotel for three days. Sheila was greeted and welcomed to the hotel on arrival and was escorted to room number 512 by the bellboy.

On the second day of her stay at the hotel, at around 8 p.m., Sheila called the front office and complained about her missing diamond necklace. She had left her room for official work at 9 a.m. and had returned to the hotel at around 7 p.m. The audit trail report of the electronic door showed that the room had been opened only once for routine cleaning.

Housekeeper Sudeep had serviced the room at around 11 a.m. Since Sudeep's shift was over, he had left for the day. Sheila had not kept the necklace in the electronic safe provided in the closet, but had instead left it on the bedside table provided. Duty manager Somjit Chakraborty and security officer Amit Das were called in, and after listening to Sheila's complaint about the missing necklace, said they would look into the matter very seriously.

(a) What went wrong in this case?
(b) If you were Somjit, what would you have done?
(c) Sheila had kept the diamond necklace on the bedside table. Could she claim damages? Justify your answer.

Answers to Multiple Choice Questions

| 1. (d) | 2. (d) | 3. (d) | 4. (c) | 5. (d) | 6. (c) | 7. (a) | 8. (a) | 9. (a) | 10. (c) |

CHAPTER 11

Management Information System

LEARNING OBJECTIVES

After reading the chapter, you will be able to understand the following:

- Meaning and concept of management information system (MIS)
- Usage of MIS in hotels
- Design and functions of MIS
- Software development life cycle
- Evaluation of MIS
- Security issues of MIS

A management information system (MIS) is an integrated man-and-machine system that provides the basic information that is necessary for supporting, planning, and controlling an organization. It is a combination of human and computer-based resources that results in collection, storage, retrieval, communication, and use of data for efficient management operations. Management information system is used for supplying information that can be used for arriving at an effective

---

EXHIBIT 11.1 Creating an Effective Information System

Abhimanyu Seth had joined a hotel as a general manager. The director of the hotel, Praveen Raheja, had asked him to look into the hotel's operations and increase the sales of the hotel.

Abhimanyu called the departmental managers to his room for a meeting and asked for sales-related details. Though the hotel had computers, the required reports of each department, as requested by Abhimanyu, were not available.

He then called his friend Girish Kumar, a software expert, to his room. Abhimanyu wanted a management information system (MIS) for the hotel, which he discussed with Girish and all the department heads. After listening to

the heads' requirements, Girish came up with a feasible integrated MIS for the hotel. The MIS was designed to cater to the needs of the various levels of management in the hotel. As the information was confidential, proper security measures were required so that no information would be accessible to anyone other than the designated person.

Computer systems in hotels and restaurants can produce reports for the managers. However, simply preparing and distributing routine reports does not in itself ensure a very effective information system. To achieve the maximum benefit of an automated information system, the computer functions should be integrated with the management's information requirements.

decision. Computerization has helped in increasing the speed and accuracy of processing data, which in turn has led to better decision-making.

The term 'management information system' is made up of three words—management, information, and system. We will discuss the three components individually and then discuss MIS in detail.

MANAGEMENT

Management is both a science and an art. It is the process of achieving organizational goals and objectives by using manpower, materials, machines, money, and methods. The various functions of management are planning, organizing, staffing, controlling, and directing.

Planning   It is the process of deciding the course of action well in advance. The following processes are needed for efficient planning: forecasting sequence of events, arriving at objectives, evolving strategies, deciding programmes, setting budgets, planning procedures, and developing policies.

Organizing   It refers to the formal grouping of people and activities to achieve a firm's objectives. Organizing is also necessary to establish a hierarchy for production of a new or existing product.

Staffing   It involves the selection of people to accomplish specifically allocated tasks. These people have to be oriented towards planning the objectives, strategies, procedures, and policies. People have to be trained to perform the necessary tasks.

Controlling   It is the means of checking the progress of plans and correcting deviations that may occur along the way. Control is used to measure physical quantities such as volume of output (i.e., results or outcome) and number of guests checking in per hour, or monetary results such as sales in a coffee shop and return on investments.

Directing   It is the process of activating plans and structuring group efforts in the desired direction. It is needed for implementation of plans by providing leadership, motivation, and proper communication. It involves the steps associated with getting things done through people.

The management can also be grouped into three hierarchical levels—top, middle, and junior levels. The top (strategic management) establishes the policies, plans, and objectives of the organization as well as a budget framework for the organization. The middle (tactical) management is responsible for the implementation of policies and plans of the top management. The junior (operational) management is responsible for the implementation of the regular operational decisions of the middle management, such as to produce goods and services, and to meet the revenue, profit, and goals of the organization.

INFORMATION

It is defined as data that is organized and presented at a particular time and place so that the decision-maker can take the necessary action. It can also be said that information is the result or product of processing data. Information is of value to only those who understand its use in taking decisions. In MIS, the value of information is used to find out the benefit of disseminating perfect information.

SYSTEM

It is a group of elements or components that are joined together to fulfill certain functions. It is an assemblage of procedures, processes, methods, and routine techniques that are united in some form to form an organized whole. A system has three basic parts, which are set in an orderly manner. These parts are shown in Fig. 11.1.

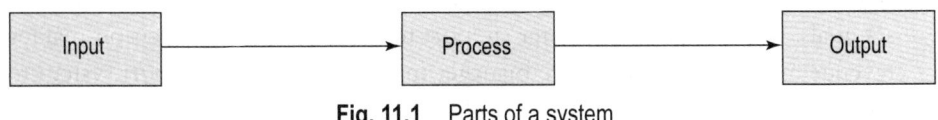

Fig. 11.1   Parts of a system

A system can have a single input and multiple outputs or many inputs and outputs. A business organization's system could have many inputs and multiple objectives—sales, profit, service, and growth.

A system can operate in any environment. The environment, in turn, can also influence the system's design and performance. As a system is designed to obtain certain objectives, it creates a boundary for itself. Knowing the boundaries of the system is vital to bring about precision in the system components and their arrangement.

A system that functions in isolation from the environment, does not exchange information with the environment, nor is influenced by environmental changes, is said to be a closed system. A system that exchanges with the environment and is also influenced by the environment is said to be an open system. Accounting systems such as stock and attendance of employees are closed systems. On the other hand, marketing, communication, and forecasting systems have to respond to changes in the environment and are hence open systems.

A system can be classified into different categories based on the certainty of the output and degree of information exchanged within the environment. A system is known as deterministic if the input, process, and output are certain. A system is said to be probabilistic when its output can be predicted only in probabilistic terms.

The aim of an information system is to support organizational processes to achieve organizational goals. A system helps the management in various planning, controlling, and decision-making processes. The following information systems support management processes:

- Transaction processing system (TPS)
- Decision support system (DSS)

Transaction processing system   The main function of a transaction processing system is to record, process, validate, and even store transactions that have taken place during business, and can be obtained later and used. Such a system processes transactions in two basic ways. In a batch process, the transaction data is accumulated over a period of time and then processed periodically. A transaction support system puts forward a set of rules and guidelines, which specify how we record, process, and store a particular transaction. The main functions of a transaction processing system are bookkeeping, issuance, and control reporting. Bookkeeping involves maintaining a record of an organization's financial statements. Issuance involves the maintenance of business documents such as invoices, vouchers, and payables, whereas, control reporting involves preparation of reports on various transactions for controlling accounts.

Some examples of transaction processing systems are payroll processing, account receivable/payable systems, and inventory processing systems.

Decision support system   It is an interactive and integrated system, which provides managers with data, tools, and models to facilitate semi-structured decisions. This system is suited for problems such as location selection, identification of new products to be marketed, and scheduling of personnel. This system helps managers to perform resourcefully to accomplish set goals. It gives direct computer support to managers during decision-making processes. A revenue manager or front-office manager may use a decision support system to decide on an appropriate rate at which rooms should be sold, taking into account the forecasted booking and the number of rooms sold on the same date, the previous year.

MANAGEMENT INFORMATION SYSTEM

The definition of management information system varies from individual to individual. It is defined by Schwartz as, 'MIS is a system of people, equipment, procedures, documents, and communication that collects, validates, operates, stores, retrieves, and presents data for use in planning, budgeting, accounting, controlling, and other management process.' An MIS is designed to present managers with definite information that is required to plan, organize, staff, direct, and control operations. An effective MIS increases the capabilities of the computer and provides the managers with necessary information, as mentioned here:

- Monitors the progress made in achieving organizational goals
- Measures performance
- Identifies new trends
- Assesses the alternatives
- Supports decision-making
- Assists in remedial action

Management Information System in Hotels

Management information system is used in hotels to understand the needs and expectations of the guests, and fulfilling them in the best manner. It is used for providing the following information:

Guest profiling   A guest database containing the following information is required:

- Type of guest
- Purpose of visit
- Duration of stay
- Services requested
- Country of origin
- Socio-economic status

This information helps the management in understanding guest requirements as well as his/her service expectations, which might lead to additional and better guest-supporting infrastructure, such as spas, speciality restaurants, and even better equipped rooms.

Occupancy status   A hotel's room inventory is fixed. Hence, the hotel would try its level best to sell all its room every single day. To meet this, the reservation and occupancy status of the rooms should be known to the management. This helps them to plan various strategies in advance, for example, different room rates for groups, corporates, and travel agents. Knowing the reservation status also helps in planning special discount rates for guests with advanced reservations and separate rates for guests without prior reservations. This leads to higher guest occupancy and revenue generation.

Future needs   The MIS should be able to forecast the needs of the future. Though earlier, hotels were thought to be a place for providing accommodation to guests, now, they are also a place for meetings, incentives, conferences, and exhibitions (MICE). Therefore, the management has to provide the necessary infrastructure and logistics.

Meeting customer expectation levels   Customer expectation about services depends on the class of the hotel and service provided. A guest may have a set idea about the manner or the process of how a particular service is to be offered. The process involved in offering services is a function of training and understanding of the capability of the manpower, whereas, the speed of the process would depend upon the strength and competency of the manpower. For this, a proper manpower network has to be created to take care of the quality of service.

Guest history   Hotels have a guest database, usually known as guest history, which is filled after a guest checks out from the property. The guest history contains personal information about the guest, his/her room choice, food choice, special likings, etc. This record helps in building a good customer relationship, thus creating a sense of care and concern.

Usage

A manager who is not properly informed about a particular situation could sometimes take a hasty decision, which may lead to a guest opting for a competing hotel. In this regard, MIS is a timely decision-making tool for an organization to optimize operations and bring about growth.

MIS can also be used in a property management system, sales and marketing, quality and human resource management, revenue generation and accounts, room management, and also in customer relationship management (CRM).

Though MIS was originally preferred by hotel chains, now, stand-alone properties have also started using it. MIS in hotels can be used for project scheduling, yield management, and generating revenue for the food and beverage (F&B) department. It is also used for customer–relationship management, which could help in improving the standards of service and removing unpopular items from the menu. It also helps in understanding guest satisfaction and spotting dissatisfied guests, who may feel so due to misbehaviour by the staff.

Design and Function

In designing an MIS, a particular approach has to be followed so that a suitable system can be devised to cater to the needs of different organizations, as per their functions and decision-making requirements.

The design of an effective MIS is prepared after taking into consideration the information needs of managers. The managers have to describe and organize their information needs. Bearing this in mind, the information system is designed to organize computer applications so that they support decision-making activities at various levels within the organization.

An MIS supports the following levels of decision-making: strategic planning, tactical decision-making, and operational decision-making. Strategic planning refers to decision-making activities through which the future goals and purposes of an organization are established.

Tactical decisions include the various steps that are required to implement strategic planning decisions. Operational decisions address specific tasks that normally follow previously established rules and patterns.

As and when the information needs of managers are met, an MIS is designed to perform further functions as mentioned here:

- Allow the managers to monitor and direct business transactions and activities in a better way.
- Offer advanced functional and internal control over business resources.
- Prepare timely and completed reports that are formatted to the specific needs of managers.
- Lessen managerial paperwork and other operational expenses by removing unnecessary source documents, organizing data transfers, and recording procedures.

To perform these functions effectively, an MIS uses a variety of information processing technologies and decision support systems. The following steps are generally taken into account while designing an MIS.

Identifying needs at all levels of management   There are problems in every business organization and many a time the reasons for the problems and solutions are unknown. Hence, the first step in designing an MIS is to identify the problem and a tentative solution for the same. The following steps are necessary to initiate the design process, and should be repeated until all the information that is required, and the problems to be solved are understood.

- Initiating the need for information
- Asking a query about the need
- Interpreting the need
- Detailing the original statement
- Reviewing with management, the detailed statement of the need

Defining objectives and benefits of MIS   A user must define the system objectives in terms of information demands and not in terms of satisfaction demands, which are without an objective. The system objective should be defined in terms of what the decision-maker can do and how effectively an organization will be able to function after the information requirements have been compiled. The basic questions that are asked while listing down the objectives of an MIS design are given here:

- Purpose of the system
- Need
- Expectations
- Users and their objectives

SOFTWARE DEVELOPMENT LIFE CYCLE

The software development life cycle (SDLC) for a particular business is not an easy task. Depending upon the software need, it may require many people and sometimes months for it to be prepared. However, a small application can be developed within a few weeks by an individual or a few programmers. The following are the various steps involved in the development of software (Fig. 11.2).

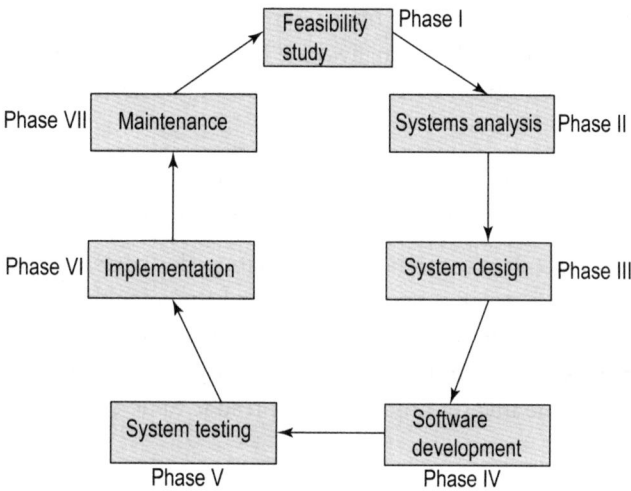

Fig. 11.2   Software development life cycle

Feasibility study   It is usually the first phase in the development of a new system. This phase starts when the user faces problems in the current system. On receiving the request, system analysts first try to determine if the requested system is feasible or not.

System analysis   When the system analyst is assured that the requested system is feasible and the management also agrees to develop it further, the next phase begins. In this phase, an existing system is studied and data collection is done to find the needs of the users.

System design   After the data has been collected and user requirements studied, the system is designed. In the identification stage, data inputs, output reports, and the procedures to process the data are made.

Software development   After the design has been accepted by the department that requested for it, the programmers start designing the data structure and write the program. The programmers slowly test each program and try to integrate them into a single unit.

System testing   It is a vital phase of the SDLC. The whole system is tested with numerous techniques to ensure that it is error free.

Implementation   The tested system is installed at the user's place and then implemented. Though it is generally considered as the last phase of the SDLC, the development work continues till the user department accepts the system.

Maintenance   After its implementation, the system needs to be maintained so as to adapt to the changing business needs of an organization.

Management Information System Evaluation

The feasibility study actually begins when the managers and the users of the MIS department identify the problem. Both groups of people discuss the major problems in the existing system and their main requirements. The objectives of the feasibility study is to identify the deficiencies in the present system, prepare the objectives of the proposed system, identify the users, and finally analyse if it is feasible to develop a new system. After the feasibility study has been done, a report is prepared.

A cost–benefit analysis is a major part of the feasibility study to find out the economic feasibility of the project. It is usually done by the identification and classification of costs and benefits, and the selection of evaluation methods. The following are the common methods of evaluation:

- Payback method
- Present value method
- Net benefit method
- Break-even method

Payback method   It is a method of evaluating the costs and benefits, and determining the time when the accumulated benefits will equal the initial investment. Using this method, the system analyst can know when the money spent on the project will be recovered.

Present value method   In this method, the project cost is planned keeping in mind the time-related value of the invested money. The present value of the money would not be the same as that later.

Net benefit method   It is a method by which the net benefit is calculated by deducting the total approximate cost from the total approximate benefit.

Break-even method   It is a method in which the costs of the current and candidate systems are compared to find out when both will become equal. The point at which this happens is said to break-even.

MANAGING MANAGEMENT INFORMATION SYSTEMS

In a fully automated hotel, the MIS management staff may either have both, a property systems manager and designated department systems supervisors, or only the systems manager. The property systems manager plays a major role in the evaluation, selection, and installation of computer system hardware and is also trained in software applications that are used throughout the property. The property systems manager provides on-premises systems support as and when necessary. In some hotels, other than the property systems manager, there are also department systems supervisors. They are individuals who are employed for a particular department and receive extensive training in operation of software applications pertaining to the department. These supervisors train other staff members within their departments and also provide technical support if required.

The property system manager has a wide range of responsibilities. He/She should not only have knowledge of computer systems, but also be familiar with information processing techniques and principles of management. The manager must also be aware of the operations of each department of a hotel and understand the interrelationship between all functional areas within the property. Without proper understanding, it becomes very difficult to direct the MIS to meet the specific information needs of managers throughout the property. Other duties of a property systems manager are given here:

- Planning and controlling various MIS activities, which include identifying and processing the priorities within the system
- Selecting department systems managers and establishing training programmes as required
- Developing system configuration and design alternatives taking into account the placement and processing capabilities of both—computer hardware and software
- Designing and implementing information security controls

SECURITY ISSUES OF MANAGEMENT INFORMATION SYSTEMS

There are three major areas of MIS security—power backup systems, information backup systems, and information protection. The management of an organization focuses on the risks involved during power failure, information backup procedures, and unprotected information that might pose a potential threat to the hospitality operations. It may so happen that a competing hotel might gain access to the guest history files, which could lead to a hotel losing revenue. However, these can be avoided with proper security systems.

Power Backup Systems

Fluctuation or interruption of power supply can lead to problems in working with computers. This problem can be significantly reduced by using an uninterruptible power supply (UPS). The UPS is equipped with a battery that is placed on the computer's electric line so that if there is any fluctuation in the power line of the computer, it would activate the battery, which would compensate for the energy deficiencies. This battery backup gives the computer continuous power supply. The system automatically recharges the battery as and when a normal power source becomes available.

A preventive maintenance programme is also necessary to protect against system breakdown. There should also be a predetermined emergency maintenance plan, outlining the steps to be followed during a crisis. For example, in addition to power backup systems, there should also be plans for hardware backup availability, that is, sources for quickly obtaining essential parts such as a printer or keyboard.

Information Backup Procedures

Information backup should be a standard operating procedure to ensure that no data is ever lost. There are three main ways of maintaining information backup:

- Redundant copy
- Duplicate copy
- Hard copy

Though many computer manufacturers suggest using more than one backup procedure, it is for the management to ensure that at least one practice is being followed regularly. We will now discuss each procedure in detail.

Redundant copy   This kind of copy is simultaneously prepared in two storage devices as and when transactions are carried out (the copies are saved in two external storage devices). Since an accounting system employs a disk drive as the base for one external storage device, it also requires a second disk drive. The data is stored in two separate disks, as and when data input is performed. It could be an expensive hardware configuration and requires more attention than other backup methods.

Duplicate copy   This kind of copy is one of the most popular and efficient means of obtaining information backup. In a duplicate copy, the computer system records data in only one storage device; hence a second disk drive (as is necessary for a redundant copy) is not required. A copy of the data on the single drive can be later made on the same device using a blank disk. The backup tapes are stored and only used in case of disk error or failure; the orderly access to data does not reduce the property's ability to maintain efficient computer operations.

Hard copy Printouts from disk files can be used as a backup technique only in combination with either redundant or duplicate copy procedures. A user who relies on only hard copy information backup will face the difficult task of maintaining data files. In this situation, all the information stored on the hard copy has to be manually re-entered to restore the system's database. When a hard copy is used to supplement one of the other two methods of data storage, it usually provides the means for troubleshooting any missing or incorrect transactional recording. A hard copy backup is usually taken by the hotels at the end of a day's procedures or after the night audit has been completed. The hard copies are usually filed and a copy is sent to the department heads.

Information Protection

Information protection is much more complicated than power backup procedures or information backup procedures, and should involve strategic considerations. Information must be protected from two major threats—external and internal.

External threats Since an organization is connected to external devices through the Internet, there could be instances of some of the data being transferred from the hotel's system to an individual's system. Cybercrime is now a huge threat to society and is caused by criminal or irresponsible action by individuals, making the Internet and other networks vulnerable to its effects. Hacking can be carried out by either an outsider or an employee of the company, who uses the Internet and other networks to steal or damage data and programs.

Internal threats To reduce internal threats to the security of information, organizations have passwords and different authorization levels for accessing the data. Security codes (e.g., a multi-level password system) are being used for security management. In this procedure, passwords are allotted to every individual using the system. Data would be provided as per the authorization level accessibility of the password. Hence, a waiter's password will not have the ability to void a sale or view the sales figures of an outlet, but the manager's password would have access to such data. Routine maintenance of the software and software updates are necessary to avoid transfer of information to unauthorized individuals.

A computer system may also be affected by a virus from either external or internal sources. A virus is an unauthorized programmed code that attaches itself to other programs. Viruses usually enter computer systems through external programs or files.

SUMMARY

The chapter describes the function of an management information system (MIS). The MIS provides managers with the necessary information to plan, organize, staff, direct, and control operations. An effective MIS extends the power of the computer beyond routine reporting and provides managers with information that is useful for strategic planning, tactical decision-making, and many a times in operational decision-making.

We also discussed the three components of MIS—management, information, and system—in detail. MIS accomplishes functions through information processing technologies that include decision support systems and transaction processing systems. These technologies establish a communication process in which data is transferred from independent computer systems; processed according to pre-established decision-making rules, financial models, or other analytical methods; and stored in information formats tailored to the needs of individual managers.

We also talked about the various steps in a software development life cycle. The use of MIS in hotels, its design, functions, and evaluation have also been explained. In the last part of the chapter, we described the three major areas of MIS security—power backup systems, information backup procedures, and information protection from various threats.

KEY TERMS

Controlling   It is the process by which it is ensured that the organization is proceeding in the desired direction and progress is being made towards the achievement of goals.

Customer relationship management   CRM is an accepted model for managing a company's interactions with customers, clients, and sales prospects.

Decision support system   This is an interactive and well integrated system that provides managers with data, tools, and models to facilitate semi-structured or tactical decisions.

Directing   It is the managerial function of guiding, supervising, motivating, and leading people towards attaining planned performance targets.

Feasibility study   The aim of this study is to objectively and rationally uncover the strengths and weaknesses of existing businesses, proposed ventures, opportunities, threats as presented by the environment, resources required to carry out the businesses, and ultimately the prospects for success.

Information   It is defined as data that is organized and presented at a particular time and place so that the decision-maker can take necessary action.

Management information system (MIS)   It is a refined arrangement of information sources, which help managers in planning and controlling methods for operational system implementation.

Organizing   It is the process of establishing a harmonious authority–responsibility relationship among the members of an enterprise.

Planning   It is the process of deciding the course of action in advance. It also involves determining the objectives.

Staffing   It is the process of building an organizational structure through proper and effective selection of individuals, appraisals, and development of personnel to fill the roles designed in the structure.

System   It is a group of elements or components that are joined together to fulfill certain functions. A system is an assemblage of procedures, processes, methods, and routine techniques, which are united in some form of regulated interaction to form an organized whole.

Transaction processing system   It is a system to record, process, validate, and store transactions in various functional areas of business for future retrieval and use.

REFERENCES

Arora, A. and A. Bhatia, *Information Systems for Managers*, First edition, Excel Books, New Delhi, 2001.

Brien O', J. and G.M. Marakas, *Management Information Systems*, Seventh edition, Tata McGraw Hill Education Pvt. Ltd, New Delhi, 2006.

Gupta, A.K., *Management Information Systems*, First edition, S. Chand & Company Ltd, New Delhi, 2000.

Jawadekar, W.S., *Management Information Systems*, Second edition, Tata McGraw-Hill Publication, New Delhi, 2002.

Web References

http://www.scribd.com/doc/29919138/Hotel-Property-Management-Systems, last accessed on 4 October 2011.

www.expresshospitality.com/20060715/edge01.shtml, 1–15 July 2006, last accessed on 8 May 2012.

EXERCISES

Concept Review Questions

1. What is a management information system (MIS)? What are its components?
2. What is the use of MIS in hotels?
3. Explain the feasibility study of a project.
4. How is transaction processing system (TPS) different from decision support system (DSS)?
5. What are the steps in MIS design?
6. How is evaluation of MIS done?
7. What are the various MIS security issues?
8. Write down the steps involved in software development life cycle.
9. What are the qualities of a system analyst?

Multiple Choice Questions

1. Cost benefit analysis _____.
   (a) compares the cost with the benefits of introducing a computer-based system
   (b) estimates the cost of hardware and software
   (c) evaluates the tangible and non-tangible factors
   (d) all of these
2. A feasibility study _____.
   (a) is partly computerized and partly manual
   (b) refers to conceptual solutions of the problem
   (c) is an assessment of the validity of the project
   (d) none of these
3. A system that does not interact with the external environment is called _____.
   (a) a closed system
   (b) a logical system
   (c) an open system
   (d) a hierarchical system
4. Which of the following is a part of the software development life cycle?
   (a) Requirement analysis
   (b) Program specification
   (c) Bench marking
   (d) All of these
5. The evaluating method for a cost–benefit analysis is _____.
   (a) pay back method
   (b) break-even method
   (c) both (a) and (b)
   (d) none of these

Project Work

1. You have been assigned to prepare an MIS for guest profiling. What steps would you take for the same?

CASE STUDY 1

Hotel Park Residency has installed a centralized call management system. Whenever a guest makes a call to the hotel, the call is directed to this call centre. The system in place is for both, in-house guests as well as guests who would call for room reservations. It is the single point of contact for all the guests.

This method was put in place as there was dissatisfaction and complaints from guests. It was also found that the maximum number of guests who were not satisfied did not register a complaint but never returned to the hotel. Whenever a guest had to register a complaint or wished for an additional service, he/she had to inform the front office or the concerned department. As the complaint was registered at individual departments or was sometimes not conveyed by the front office, the complaint was not passed on to the concerned department and if passed, was sometimes exaggerated. At times, this led to differences between the departments, ultimately leaving the guest dissatisfied.

However, now, with the new system, the in-house guest had to just dial a single contact number from their console and register their grievance. A guest complaint/suggestion was registered in the call centre and immediately forwarded to the concerned department—housekeeping desk, front office, or the food and beverage department. If there was any maintenance-related problem, the same would be forwarded to the department by the call centre. As and when the guest complaint was resolved, the same would be informed to the call centre, who reverted to the guest to obtain their feedback and satisfaction levels. A standard operating procedure was laid out for handling complaints. If the complaint could not be attended to or resolved by the staff within the first 30 minutes, it was immediately forwarded to the next higher authority, and the problem was taken up seriously until it was resolved.

The hotel had even allotted a few department managers at the various guest floors of the hotel. Depending upon the guest's complaint, the manager would sometimes personally speak to or interact with him/her to resolve the problem and apologize for the situation. This extra care led to guests being satisfied, customers coming back, and hence increased hotel revenues.

(a) What are the benefits of the centralized call monitoring system?
(b) How does this system help in improving guest satisfaction and give the hotel an edge over competitors?
(c) How will the call monitoring system help the hotel in the future?

CASE STUDY 2

Vikram Foods Pvt. Ltd has a chain of restaurants by the name, Quality Restaurant. Entrepreneur Vikram Raheja, started the company in early 2004. Now the company has over 10 outlets in the city and its outskirts. However, the expansion of the restaurant took place without much planning. The menu was different at different locations and there was lack of standardization in the food being prepared and served to the guests. All the restaurants purchased their raw materials separately; if the purchases had been done from one location, ingredients could have been purchased at competitive prices.

The restaurant purchased raw materials from various vendors and due to lack of proper storage and inventory procedures, some of the items were either wasted or pilfered. Since the stocks were manually maintained, it was difficult to find out if any ingredient was out of stock; as a result of this, sometimes, dishes would not be available. This led to cancellation of orders and dissatisfaction of the guests. Vikram wanted to have a centralized purchasing and receiving system. He also wished to have a similar menu in all the restaurants so that, if for some reason, an item ran short in one outlet, the same could be delivered by another outlet. Vikram thought about computerizing the inventory system and called Kulkarni, a system analyst, and discussed his problem and intention to have a centralized system. Kulkarni advised him to go in for computerization. Vikram asked Kulkarni to conduct preliminary investigations regarding the feasibility of the proposed system.

Feasibility Study

Kulkarni visited all the outlets and observed the manual procedures being followed. The analyst then finally prepared a feasibility study report bearing in mind Vikram's requirements, and presented it to him. The analyst noted the following problems in the present system of inventory control:

- No proper inventory was made on a regular basis.
- A dish was not available due to unavailability of raw ingredients.
- No purchase department existed, hence, no re-order level or par stock of an ingredient existed.

Requirement Analysis

After the feasibility analysis, the system analyst submitted a project proposal to Vikram. After the proposal was approved by him, Kulkarni studied the present system in detail. Kulkarni understood the problem of the current system and came up with various inputs and outputs of the system such as those mentioned here:

Input to the system

- Item/Ingredient codes and names
- Item quantity, unit, and opening stock
- Details of items received from supplier/vendor
- Details of items issued to various outlets
- Details of items rejected, if any
- Output to the system
- Purchase order
- Stock status report
- Stock summary report
- Purchase analysis report

Design and development of the system   The system analyst designed the system using a structured designed methodology. The input forms for purchase orders, goods received notes, etc., were designed.

Testing and implementation   The system was tested by preparing a test plan and test data. After successful testing, the system was installed in the proposed central store of the restaurant. A user manual was prepared for the users and the staff was trained to operate the system. Though there were problems initially, later, the system was successfully implemented.

(a) How did the central purchase system help Quality Restaurant in its operations?

(b) What are the various inputs and outputs of the proposed inventory system?

(c) Would a commissary or central kitchen be a better alternative?

(d) If you were appointed as an analyst for Quality Restaurant, what would you have done? Justify your action.

Answers to Multiple Choice Questions

1. (a)   2. (b)   3. (a)   4. (d)   5. (c)

Index

Related Titles

Food Production Operations (with DVD)
9780198061816
Parvinder S. Bali, Programme Manager, Culinary Services, Oberoi Center of Learning and Development, Delhi

Key Features

- Discusses the various methods of cooking such as sautéing, steaming, braising, microwave cooking, etc.
- Includes chapters on stocks, soups, sauces, fish, eggs, bread making, etc.

In the DVD

- 55 videos showcasing various food production procedures and techniques to operate complex kitchen equipment
- 365 recipes, which include methods to prepare Indian, western, and pastry food items

Food and Beverage Service
9780198065272
R. Singaravelavan, Head, Department of Catering Science and Hotel Management at SNR Sons College, Coimbatore

Key Features

- Illustrates the key concepts with the help of photographs of various table layouts and other services, colour plates, sample menus, and side bars
- Discusses the duties and responsibilities of the F&B staff
- Provides a detailed description of the various types of wines, non-alcoholic beverages, guéridon service, and specialized service skills for breakfast, afternoon, tea, brunch, and so on
- Includes the French terms used for the staff members, menu, and dishes

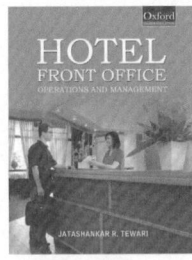

Hotel Front Office: Operations and Management
9780195699197
Jatashankar R. Tewari, Assistant Professor, Hotel Management, Uttarakhand Open University

Key Features

- Gives an overview of the hospitality industry
- Discusses the functioning of front office operations, and suggests ways and means to make them more effective
- Includes well-illustrated chapters with numerous photographs, flowcharts, illustrations, tables, and examples
- Provides end-chapter review questions to reinforce concepts
- Contains mini case studies to enhance critical thinking and relate the concepts to real-life situations

OTHER RELATED TITLES

Raghubalan and Raghubalan: *Hotel Housekeeping 2/e*
Bali: *Quantity Food Production Operations and Indian Cuisine*
Bali: *International Cuisine and Food Production Management*
Biswas: *Human Resource Management in Hospitality*
Bansal: *Hotel Facility Planning*
Ghosal: *Hotel Engineering*

Iyengar: *Hotel Finance*
Swain and Mishra: *Tourism: Principles and Practices*
Roday, Biwal, and Joshi: *Tourism Operations and Management*
Chaudhary: *Tourism Marketing*
Roday: *Food Science and Nutrition 2/e*